CHEMICAL AND FUNCTIONAL GENOMIC APPROACHES TO STEM CELL BIOLOGY AND REGENERATIVE MEDICINE

CHEMICAL AND FUNCTIONAL GENOMIC APPROACHES TO STEM CELL BIOLOGY AND REGENERATIVE MEDICINE

Edited by

SHENG DING
Departments of Chemistry and Cell Biology
The Scripps Research Institute

A JOHN WILEY & SONS, INC., PUBLICATION

Published by John Wiley & Sons, Inc., Hoboken, New Jersey
Published simultaneously in Canada

Library of Congress Cataloging-in-Publication Data:

Chemical and functional genomic approaches to stem cell biology and regenerative medicine / [edited by] Sheng Ding.
p. ; cm.
Includes bibliographical references and index.
ISBN 978-0-470-04146-8 (cloth)
1. Stem cells–Research–Methodology. 2. Physiological genomics.
3. Molecular genetics. I. Ding, Sheng, 1975-
[DNLM: 1. Genomics. 2. Regenerative Medicine–methods. 3. Stem Cells.
QU 58.5 C5165 2008] QH588.S83C455 2008
616′.02774–dc22 2007039335

Printed in the United States of America

10 9 8 7 6 5 4 3 2 1

CONTENTS

Contributors vii

1 **Embryonic Stem Cells** 1
Crystal L. Sengstaken, Eric N. Schulze, and Qi-Long Ying

2 **Adult Stem Cells** 27
Lief Fenno and Chad A. Cowan

3 **Genomewide Expression Analysis Technologies** 59
John R. Walker

4 **Genomic cDNA and RNAi Functional Profiling and Its Potential Application to the Study of Mammalian Stem Cells** 83
Jia Zhang, Myleen Medina, Genevieve Welch, Deanna Shumate, Anthony Marelli, and Anthony P. Orth

5 **Chemical Technologies: Probing Biology with Small Molecules** 109
Nicolas Winssinger, Zbigniew Pianowski, and Sofia Barluenga

6 **Protein Characterization by Biological Mass Spectrometry** 145
Venkateshwar A. Reddy and Eric C. Peters

7 **Large-Scale Genomic Analysis of Stem Cell Populations** 169
Jonathan D. Chesnut and Mahendra S. Rao

8 Exploring Stem Cell Biology with Small Molecules and Functional Genomics **187**
Julie Clark, Yue Xu, Simon Hilcove, and Sheng Ding

9 Regeneration Screens in Model Organisms **207**
Chetana Sachidanandan and Randall T. Peterson

10 Proteomics in Stem Cells **223**
Qiang Tian and W. Andy Tao

Index **243**

CONTRIBUTORS

Sofia Barluenga, Organic and Bioorganic Chemistry Laboratory, Institut de Science et Ingénierie Supramolécularies, Université Louis Pasteur, 8 allée Gaspard Monge 67000, Strasbourg, France

***Jonathan D. Chesnut**, Stem Cells and Regenerative Medicine, Invitrogen Corporation, Carlsbad, CA 92008 (ion.chesnut@invitrogen.com)

Julie Clark, Department of Chemistry and the Skaggs Institute for Chemical Biology, The Scripps Research Institute, 10550 North Torrey Pines Road, La Jolla, CA 92037

***Chad A. Cowan**, Stowers Medical Institute, Center for Regenerative Medicine and Technology, Cardiovascular Research Center, 185 Cambridge St., Boston, MA 02114 (ccowan1@partners.org)

***Sheng Ding**, Department of Chemistry and the Skaggs Institute for Chemical Biology, The Scripps Research Institute, 10550 North Torrey Pines Road, La Jolla, CA 92037 (sding@scripps.edu)

Lief Fenno, Stowers Medical Institute, Center for Regenerative Medicine and Technology, Cardiovascular Research Center, 185 Cambridge St., Boston, MA 02114

Simon Hilcove, Department of Chemistry and the Skaggs Institute for Chemical Biology, The Scripps Research Institute, 10550 North Torrey Pines Road, La Jolla, CA 92037

Anthony Marelli, Genomics Institute of the Novartis Research Foundation, 10675 John Jay Hopkins Drive, San Diego, CA 92121

Myleen Medina, Genomics Institute of the Novartis Research Foundation, 10675 John Jay Hopkins Drive, San Diego, CA 92121

***Anthony P. Orth**, Genomics Institute of the Novartis Research Foundation, 10675 John Jay Hopkins Drive, San Diego, CA 92121 (aorth@gnf.org)

***Eric C. Peters**, Genomics Institute of the Novartis Research Foundation, 10675 John Jay Hopkins Drive, San Diego, CA 92121(epeters@gnf.org)

***Randall T. Peterson**, Cardiovascular Research Center, Massachusetts General Hospital and Harvard Medical School, Boston, MA (peterson@cvrc.mgh.harvard.edu)

Zbigniew Pianowski, Organic and Bioorganic Chemistry Laboratory, Institut de Science et Ingénierie Supramolécularies, Université Louis Pasteur, 8 allée Gaspard Monge 67000, Strasbourg, France

Mahendra S. Rao, Stem Cells and Regenerative Medicine, Invitrogen Corporation, Carlsbad, CA 92008

Venkateshwar A. Reddy, Genomics Institute of the Novartis Research Foundation, 10675 John Jay Hopkins Drive, San Diego, CA 92121

Chetana Sachidanandan, Cardiovascular Research Center, Massachustetts General Hospital and Harvard Medical School, Boston, MA

Eric N. Schulze, Center for Stem Cell and Regenerative Medicine, Department of Cell and Neurobiology, Keck School of Medicine, University of Southern California, 1501 San Pablo Street, Los Angeles, CA 90033

Crystal L. Sengstaken, Center for Stem Cell and Regenerative Medicine, Department of Cell and Neurobiology, Keck School of Medicine, University of Southern California, 1501 San Pablo Street, Los Angeles, CA 90033

Deanna Shumate, Genomics Institute of the Novartis Research Foundation, 10675 John Jay Hopkins Drive, San Diego, CA 92121

W. Andy Tao, Department of Biochemistry, Purdue University, 175 S. University Street, West Lafayette, IN 47907

***Qiang Tian**, Institute for Systems Biology, 1441 North 34th Street, Seattle, WA 98103 (qtian@systemsbiology.org)

***John R. Walker**, Group Leader, RNA Dynamics, Genomics Institute of the Novartis Research Foundation, 10675 John J. Hopkins Drive, San Diego, CA 92121 (walker@gnf.org)

Genevieve Welch, Genomics Institute of the Novartis Research Foundation, 10675 John Jay Hopkins Drive, San Diego, CA 92121

***Nicolas Winssinger**, Organic and Bioorganic Chemistry Laboratory, Institut de Science et Ingénierie Supramolécularies, Université Louis Pasteur, 8 allée Gaspard Monge 67000, Strasbourg, France (winssinger@isis-u.strasbg.fr)

Yue Xu, Department of Chemistry and the Skaggs Institute for Chemical Biology, The Scripps Research Institute, 10550 North Torrey Pines Road, La Jolla, CA 92037

***Qi-Long Ying**, Center for Stem Cell and Regenerative Medicine, Department of Cell and Neurobiology, Keck School of Medicine, University of Southern California, 1501 San Pablo Street, Los Angeles, CA 90033 (qying@keck.usc.edu)

Jia Zhang, Genomics Institute of the Novartis Research Foundation, 10675 John Jay Hopkins Drive, San Diego, CA 92121

*Corresponding author.

1

EMBRYONIC STEM CELLS

CRYSTAL L. SENGSTAKEN, ERIC N. SCHULZE, AND QI-LONG YING

Center for Stem Cell and Regenerative Medicine, Department of Cell and Neurobiology, Keck School of Medicine, University of Southern California, Los Angeles, California

Embryonic stem (ES) cells are pluripotent cell lines derived from the inner cell mass (ICM) of preimplantation embryos [1]. ES cells are invaluable tools to create genetic modifications in mice for the study of gene function and disease. The first true ES cell lines were derived from the 129 strain of mouse in 1981 by two groups simultaneously [2,3]. In 1998, Thomson and colleagues reported the isolation of the first human ES cell lines from human blastocysts left over from *in vitro* fertilization [4]. Putative ES or ES-like cells from other species, such as, bird, fish, monkey, dog, cow, and rat, have also been reported. However, only ES cells from mice have proved to be able to efficiently contribute to chimeras and reenter germline. There are other pluripotent cell lines including embryonic germ (EG) cells, which are derived from gonadal ridges of the embryo, and embryonic carcinoma (EC) cells, which are isolated from spontaneously arising teratocarcinomas and are therefore karyotypically abnormal. In this chapter, we will focus on mouse and human ES cells.

1.1 THE ORIGIN OF EMBRYONIC STEM (ES) CELLS

There appears to be a very limited period of embryonic development during which pluripotent ES cells can be established in culture. This period is termed the *preimplantation blastocyst stage* (Figure 1.1). From the one-cell to early eight-cell stage, the blastomeres of the embryo are equipotent. Each blastomere is considered to be *totipotent*; that is, they have the ability to differentiate into all cell types in an

Chemical and Functional Genomic Approaches to Stem Cell Biology and Regenerative Medicine
Edited by Sheng Ding

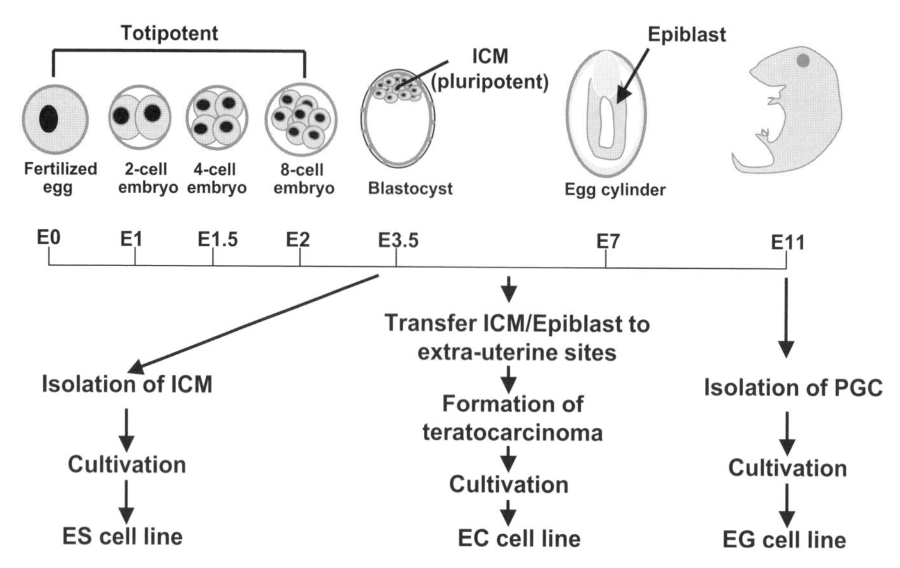

FIGURE 1.1 Origin of pluripotent stem cells of the mouse. There are three main types of pluripotent stem cells: embryonic stem (ES) cells, embryonic carcinoma (EC) cells, and embryonic germ (EG) cells. ES cells are derived from ICMs isolated from preimplantation blastocysts. EC cells are isolated from embryo-derived teratocarcinomas. Teratocarcinomas can be produced from ICMs or epiblasts grafted into adult mice. EG cells are derived from primordial germ cells (PGC) isolated from the genital ridges between embryonic days 9 and 12.5. In most respects, EG cells are indistinguishable from ES cells, and can contribute to chimeras and give germline transmission.

organism, including the extraembryonic tissue associated with that organism, and are able to form an entire organism. However, blastomeres lack the capacity for self-renewal and therefore are not considered stem cells. As cleavage proceeds, there is a gradual restriction in the developmental potency of the cells, eventually resulting in the generation of the first two distinct lineages: the trophectoderm and the ICM, followed by the formation of a fully expanded blastocyst consisting of a hollow vesicle of trophectoderm surrounding a fluid-filled cavity and a small group of ICM cells. The preimplantation blastocyst is formed after about six cleavage divisions, which occur about 3.5 or 5.5 days after fertilization in the mouse and human, respectively. The trophectoderm cells are required for implantation and development of the placenta. The ICM is the foundation of all somatic tissues and germ cells in adults. The ICM comprises the inner pluripotent primitive ectoderm cells and the outer primitive endoderm cells. ES cells are derived from the primitive ectoderm cells when ICM is isolated and cultured *in vitro* under proper conditions. Little is known, however, about how pluripotent primitive ectoderm cells in the ICM are transformed into pluripotent ES cells in culture. It is thought that the pluripotent ES cells remain in a state *in vitro* that may occur only transiently *in vivo*, since the primitive ectoderm cells progress rapidly through development to form the epiblast cells. Although these epiblast cells are still pluripotent and can give rise to all three primary germ layers, ectoderm, mesoderm, and endoderm, there is still no evidence suggesting that ES cells can be derived from the pluritpotent epiblast cells of the implanted embryo. However, cells resembling ES cells in both morphology and growth characteristics have been obtained previously from preblastocyst stages [5,6]. Since attached embryos from progressively earlier developmental stages require longer periods in culture before they reach a state that yield ES cells, the possibility exists that preblastocyst embryos progress to an equivalent stage of blastocyst embryos in culture before ES cells can be obtained. It remains unclear, however, whether this state represents the sole timepoint from which ES cells can be obtained or whether it demarcates the beginning of a window of time during which the derivation of ES cells is possible.

1.2 DERIVATION OF ES CELLS

The research on embryonic carcinoma (EC) cells in the 1970s eventually paved the way for the establishment of the first ES cell lines [7–10]. When epiblast cells from early postimplantation embryos were grafted into adult mice, they produced teratocarcinomas. Teratocarcinomas are malignant multidifferentiated tumors containing a proportion of undifferentiated cells. These undifferentiated cells could be propagated in culture and established as cell lines termed *EC cells*. Clonally isolated EC cells retained the capacity for differentiation into derivatives of all three germ layers. EC cells could also participate in embryonic development when introduced into the ICM of blastocysts to generate chimeric mice [11]. Maintenance of the undifferentiated state of EC cells relied on cocultivation with feeder cells, usually mitotically inactivated mouse embryonic fibroblasts (MEFs) [12]. It was reasoned that these feeder cells were providing some critical factors to sustain the pluripotency of the EC cells. Since EC cells have undergone transformation and karyotypic changes prior to establishment as cell lines, they are almost always aneuploid and are not capable of proceeding through meiosis to produce mature gametes. Following the discovery of EC cells, the next logical step was to attempt to directly isolate pluripotent cells from embryos. In 1981, two groups succeeded in establishing pluripotent ES cell lines from mouse embryos [2,3]. The protocol for the derivation of ES cells is relatively simple, and most labs are still using the original protocol developed by Evans and Martin. In brief, the protocol involves plating intact embryos at the expanded blastocyst stage onto a mitotically inactivated feeder layer with either DMEM or GMEM as the basal culture medium supplemented with 10–20% fetal calf serum, 2-mercaptoethanol, nonessential amino acids, L-glutamine, and sodium pyruvate. After several days of culture, the cells from the ICM will expand to form a cell mass. After being disaggregated and replated onto fresh feeders, various types of differentiated colonies as well as colonies of a characteristic undifferentiated morphology will appear. The undifferentiated colonies can generally be expanded further to establish cell lines, now known as *ES cells*. The first human ES cell line was established in essentially the same method by Thomson and colleagues in 1998 [4]. Since ES cells were first created, it has become clear that different strains of mice vary considerably in their facility for derivation [13]. The inbred 129 and C57Bl/6 strains are the most permissive;

approximately 30% of 129 or C57Bl/6 embryos can be expected to give rise to ES cell lines. Most other strains rarely produce ES cell lines at all. The reason for this intriguing variability is not yet understood. The ES cell derivation efficiency can be increased by several modifications of the protocol. For example, subjecting the embryos to delayed implantation or diapause can improve the efficiency [14]. Removal of the extraembryonic tissues from blastocysts mechanically or by immunosurgery can also facilitate the derivation of ES cells. It has been reported that, by combining the two techniques, the ES cell yield rates can be increased up to 100% for 129 and to over 50% for CBA embryos [1]. In addition to removing the differentiative signals from the surrounding tissues of ICM, it is possible to encourage the self-renewal aspect of stem cell proliferation by preferentially inhibiting signaling pathways that promote differentiation. For example, commercially available inhibitors of the Erk-activating MEK pathway such as PD98059 have been used to promote ES cell self-renewal and improve the efficiency of ES cell derivation [15]. ES cell lines from 129 and C57Bl/6 strain mice can also be efficiently derived under serum and feeder-free conditions with the addition of both the leukemia inhibitory factor (LIF) and bone marrow morphogenetic protein 4 (BMP4) [16,17].

1.3 KEY PROPERTIES OF ES CELLS

Embryonic stem cells are a remarkable cell type, mainly because of the two key properties they possess: unlimited proliferation and unlimited differentiation. ES cells can be maintained in culture for an extended amount of time, perhaps even indefinitely. Additionally, even after many passages in culture, ES cells continue to maintain the ability to differentiate into any type of cell in the body. For this reason, ES cells are considered as pluripotent. Although there are some similarities between the various types of adult stem cells, pluripotent ES stem cells clearly differ from other adult stem cells in several ways. Adult stem cells are generally limited to differentiation into the cell types of their tissue of origin, making them only multipotent, and can be maintained in culture for only a very limited number of passages before they differentiate. However, pluripotent ES cells, when reintroduced into early-stage embryo, have the ability to reenter developmental processes and contribute to all cell lineages, including germ cells [14,18–20]. The capacity to generate all fetal and adult cell lineages *in vitro* and *in vivo*, combined with the facility of genetic manipulations, makes ES cells a very powerful tool for molecular dissection of tissue differentiation and cellular (patho)physiology [21]. Human ES cells also create a platform for a renewable source of differentiated cells for applications in pharmacogenomics and cell transplantation therapies [22,23]. ES cells have other functionally important or unique properties, including derivation without transformation or immortalization, stable diploid karyotype (prolonged culture will increase genetic or epigenetic abnormality), clonogenic (can be grown from single cells; it has proved difficult for human ES cells), absence of G1 cell cycle checkpoint, and absence of X inactivation (in XX lines).

1.4 MAINTENANCE OF ES CELL SELF-RENEWAL

When an ES cell divides, it has to decide whether to produce identical copies of itself (self-renewal) or to differentiate into specific cell types. This ES "cell fate" decision, that is, self-renewal versus differentiation, is greatly influenced by both extrinsic and intrinsic factors (Figure 1.2).

1.4.1 Extrinsic Factors in Mouse ES Cell Fate Determination

LIF When ES cells are cultured on feeders in medium supplemented with fetal calf serum, they can be sustained in an undifferentiated state indefinitely while retaining the ability to give rise to all the cell types of the embryo and adult. It has been shown that conditioned media from the feeders has the same effect on ES cell self-renewal as

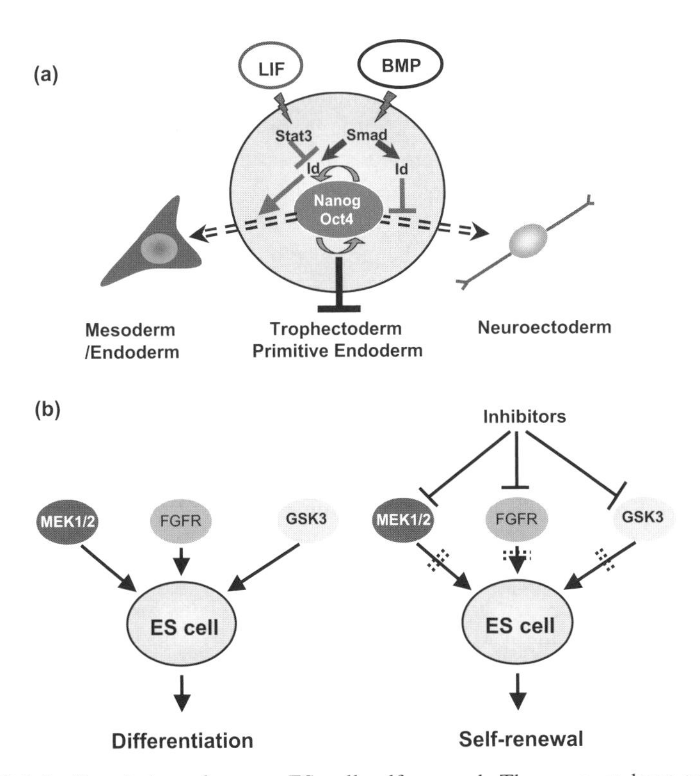

FIGURE 1.2 Regulation of mouse ES cell self-renewal. There are at least two known independent pathways regulating mouse ES cell self-renewal.(**a**) Mouse ES cell self-renewal can be driven by the combined action of LIF/Stat3 and BMP/Smad/Id pathways. LIF activates Stat3 and blocks nonneural differentiation. BMP blocks neural differentiation by induction of Ids.(**b**) Mouse ES self-renewal can also be sustained by simple elimination of differentiation cues associated mainly with MAPK, GSK3, and FGF receptor signaling pathways. In both conditions, the proper function of Oct4 and Nanog remains important for the maintenance of ES cell identity.

the feeder layers. This observation eventually leaded to the identification of leukemia inhibitory factor (LIF) as the key cytokine secreted by feeder cells in supporting mouse ES cell self-renewal [24,25]. Feeders lacking functional LIF gene do not support ES propagation effectively [26]. LIF is a soluble glycoprotein and is a member of the interleukin-6 (IL6) family. LIF was initially identified by its activity to induce differentiation of M1 leukemia cells [27,28]. Other members of the IL6 family, including oncostatin M (OSM) [29], ciliary neurotrophic factor (CNTF) [30] and cardiotrophin 1 (CT1) [31], also have an effect similar to that of LIF in suppressing differentiation and sustaining self-renewal of mouse ES cells. LIF works by binding directly to a heterodimeric receptor complex containing two transmembrane glycoproteins, gp130 and the LIF-specific receptor subunit LIFRβ, bringing the associated JAKs into close proximity and causing their cross-phosphorylation and activation [32]. A wide range of downstream effectors can be activated via gp130. These include the signal transducer and activator of transcription (STAT) 1, 3, and 5, the mitogen-activated protein kinases (MAPK), extracelluar regulated kinases (ERK) 1 and 2, and phosphoinositol-3 kinase (PI3K). In mouse ES cells LIF and other related cytokines sustain ES cell self-renewal mainly through activation of Stat3 [33]. Inhibition of Stat3 signaling directly, through the use of interfering mutant forms of the transcription factor Stat3F, provides the best evidence for an essential role of Stat3 in ES cell self-renewal mediated by LIF or related cytokines. Stat3F is a dominant-negative mutant of Stat3. Overexpression of Stat3F in ES cells resulted in induction of differentiation even in the presence of LIF, indicating that activation of Stat3 is essential to the LIF-mediated ES cell self-renewal. Studies using a chimaeric Stat3 molecule that can be activated directly by estradiol indicate that Stat3 activation is not only necessary but might be sufficient to block differentiation [34]. Several groups have tried to identify LIF/Stat3 target gene(s) that are responsible for ES cell self-renewal [35–37], but so far it has not been very successful. Downstream target genes of Stat3 in ES cells that mediate self-renewal are still largely unknown. In mouse ES cells, LIF/gp130 signaling also activates MAPK pathway. The ERK MAPKs Erk1 and Erk2 are strongly phosphorylated on stimulation of LIF. Phosphorylated ERKs then undergo nuclear translocation and modulate the activities of transcriptional regulators such as Elk, Myc, and the serum response factor (SRF). Inhibition of MAPK/ERK signaling by small molecule inhibitors PD98059 and U0126 limits the differentiation of ES cells and promotes self-renewal [15]. This strongly suggests that while LIF/Stat3 pathway sustains ES cell self-renewal, the LIF/gp130/ERK pathway is antagonistic to it. The overall balance of conflicting activation of Stat3 and ERKs might well determine the efficiency of mouse ES cell self-renewal.

Serum/BMP Feeder layers or LIF can support mouse ES cell self-renewal. However, these observations were made in the presence of fetal calf serum. In the absence of serum, LIF is not sufficient to sustain mouse ES cell self-renewal (Figure 1.2a). Instead, the ES cells will differentiate predominantly into neural phenotypes. Hence, there is another factor or factors needed in combination with LIF to achieve self-renewal. It has been demonstrated that bone morphogenetic protein 4 (BMP4) can replace the requirement for serum both during clonal propagation of mouse ES cells and during their de novo derivation [17]. BMPs were originally isolated from demineralized bone matrix and identified as factors responsible for inducing bone formation in muscular tissues [38]. The critical contribution of BMP4 to self-renewal is to induce expression of the negative helix–loop–helix factors, inhibitors of differentiation (Ids). Id proteins lack a DNA binding domain and are not thought capable of inducing gene transcription. They act by binding and sequestering E proteins, thereby inhibiting the E–protein–dependent transcriptional activity of basic helix–loop–helix (bHLH) factors. Constitutive Id gene expression in ES cells replaces the need for BMP4 in the media. Serum also induces Id genes via multiple pathways, including integrin engagement by extracellular matrix molecules such as fibronectin [39,40]. A BMP-like factor was purified from serum that is responsible for both inhibition of myogenesis and stimulation of osteoblast differentiation *in vitro*. This BMP-like factor was identified as BMP4. In addition, BMP4 in serum was found to form a large complex with other molecules, resulting in potentiation of its activity [41]. Therefore it is not surprising that ES cells cultured in serum also show appreciable levels of Id gene expression [17]. These findings together suggest that the biological activities of serum in suppressing differentiation and sustaining self-renewal of mouse ES cells might be mediated at least in part by BMPs. Significantly, in the absence of LIF, both serum and BMP4 drive ES cells differentiation into nonneural fates (Figure 1.2a). Thus serum or BMP4 stimulation has a dual potential in ES cells, and the outcome is dictated by the presence or absence of LIF/Stat3 pathways. The ability of BMP4 to suppress differentiation and maintain ES cell self-renewal in collaboration with LIF/Stat3 is shared by other BMP family, BMP2 and GDF6, but not other transforming growth factor β (TGFβ) superfamily members, such as TGFβ1 and activin A. Mouse ES cells can also be maintained in serum-free medium supplemented with LIF and knockout serum replacement without addition of serum or BMPs. Under these culture conditions, it was reported that activin or Nodal, but not TGFβ1 or BMP4, can significantly enhance ES cell proliferation without affecting pluripotency [42]. Although serum replacement is thought to be better defined than serum, it is a proprietary product that cannot be regarded as fully defined. In fact, it has been shown that BMP-like activity is present in serum replacement. Indeed, both BMP2 and BMP4 proteins were detected in serum replacement [43].

Inhibitors ES cells exist in the artificial milieu of tissue culture, therefore the composition of the *in vitro* environment can obscure the actual requirements for maintaining pluripotency. Pathways that appear obligatory for ES cell propagation in culture may not, in fact, be core components of the pluripotent state, but accessories demanded by unrefined culture settings. It has been suggested that mouse ES cell self-renewal is not driven by activating signals but by the elimination of differentiation cues and consolidation of a basal program of cell growth and proliferation [44]. MAPK, FGFR, and GSK3 signaling pathways are involved in numerous cell functions including cell proliferation and differentiation. Under serum-free basal conditions mouse ES cells can be efficiently derived and maintained in an undifferentiated, proliferative, and pluripotent state through the actions of chemically defined inhibitors of these three signaling pathways without added growth factors or cytokines [44]. By blocking these three pathways, it is even possible to derive Stat3-/- mouse ES cell lines, authentic ES cell lines from rats and nonpermissive strains of mice such as CBA, all at high efficiency in defined feeder-free conditions without adding any growth factors or cytokines [44]. Through a cell-base screen of chemical libraries, Chen et al. identified a small molecular inhibitor, pluripotin, that can maintain mouse ES cell self-renewal in the absence of feeders and exogenous factors [45]. Pluripotin may act by inhibition of both Erk1 and RasGAP-dependent signaling pathways. These findings indicate that mouse ES cell self-renewal may not be dependent on extrinsic stimulation and that elimination of differentiation cues is sufficient to sustain self-renewal (Figure 1.2b).

1.4.2 Extrinsic Factors in Human ES Cell Fate Determination

Currently human ES cells are routinely grown on mitotically inactivated feeder layers with medium containing either serum or serum replacements, both of which are ill-defined. Although progress has been made in delimiting culture environments for the derivation and propagation of human ES cells, it is still largely unclear what extrinsic factors are really required for sustaining human ES cell self-renewal. Human and mouse ES cells appear to be very different in terms of their requirements for sustaining self-renewal. For example, while the activation of LIF/Stat3 pathway has long been considered necessary for maintenance of mouse ES cell self-renewal, it has no obvious effect on human ES cells [46,47]. Instead, human ES cells require bFGF, unknown factor(s) secreted by feeder cells and cell–cell adhesion to sustain pluripotency in serum-free conditions. BMP4 can replace serum in supporting mouse ES cell self-renewal in collaboration with LIF/Stat3 pathway. However, BMP4 induces human ES cell differentiation toward trophectoderm lineage [48]. The question that still remains is why they are so different. Is it because of the mechanisms that regulate human and mouse ES cell self-renewal are fundamentally different, or is it because of subtle variations in the machinery regulating self-renewal modulating the response of the cell? It is worth noting that despite these differences, feeder cells support both mouse and human ES cell self-renewal. In fact, the first mouse and human ES cell lines were both derived in essentially the same culture conditions using mitotically inactivated MEFs as feeder layers in serum-containing media [2–4]. For mouse ES cells, the requirement for feeders can be partially replaced by LIF or related cytokines that activate the gp130 signal transducer and the downstream transcriptional effector Stat3 [24,25,33]. However, LIF has no obvious effect on human ES cell self-renewal. When mouse ES cells are "weaned off" feeder cells, they go through a "crisis" in which most ES cells die or differentiate, even in the presence of LIF. Although sublines of feeder-independent cells can be derived, they generally lose the hallmark of normal ES cells, which is competence for germline transmission. This suggests that feeder cells must produce a factor or factors, other than LIF, that may not yet be identified, that support both human and mouse ES cell growth. It may or may not, however, be the same factor (s) for human and mouse ES cells. Another major difference between mouse and human ES cells is that mouse ES cells can easily adhere to substrates such as gelatin, fibronectin, or laminin, while most human ES cells do not adhere to gelatin, and adhere only poorly to fibronectin or laminin. Adherent growth appears to be important for maintenance of ES cell self-renewal, since ES cells aggregate in suspension culture, a state generally associated with differentiation into structures known as *embryoid bodies*. Feeder layers help to support the adhesion of human ES cells, which is possibly one of the major contributions of feeder layers in sustaining human ES cell self-renewal. There have been several reports that have investigated the effects of a broad range of factors, including Wnts, Noggin, Nodal, activin A, neurotrophins, or TGFβ1 [49–53], all of which have been claimed to be involved in maintenance of human ES cell self-renewal. However, bFGF is still the single most potent extrinsic factor identified for human ES cell self-renewal under serum-free conditions [54]. Why do human ES cells require bFGF? While bFGF has no obvious effect on wild-type mouse ES cells, it was found that Eras-null mouse ES cells need bFGF for efficient propagation in serum-free conditions (unpublished data). This finding is interesting because bFGF and Eras both activate PI3-kinase/Akt pathway and Eras is not expressed in human ES cells [55]. Eras is a member of the ras family that is expressed specifically in undifferentiated mouse ES cells [56]. It is constitutively active and is important for ES cell proliferation; therefore it would be interesting to see if lack of Eras function in human ES cells underlies the requirement for bFGF to sustain self-renewal.

1.4.3 Intrinsic Determinants

Oct4 Oct4 is a POU family transcriptional factor that is encoded by *Pou5f1*. Oct4 plays a pivotal role in both human and mouse ES cell fate determination. Expression of Oct4 is restricted to pluripotent cell types, such as cells in early embryos; germline cells; and undifferentiated EC, EG, and ES cells [57]. *In vivo*, zygotic expression of Oct4 is essential for the initial development of pluripotential capacity in the ICM [58]. Oct4 deficient embryos fail to initiate fetal development because the prospective founder cells of the ICM do not acquire pluripotency and differentiate into the trophectoderm lineage. Investigation via conditional repression/expression in ES cells has suggested that the level of Oct4 is critical for ES cell fate determination. Artificial repression of Oct4 in mouse or human ES cells induces differentiation along the trophectodermal lineage [59–61], which suggests that Oct4 is continuously required by ES cells in order to maintain their pluripotent identity and may act as a lock that prevents differentiation into the trophoblast stage [61]. It has been shown that Oct4 directly prevents differentiation toward trophectoderm by forming a repressor complex with Cdx2, an inducer for trophectoderm differentiation [62]. This complex interferes with the autoregulation of these two factors, giving rise to a reciprocal inhibition system that establishes their mutually exclusive expression. The downregulation of Oct4 results in an upregulation of Cdx2, and vice versa. This might account for the segregation of the first two cell lineages in early embryonic development: pluripotent stem cells in ICM and trophectoderm cells. When the level of Oct4 was artificially raised by more than 50% of wild-type levels, with ES cells differentiated into endoderm and mesoderm, indicating that in ES cells, continuous Oct4 function at appropriate levels may be crucial to maintain pluripotency. However,

Oct4 cannot act alone to maintain pluripotency, and interaction with other factor(s) and signaling pathways like LIF/Stat3 and BMP/Smad is required. Many of the genes that are regulated by Oct4 signaling also contain Stat binding sites, suggesting that these Oct4 and LIF/Stat3 could act cooperatively in regulating the expression of ES cell-specific genes.

Nanog Nanog is a homeodomain-containing protein, and is expressed in a restricted range of cell types [63,64]. Nanog mRNA is present in pluripotent mouse and human ES cells, and absent from differentiated cells. In embryo, Nanog expression is first detected in morula, increases in the early blastocyst, and declines prior to implantation. Following implantation, Nanog is expressed in only a subset of epiblast cells and rapidly downregulated on entry into the primitive streak. Nanog is also expressed in developing germ cells. Deletion of Nanog prevents acquisition of pluripotency in the ICM of preimplantation mouse blastocyst. Downregulation of Nanog via siRNA in human ES cells leads to a significant downregulation of Oct4 and loss of ES cell surface antigens, and differentiation toward extraembryonic endodermal lineages [65]. Endogenous Nanog acts in parallel with cytokine stimulation of Stat3 to drive ES cell self-renewal. Nanog overexpression from transgene constructs is sufficient to maintain constitutive self-renewal in mouse ES cells, bypassing LIF/Stat3 and BMP/Smad/Id pathways [66]. It has been shown that overexpression of Nanog can also enable both human and primate ES cell self-renewal in the absence of feeder layers [67,68]. Perhaps appropriately, Nanog is considered a core element of the pluripotent state. However, Nanog function still requires the continued presence of Oct4. Elevated Nanog expression is not sufficient to prevent ES cell differentiation into trophectodermal lineage when Oct4 expression was repressed. Using a genome-scale analysis to identify a cohort of genes that respond to Oct4, Sox2, and Nanog in human ES cells, it has been observed that a majority of genes coregulated by Oct4 and Sox2 are also targets of Nanog. It will be interesting to see whether this regulatory circuitry is functionally important in both mouse and human ES cell fate regulation. More recently, the same group that initially identified the Nanog found that transient downregulation of Nanog in ES cells appears to predispose cells toward differentiation, yet is fully reversible. Permanent deletion of Nanog gene in ES cells reduces but does not eliminate clonogenic self-renewal. Nanog-null ES cells can reenter embryo development and contribute extensively to all three germ layers. Interestingly, Nanog-null ES cells can also be recruited to the germline. However, primordial germ cells lacking Nanog fail to mature on reaching the genital ridge. This suggests that Nanog also has a specific role in the formation of germ cells in addition to epiblast. Rather than being part of the integral machinery of pluripotency, it was proposed that Nanog may act primarily in establishing the unique states of epiblast and germ cells.

Sox2 and FoxD3 Sox2 and FoxD3 are the two transcription factors that have been suggested to interact with Oct4 and contribute to pluripotency. Sox2 is a HMG DNA-binding domain containing transcription factor. ICMs from Sox2 knockout embryos cannot give rise to ES cell lines. Instead, they differentiate into both trophectodermal and primitive endodermal cell types. This suggests that Sox2 also play a pivotal role in the establishment of ES cell identity [69]. Sox2 is one of the four transcription factors that were reported to be sufficient to establish pluripotency in the nuclei of fibroblasts when they were forcibly expressed [70]. The other three factors are Oct4, Klf4, and c-Myc. Sox2 seems to be essential in the regulation of several Oct4 target genes at the transcriptional level [71]. However, it has not yet been proved that Sox2 is required in order to enable Oct4 to block trophectodermal differentiation. FoxD3 is a forkhead transcription factor. Its expression is detectable in the blastocyst and later in the postimplantation egg cylinder epiblast. FoxD3 knockout embryos survive until about E6.5 [72], suggesting that Foxd3 is required at a stage beyond establishment and maintenance of a pluripotent ICM.

1.4.4 Negative Regulators of ES Cell Self-Renewal

Socs3 Socs3 is a member of the family of suppressors of cytokine signaling, which act as classical negative regulators to attenuate the signal leading to their induction. In ES cells, Socs3 is an immediate target gene of the LIF/Stat3 pathway. Overexpression of Socs3 has been suggested to have an apoptotic effect on ES cells [73]. When ES cells are transfected with Socs3 transgenes, there is a significant decrease in colony formation. Among the formed colonies, the majority are morphologically differentiated even in the presence of LIF [66]. Therefore induction of Socs3 is considered a negative feedback of the LIF/Stat3 pathway in ES cell self-renewal regulation. However, a more recent study on Socs3-null ES cells suggested otherwise [74]. When cultured in LIF levels that sustain self-renewal of wild-type cells, Forrai and coworkers found that Socs3-null ES cell lines actually exhibited less self-renewal and greater differentiation into primitive endoderm. The absence of Socs3 enhanced JAK-STAT and Erk1/2 signal transduction via gp130 in response to LIF stimulation. Attenuation of ERK signaling by the addition of MAPK/ERK kinase inhibitors to Socs3-null ES cell cultures rescued the differentiation phenotype, but did not restore proliferation to wild-type levels. These data suggest that the level of Socs3 expression might be critical in the regulation of mouse ES cell self-renewal mediated by the LIF/ Stat3 signaling pathway.

Gata Factors Gata4 and Gata6 are zinc finger-containing transcription factors that have been shown to play a pivotal role in the initiation and promotion of differentiation of extraembryonic endoderm. They are expressed in the primitive endoderm and its derivatives, the visceral endoderm and parietal endoderm [75]. Gata4-null mice die between E8 and E9 as a result of defects in heart morphogenesis [76,77]. Gata6 null mice die at E5.5 because of defects in visceral endoderm formation and subsequent extraembryonic development [78]. Gata6 is considered an upstream regulator of Gata4 because loss of Gata6 expression results in the absence of Gata4, whereas loss of Gata4 leads to the upregulation of Gata6 [76,78]. Gata6 and Gata4 mRNAs are detectable in undifferentiated ES cells by RT-PCR, but not by Northern blot analysis, indicating very weak expression of these genes. Gata4 mRNA is increased during ES cell differentiation induced by elevation of Oct4 or withdrawal of LIF [61]. Forced expression of either Gata6 or Gata4 in mouse ES cells causes ES cells to differentiate uniformly into primitive endoderm even in the presence of LIF [79]. Interestingly, ES cells lacking Nanog also tend to differentiate into primitive endoderm [63,64].

1.5 DIFFERENTIATION OF ES CELLS

Embryonic stem cells have the capacity to produce every type of fetal and adult cell both *in vitro* and *in vivo*. In culture, when factors that sustain ES cell self-renewal are removed, ES cells will differentiate and, under appropriate conditions, will generate progeny consisting of derivatives of the three germ layers: mesoderm, endoderm, and ectoderm. In order to use ES cell-derived cells therapeutically, it is critical to know whether they are functional *in vivo*. There is increasing evidence suggesting that these *in vitro*–generated cells can integrate and function when transplanted into adult tissue. For example, cardiomyocytes from differentiating ES cells have been shown to form stable intracardiac grafts when injected into ischemic rat hearts [80]; and glia precursors derived from mouse ES cells interacted with host neurons and efficiently myelinated axons in brain and spinal cord when transplanted into a rat model of acute demyelination [81]; dopaminergic neurons from ES cells can show electrophysiological and behavioral properties expected of neurons from the midbrain and can restore cerebral function and behavior in an animal model of Parkinson's disease [82,83].

Two general approaches are used to initiate ES cell differentiation (Figure 1.3). The primary approach is to allow ES cells to grow in suspension and form three-dimensional aggregates known as *embryoid bodies* (EBs) [84]; another approach is called *monolayer differentiation*. ES cells differentiate readily in monolayer culture when deprived of LIF or feeder support [85,86]. Both approaches have been successful in the generation of certain cell types from ES cells.

Within the EBs, cellular differentiation proceeds on a schedule similar to that in the embryo but in the absence of proper axial organization or elaboration of a body plan [87]. Each EB develops multiple cell types and further differentiation is elaborated on subsequent attachment and outgrowth. It is possible to bias the differentiation for or against certain cell types by addition of different factors such as retinoic acid [88]. However, in the absence of understanding how to instruct ES cells uniformly to enter a lineage of choice, it is still a challenge to direct ES cells into specific pathways and then to support the viability and maturation of individual differentiated phenotypes. As a result, the differentiation products from ES cells by different protocols remain a mixture of cell types. Several strategies have been developed to isolate cells of interest from the mixed cell populations. One cell type from each germ layer will be chosen to elucidate such strategies.

1.5.1 Neural Differentiation (Ectoderm)

Both mouse and human ES cells differentiate efficiently into neural precursor cells on withdrawal of serum in adherent monolayer culture [86,89,90] or via treatment of embryoid bodies with retinoic acid [91–93]. Under appropriate conditions, each of the

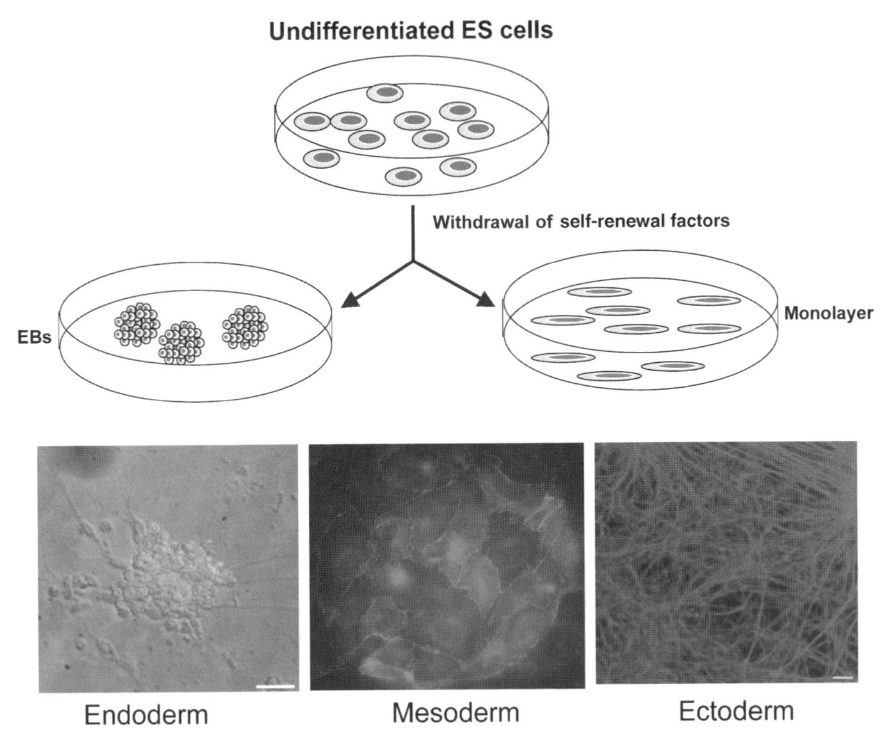

FIGURE 1.3 Differentiation potential of ES cells. ES cells can be induced to differentiate into all somatic cell lineages via the formation of three-dimensional EBs or monolayer culture. (See insert for color representation.)

three major neural cell types of the central nervous system, neurons, astrocytes, and oligodendocytes, as well as subtypes of neurons, can be generated [94,95]. For instance, midbrain dopaminergic neurons were derived at relatively high efficiency from Nurr1 overexpressed ES cells or ES cells grown on stromal cell layers [83,96]. Addition of SHH and FGF8 seemed to dramatically increase the dopaminergic neuron production [83]. Nurr1, SHH, and FGF8 are required for the development of this class of neurons in the early embryo [97,98]. Oligodendrocytes can also been derived and enriched from both mouse and human ES cells and proved to have biological functionality [99,100].

Embryonic stem cell–derived neural cells provide an unlimited source for potential application in cell replacement therapies for neurodegenerative diseases such as Parkinson's disease and Alzheimer's disease. However, before ES cell–derived neural cells can be used clinically, it is important to develop strategies to purify neural cell population of interest, as the resulting cell populations from ES cells are always heterogeneous. Neural stem cells or specific differentiated neural cells can be purified with restricted expression of marker genes, a process termed the *lineage selection strategy* [92,101], or through selective culture conditions and/or clonal expansion [102,103]. Both Sox1 and Sox2 have been used as markers to purify neural precursors from ES cell culture [101]. Sox1 and Sox2 are members of the B group of Sox family transcription factors [104]. Sox2 is expressed in neural precursors as well as in undifferentiated ES cells. In order to enrich Sox2-positive neural precursors and eliminate any residual Sox2-positive ES cells after differentiation, an ES cell line OS25 has been generated in which a bifunctional β*geo* cassette was inserted into the Sox2 locus and an *hphtk* was inserted into the Oct4 locus. The β*geo* in the Sox2 locus will permits enrichment for neural precursors and the *hphtk* in the Oct4 locus will allow to eliminate any residual ES cells by the addition of gancyclovir [101]. Sox1 is one of the best markers for neural stem cells. It is expressed specifically in neuroectoderm at the formation of the neural plate and downregulated at the onset of both neuronal and glial differentiation [105], and importantly, it is not expressed in undifferentiated ES cells. A Sox1-GFP ES cell line was generated by introducing the eGFP into the Sox1 locus, so the Sox1-positive neural precursors derived from ES cells can be purified by FACS sorting. The Sox1-GFP ES cells also harbor a *pac* gene that is coupled to Sox1-GFP via an internal ribosome entry site and is therefore co-expressed with Sox1. After neural differentiation, brief exposure to puromycin results in pure Sox1-positive neural precursors by rapid eliminating on of Sox1-negative cells [86].

Conti's group reported that neural precursors generated from both mouse and human ES cells can be maintained in culture for a prolonged time while retaining the capability to differentiate into all three types of neural cells [102]. Neural precursors are expanded readily in adherent conditions using a combination of EGF and bFGF in serum-free medium. Interestingly, they found that nonneural cell types and differentiated neural cells cannot grow under these conditions and thus are able to generate pure neural stem cell population without any genetic interventions. These cells are remarkably homogeneous and show similarities to radial glia.

1.5.2 Hematopoietic Differentiation (Mesoderm)

Embryonic stem cells can be induced to differentiate into several mesodermal cell lineages, including hematopoietic, cardiac [106], adipogenic [107], osteoblast [108], and myogenic cells [109]. Specific differentiation factors are required for the efficient generation of these cell types from pluripotent ES cells. Of all mesoderm lineages derived from ES cells, hematopoietic differentiation has been studied in most detail.

Under appropriate culture conditions, development of the hematopoietic lineages has been demonstrated to be highly reproducible and efficient [84]. Hematopoietic commitment within these cultures can be monitored by gene expression pattern, the appearance of specific cell surface markers, and the development of clonable progenitors. With these assays, several groups have analyzed in details the early stages of hematopoietic commitment within EBs. Application of serum and cytokines such as IL3, IL1, and granulocyte macrophage colony stimulating factor (GM-CSF) to ES cells generates early hematopoietic precursor cells expressing both embryonic *z* globin (βH1) and adult β major globin RNAs. Different hematopoietic cell types including erythroid, myeloid, and lymphoid lineages and natural killer cells have also been generated from ES cells when cocultured with OP9 cells. In addition to phenotypic characterization of ES cell-derived hematopoietic cells by specific gene expression patterns and by surface antigens, it is of even greater importance to demonstrate that these cells are indeed functional. ES cell–derived hematopoietic progenitors have been shown to have a long-term multilineage hematopoietic repopulating potential [110,111]. There is also strong evidence suggesting that regulation of hematopoietic development in the EBs is similar to that in the early embryo.

Several studies have documented hematopoietic development in human ES cells. By applying a combination of different factors, including BMP4, stem cell factor, and FLT3L, human ES cells have been induced to differentiate into hematopoietic lineages robustly [112–114]. Differentiation of hematopoietic colony-forming cells from human ES cells can also be achieved by coculture with mouse bone marrow stromal cells in the presence of serum [115].

1.5.3 Insulin-Producing Pancreatic Cells (Endoderm)

Efficient and reproducible generation of endoderm derivatives from ES cells, such as pancreatic β cells and hepatocytes, holds great promise for potential clinical treatment of type I diabetes and liver disease. Several groups have claimed to have derived endoderm lineages from ES cells, including pancreatic islets [116,117], hepatocytes [118–120], thyrocytes [121], lung [122], and intestinal cells [123]. Compared to the generation of mesoderm and ectoderm from ES cells, the progress in endoderm cell differentiation from ES cells has been slow. This is in part because there are no specific inducers of endoderm lineage and there are no good markers to identify or separate definitive endoderm from other lineages at different differentiation stages. Most of the markers currently used for definitive endoderm have an overlapping expression patterns with visceral endoderm, an extraembryonic tissue, and do not contribute to the formation of any tissue of embryo or adult.

Lumelsky and colleagues [124] developed a five-step protocol to induce ES cells differentiation into pancreatic islet-like cells. These cells expressed insulin and other pancreatic endocrine hormones. They self-assembled to form three-dimensional clusters similar in topology to normal pancreatic islets where pancreatic cells are in close association with neurons. Glucose triggers insulin release from these cell clusters by mechanisms similar to those employed *in vivo*. When injected into diabetic mice, the ES cell–derived insulin-producing cells undergo rapid vascularization and maintain a clustered, islet-like organization. However, when tested for their ability to function *in vivo* following transplantation into streptozotocin (STZ)-induced diabetic mice, these cells failed to correct the hyperglycemia of these animals [124]. This could be due to the fact that the cells were too immature or that they were not true islet cells. Subsequently several groups have modified this protocol and reported that the islet-like cells they derived from ES cells were functional and could improve the hyperglycemia in STZ-induced diabetic mice. These were achieved by forced expression of Pax4 [125] or pancreatic duodenal homeobox 1 (Pdx1) [126] during ES cell differentiation, or by treatment of cells with the inhibitor of phosphoinositide 3-kinase (PI3K), LY294002 [127] at the final stage. Pax4 and Pdx1 are transcription factors that play an essential role during β-cell development [128].

Kahan's group [129] developed another approach to derive islet-like structure from ES cells. They found that, following growth and differentiation in nonselective medium containing serum, mouse ES cells spontaneously differentiated into cells individually expressing each of the four major islet hormones: insulin, glucagon, somatostatin, and pancreatic polypeptide [129]. This system allows the investigation of many facets of islet development since it promotes the appearance of the complete range of islet phenotypes and reproduces important developmental stages of normal islet cytodifferentiation in differentiating ES cell cultures. However, the efficiency of generation of islet-like cells from ES cells is very low. Micallef and coworkers [130] developed a strategy to select for early pancreatic cell population from the mixed ES cell differentiation culture. They generated an embryonic stem (ES) cell line in which sequences encoding GFP were targeted to the locus of Pdx1 [130]. Pdx1 is expressed in the earliest stages of pancreatic development as the organ rudiments are specified from the gut endoderm [128].

Insulin-producing β cells have also been generated from human ES cells [131,132]. The approach used to induce human ES cell differentiation into islet-like structures is similar to that of mouse ES cells [124]. First human ES cells were grown in suspension to allow the formation EBs. The EBs were then cultured and plated in an insulin–transferring–selenium–fibronectin medium, followed by medium supplemented with N2, B27, and bFGF. Next, the glucose concentration in the medium was lowered, bFGF was withdrawn, and nicotinamide was added. Reverse transcription–polymerase chain reaction detected an enhanced expression of pancreatic genes in the differentiated cells. Immunofluorescence and *in situ* hybridization analyses revealed a high percentage of insulin-expressing cells in the clusters. In addition to insulin, most cells also coexpressed glucagon or somatostatin, indicating a similarity to immature pancreatic cells. These findings validate the human ES cell model system as a potential basis for enrichment of human β cells or their precursors, as a possible future source for cell replacement therapy in diabetes.

1.6 CONCLUSIONS

Embryonic stem cells have the ability to form all three embryonic germ layers and their differentiated derivatives *in vitro* and *in vivo*. They can be manipulated by controlling their growth conditions or by introducing genetic modifications, allowing the controlled direction of their differentiation into specific cell types. Because of these qualities, there are many ways in which ES cells might be used in basic and clinical research. The most obvious potential application of human ES cells would be the generation of cells and tissues for cell-based therapies. Although significant progress has been made toward an understanding of ES cell biology, the mechanisms by which the ES cell fate is regulated are still largely unknown. In order to fully realize the potential of ES cells, several fundamental questions need to be addressed. First, while some of the pathways controlling mouse ES cell self-renewal have been identified, none of these appear to function equivalently in both humans and mice. This raises the question as to whether the fundamental mechanisms underlying ES cell self-renewal are shared or distinct among different species. We expect that cross-species comparisons using large-scale genomic analysis will provide a better understanding of the conserved and divergent pathways required for ES cell self-renewal. While numerous studies have demonstrated that ES cells have the potential to differentiate into nearly all the specialized cell types in the body, little is known about the routes by which specialized cell types arise from ES cells. Among other methods, gene expression profiling using whole-genome microarrays and epigenomics for the detection of "silenced" region of chromosomes known to be involved in cell fate decision, are likely to provide important insights into the basic mechanisms controlling ES cell differentiation.

REFERENCES

[1] Brook FA, Gardner RL: The origin and efficient derivation of embryonic stem cells in the mouse, *Proc Natl Acad Sci USA* **94**:5709–5712 (1997).

[2] Evans MJ, Kaufman MH: Establishment in culture of pluripotential cells from mouse embryos, *Nature* **292**:154–156 (1981).

[3] Martin GR: Isolation of a pluripotent cell line from early mouse embryos cultured in medium conditioned by teratocarcinoma stem cells, *Proc Natl Acad Sci USA* **78**:7634–7638 (1981).

[4] Thomson JA, Itskovitz-Eldor J, Shapiro SS, Waknitz MA, Swiergiel JJ, Marshall VS, Jones JM: Embryonic stem cell lines derived from human blastocysts, *Science* **282**:1145–1147 (1998).

[5] Delhaise F, Bralion V, Schuurbiers N, Dessy F: Establishment of an embryonic stem cell line from 8-cell stage mouse embryos, *Eur J Morphol* **34**:237–243 (1996).

[6] Tesar PJ: Derivation of germ-line-competent embryonic stem cell lines from preblastocyst mouse embryos, *Proc Natl Acad Sci USA* **102**:8239–8244 (2005).

[7] Martin GR, Evans MJ: Differentiation of clonal lines of teratocarcinoma cells: Formation of embryoid bodies in vitro, *Proc Natl Acad Sci USA* **72**:1441–1445 (1975).

[8] Martin GR, Wiley LM, Damjanov I: The development of cystic embryoid bodies in vitro from clonal teratocarcinoma stem cells, *Dev Biol* **61**:230–244 (1977).

[9] Solter D, Skreb N, Damjanov I: Extrauterine growth of mouse egg-cylinders results in malignant teratoma, *Nature* **227**:503–504 (1970).

[10] Stevens LC: The development of transplantable teratocarcinomas from intratesticular grafts of pre- and postimplantation mouse embryos, *Dev Biol* **21**:364–382 (1970).

[11] Mintz B, Illmensee K: Normal genetically mosaic mice produced from malignant teratocarcinoma cells, *Proc Natl Acad Sci USA* **72**:3585–3589 (1975).

[12] Martin GR, Evans MJ: The morphology and growth of a pluripotent teratocarcinoma cell line and its derivatives in tissue culture, *Cell* **2**:163–172 (1974).

[13] Kawase E, Suemori H, Takahashi N, Okazaki K, Hashimoto K, Nakatsuji N: Strain difference in establishment of mouse embryonic stem (ES) cell lines, *Int J Dev Biol* **38**:385–390 (1994).

[14] Robertson EJ, ed: *Teratocarcinoma and Embryo-Derived Stem Cells: A Practical Approach*, IRL Press, Oxford, 1987.

[15] Burdon T, Stracey C, Chambers I, Nichols J, Smith A: Suppression of SHP-2 and ERK signalling promotes self-renewal of mouse embryonic stem cells, *Dev Biol* **210**:30–43 (1999).

[16] Nichols J, Ying QL: Derivation and propagation of embryonic stem cells in serum- and feeder-free culture, *Meth Mol Biol* **329**:91–98 (2006).

[17] Ying QL, Nichols J, Chambers I, Smith A: BMP induction of Id proteins suppresses differentiation and sustains embryonic stem cell self-renewal in collaboration with STAT3, *Cell* **115**:281–292 (2003).

[18] Beddington RS, Robertson EJ: An assessment of the developmental potential of embryonic stem cells in the midgestation mouse embryo, *Development* **105**:733–737 (1989).

[19] Bradley A, Evans M, Kaufman MH, Robertson E: Formation of germ-line chimaeras from embryo-derived teratocarcinoma cell lines, *Nature* **309**:255–256 (1984).

[20] Nagy A, Rossant J, Nagy R, Abramow-Newerly W, Roder JC: Derivation of completely cell culture-derived mice from early-passage embryonic stem cells, *Proc Natl Acad Sci USA* **90**:8424–8428 (1993).

[21] Gorba T, Allsopp TE: Pharmacological potential of embryonic stem cells, *Pharmacol Res* **47**:269–278 (2003).

[22] Smith A: Cell therapy: In search of pluripotency, *Curr Biol* **8**:R802–R804 (1998).

[23] Solter D, Gearhart J: Putting stem cells to work, *Science* **283**:1468–1470 (1999).

[24] Smith AG, Heath JK, Donaldson DD, Wong GG, Moreau J, Stahl M, Rogers D: Inhibition of pluripotential embryonic stem cell differentiation by purified polypeptides, *Nature* **336**:688–690 (1988).

[25] Williams RL, Hilton DJ, Pease S, Willson TA, Stewart CL, Gearing DP, Wagner EF, Metcalf D, Nicola NA, Gough NM: Myeloid leukaemia inhibitory factor maintains the developmental potential of embryonic stem cells, *Nature* **336**:684–687 (1988).

[26] Stewart CL, Kaspar P, Brunet LJ, Bhatt H, Gadi I, Kontgen F, Abbondanzo SJ: Blastocyst implantation depends on maternal expression of leukaemia inhibitory factor, *Nature* **359**:76–79 (1992).

[27] Gearing DP, Gough NM, King JA, Hilton DJ, Nicola NA, Simpson RJ, Nice EC, Kelso A, Metcalf D: Molecular cloning and expression of cDNA encoding a murine myeloid leukaemia inhibitory factor (LIF), *Embo J* **6**:3995–4002 (1987).

[28] Tomida M, Yamamoto-Yamaguchi Y, Hozumi M: Purification of a factor inducing differentiation of mouse myeloid leukemic M1 cells from conditioned medium of mouse fibroblast L929 cells, *J Biol Chem* **259**:10978–10982 (1984).

[29] Yoshida K, Chambers I, Nichols J, Smith A, Saito M, Yasukawa K, Shoyab M, Taga T, Kishimoto T: Maintenance of the pluripotential phenotype of embryonic stem cells through direct activation of gp130 signalling pathways, *Mech Dev* **45**:163–171 (1994).

[30] Conover JC, Ip NY, Poueymirou WT, Bates B, Goldfarb MP, DeChiara TM, Yancopoulos GD: Ciliary neurotrophic factor maintains the pluripotentiality of embryonic stem cells, *Development* **119**:559–565 (1993).

[31] Pennica D, Shaw KJ, Swanson TA, Moore MW, Shelton DL, Zioncheck KA, Rosenthal A, Taga T, Paoni NF, Wood WI: Cardiotrophin-1. Biological activities and binding to the leukemia inhibitory factor receptor/gp130 signaling complex, *J Biol Chem* **270**:10915–10922 (1995).

[32] Taga T, Kishimoto T: Gp130 and the interleukin-6 family of cytokines, *Annu Rev Immunol* **15**:797–819 (1997).

[33] Niwa H, Burdon T, Chambers I, Smith A: Self-renewal of pluripotent embryonic stem cells is mediated via activation of STAT3, *Genes Dev* **12**:2048–2060 (1998).

[34] Matsuda T, Nakamura T, Nakao K, Arai T, Katsuki M, Heike T, Yokota T: STAT3 activation is sufficient to maintain an undifferentiated state of mouse embryonic stem cells, *Embo J* **18**:4261–4269 (1999).

[35] Ko SY, Kang HY, Lee HS, Han SY, Hong SH: Identification of Jmjd1a as a STAT3 downstream gene in mES cells, *Cell Struct Funct* **31**:53–62 (2006).

[36] Pritsker M, Ford NR, Jenq HT, Lemischka IR: Genomewide gain-of-function genetic screen identifies functionally active genes in mouse embryonic stem cells, *Proc Natl Acad Sci USA* **103**:6946–6951 (2006).

[37] Sekkai D, Gruel G, Herry M, Moucadel V, Constantinescu SN, Albagli O, Tronik-Le Roux D, Vainchenker W, Bennaceur-Griscelli A: Microarray analysis of LIF/Stat3 transcriptional targets in embryonic stem cells, *Stem Cells* **23**:1634–1642 (2005).

[38] Wozney JM, Rosen V, Celeste AJ, Mitsock LM, Whitters MJ, Kriz RW, Hewick RM, Wang EA: Novel regulators of bone formation: Molecular clones and activities, *Science* **242**:1528–1534 (1988).

[39] Benezra R: Role of Id proteins in embryonic and tumor angiogenesis, *Trends Cardiovasc Med* **11**:237–241 (2001).

[40] Norton JD: ID helix-loop-helix proteins in cell growth, differentiation and tumorigenesis, *J Cell Sci* **113(Pt 22)**:3897–3905 (2000).

[41] Kodaira K, Imada M, Goto M, Tomoyasu A, Fukuda T, Kamijo R, Suda T, Higashio K, Katagiri T: Purification and identification of a BMP-like factor from bovine serum, *Biochem Biophys Res Commun* **345**:1224–1231 (2006).

[42] Ogawa K, Saito A, Matsui H, Suzuki H, Ohtsuka S, Shimosato D, Morishita Y, Watabe T, Niwa H, Miyazono K: Activin-Nodal signaling is involved in propagation of mouse embryonic stem cells, *J Cell Sci* **120**:55–65 (2007).

[43] Xu RH, Peck RM, Li DS, Feng X, Ludwig T, Thomson JA: Basic FGF and suppression of BMP signaling sustain undifferentiated proliferation of human ES cells, *Nat Meth* **2**:185–190 (2005).

[44] Ying QL, Nichols J, Wray J, Alessi D, Cohen P, Smith A: The ground state of pluripotency: Nature (in press).

[45] Chen S, Do JT, Zhang Q, Yao S, Yan F, Peters EC, Scholer HR, Schultz PG, Ding S: Self-renewal of embryonic stem cells by a small molecule, *Proc Natl Acad Sci USA* **103**:17266–17271 (2006).

[46] Daheron L, Opitz SL, Zaehres H, Lensch WM, Andrews PW, Itskovitz-Eldor J, Daley GQ: LIF/STAT3 signaling fails to maintain self-renewal of human embryonic stem cells, *Stem Cells* **22**:770–778 (2004).

[47] Humphrey RK, Beattie GM, Lopez AD, Bucay N, King CC, Firpo MT, Rose-John S, Hayek A: Maintenance of pluripotency in human embryonic stem cells is STAT3 independent, *Stem Cells* **22**:522–530 (2004).

[48] Xu RH, Chen X, Li DS, Li R, Addicks GC, Glennon C, Zwaka TP, Thomson JA: BMP4 initiates human embryonic stem cell differentiation to trophoblast, *Nat Biotechnol* **20**:1261–1264 (2002).

[49] Beattie GM, Lopez AD, Bucay N, Hinton A, Firpo MT, King CC, Hayek A: Activin A maintains pluripotency of human embryonic stem cells in the absence of feeder layers, *Stem Cells* **23**:489–495 (2005).

[50] James D, Levine AJ, Besser D, Hemmati-Brivanlou A: TGFbeta/activin/nodal signaling is necessary for the maintenance of pluripotency in human embryonic stem cells, *Development* **132**:1273–1282 (2005).

[51] Pyle AD, Lock LF, Donovan PJ: Neurotrophins mediate human embryonic stem cell survival, *Nat Biotechnol* **24**:344–350 (2006).

[52] Sato N, Meijer L, Skaltsounis L, Greengard P, Brivanlou AH: Maintenance of pluripotency in human and mouse embryonic stem cells through activation of Wnt signaling by a pharmacological GSK-3-specific inhibitor, *Nat Med* **10**:55–63 (2004).

[53] Vallier L, Reynolds D, Pedersen RA: Nodal inhibits differentiation of human embryonic stem cells along the neuroectodermal default pathway, *Dev Biol* **275**:403–421 (2004).

[54] Levenstein ME, Ludwig TE, Xu RH, Llanas RA, VanDenHeuvel-Kramer K, Manning D, Thomson JA: Basic fibroblast growth factor support of human embryonic stem cell self-renewal, *Stem Cells* **24**:568–574 (2006).

[55] Kameda T, Thomson JA: Human ERas gene has an upstream premature polyadenylation signal that results in a truncated, noncoding transcript, *Stem Cells* **23**:1535–1540 (2005).

[56] Takahashi K, Mitsui K, Yamanaka S: Role of ERas in promoting tumour-like properties in mouse embryonic stem cells, *Nature* **423**:541–545 (2003).

[57] Pesce M, Gross MK, Scholer HR: In line with our ancestors: Oct-4 and the mammalian germ, *Bioessays* **20**:722–732 (1998).

[58] Nichols J, Zevnik B, Anastassiadis K, Niwa H, Klewe-Nebenius D, Chambers I, Scholer H, Smith A: Formation of pluripotent stem cells in the mammalian embryo depends on the POU transcription factor Oct4, *Cell* **95**:379–391 (1998).

[59] Hay DC, Sutherland L, Clark J, Burdon T: Oct-4 knockdown induces similar patterns of endoderm and trophoblast differentiation markers in human and mouse embryonic stem cells, *Stem Cells* **22**:225–235 (2004).

[60] Matin MM, Walsh JR, Gokhale PJ, Draper JS, Bahrami AR, Morton I, Moore HD, Andrews PW: Specific knockdown of Oct4 and beta2-microglobulin expression by RNA interference in human embryonic stem cells and embryonic carcinoma cells, *Stem Cells* **22**:659–668 (2004).

[61] Niwa H, Miyazaki J, Smith AG: Quantitative expression of Oct-3/4 defines differentiation, dedifferentiation or self-renewal of ES cells, *Nat Genet* **24**:372–376 (2000).

[62] Niwa H, Toyooka Y, Shimosato D, Strumpf D, Takahashi K, Yagi R, Rossant J: Interaction between Oct3/4 and Cdx2 determines trophectoderm differentiation, *Cell* **123**:917–929 (2005).

[63] Chambers I, Colby D, Robertson M, Nichols J, Lee S, Tweedie S, Smith A: Functional expression cloning of Nanog, a pluripotency sustaining factor in embryonic stem cells, *Cell* **113**:643–655 (2003).

[64] Mitsui K, Tokuzawa Y, Itoh H, Segawa K, Murakami M, Takahashi K, Maruyama M, Maeda M, Yamanaka S: The homeoprotein Nanog is required for maintenance of pluripotency in mouse epiblast and ES cells, *Cell* **113**:631–642 (2003).

[65] Hyslop L, Stojkovic M, Armstrong L, Walter T, Stojkovic P, Przyborski S, Herbert M, Murdoch A, Strachan T, Lako M: Downregulation of NANOG induces differentiation of human embryonic stem cells to extraembryonic lineages, *Stem Cells* **23**:1035–1043 (2005).

[66] Chambers I, Smith A: Self-renewal of teratocarcinoma and embryonic stem cells, *Oncogene* **23**:7150–7160 (2004).

[67] Darr H, Mayshar Y, Benvenisty N: Overexpression of NANOG in human ES cells enables feeder-free growth while inducing primitive ectoderm features, *Development* **133**:1193–1201 (2006).

[68] Yasuda SY, Tsuneyoshi N, Sumi T, Hasegawa K, Tada T, Nakatsuji N, Suemori H: NANOG maintains self-renewal of primate ES cells in the absence of a feeder layer, *Genes Cells* **11**:1115–1123 (2006).

[69] Avilion AA, Nicolis SK, Pevny LH, Perez L, Vivian N, Lovell-Badge R: Multipotent cell lineages in early mouse development depend on SOX2 function, *Genes Dev* **17**:126–140 (2003).

[70] Takahashi K, Yamanaka S: Induction of pluripotent stem cells from mouse embryonic and adult fibroblast cultures by defined factors, *Cell* **126**:663–676 (2006).

[71] Niwa H: Molecular mechanism to maintain stem cell renewal of ES cells, *Cell Struct Funct* **26**:137–148 (2001).

[72] Hanna LA, Foreman RK, Tarasenko IA, Kessler DS, Labosky PA: Requirement for Foxd3 in maintaining pluripotent cells of the early mouse embryo, *Genes Dev* **16**:2650–2661 (2002).

[73] Duval D, Reinhardt B, Kedinger C, Boeuf H: Role of suppressors of cytokine signaling (Socs) in leukemia inhibitory factor (LIF) -dependent embryonic stem cell survival, *Faseb J* **14**:1577–1584 (2000).

[74] Forrai A, Boyle K, Hart AH, Hartley L, Rakar S, Willson TA, Simpson KM, Roberts AW, Alexander WS, Voss AK et al: Absence of suppressor of cytokine signalling 3 reduces self-renewal and promotes differentiation in murine embryonic stem cells, *Stem Cells* **24**:604–614 (2006).

[75] Morrisey EE, Ip HS, Lu MM, Parmacek MS: GATA-6: A zinc finger transcription factor that is expressed in multiple cell lineages derived from lateral mesoderm, *Dev Biol* **177**:309–322 (1996).

[76] Kuo CT, Morrisey EE, Anandappa R, Sigrist K, Lu MM, Parmacek MS, Soudais C, Leiden JM: GATA4 transcription factor is required for ventral morphogenesis and heart tube formation, *Genes Dev* **11**:1048–1060 (1997).

[77] Molkentin JD, Lin Q, Duncan SA, Olson EN: Requirement of the transcription factor GATA4 for heart tube formation and ventral morphogenesis, *Genes Dev* **11**:1061–1072 (1997).

[78] Morrisey EE, Tang Z, Sigrist K, Lu MM, Jiang F, Ip HS, Parmacek MS: GATA6 regulates HNF4 and is required for differentiation of visceral endoderm in the mouse embryo, *Genes Dev* **12**:3579–3590 (1998).

[79] Fujikura J, Yamato E, Yonemura S, Hosoda K, Masui S, Nakao K, Miyazaki Ji J, Niwa H: Differentiation of embryonic stem cells is induced by GATA factors, *Genes Dev* **16**:784–789 (2002).

[80] Klug MG, Soonpaa MH, Koh GY, Field LJ: Genetically selected cardiomyocytes from differentiating embronic stem cells form stable intracardiac grafts, *J Clin Invest* **98**:216–224 (1996).

[81] Brustle O, Jones KN, Learish RD, Karram K, Choudhary K, Wiestler OD, Duncan ID, McKay RD: Embryonic stem cell-derived glial precursors: A source of myelinating transplants, *Science* **285**:754–756 (1999).

[82] Bjorklund LM, Sanchez-Pernaute R, Chung S, Andersson T, Chen IY, McNaught KS, Brownell AL, Jenkins BG, Wahlestedt C, Kim KS et al: Embryonic stem cells develop into functional dopaminergic neurons after transplantation in a Parkinson rat model, *Proc Natl Acad Sci USA* **99**:2344–2349 (2002).

[83] Kim JH, Auerbach JM, Rodriguez-Gomez JA, Velasco I, Gavin D, Lumelsky N, Lee SH, Nguyen J, Sanchez-Pernaute R, Bankiewicz K et al: Dopamine neurons derived from embryonic stem cells function in an animal model of Parkinson's disease, *Nature* **418**:50–56 (2002).

[84] Keller GM: In vitro differentiation of embryonic stem cells, *Curr Opin Cell Biol* **7**:862–869 (1995).

[85] Nishikawa SI, Nishikawa S, Hirashima M, Matsuyoshi N, Kodama H: Progressive lineage analysis by cell sorting and culture identifies FLK1+VE-cadherin+ cells at a diverging point of endothelial and hemopoietic lineages, *Development* **125**:1747–1757 (1998).

[86] Ying QL, Stavridis M, Griffiths D, Li M, Smith A: Conversion of embryonic stem cells into neuroectodermal precursors in adherent monoculture, *Nat Biotechnol* **21**:183–186 (2003).

[87] Doetschman TC, Eistetter H, Katz M, Schmidt W, Kemler R: The in vitro development of blastocyst-derived embryonic stem cell lines: Formation of visceral yolk sac, blood islands and myocardium, *J Embryol Exp Morphol* **87**:27–45 (1985).

[88] Rohwedel J, Guan K, Wobus AM: Induction of cellular differentiation by retinoic acid *in vitro*, *Cells Tissues Organs* **165**:190–202 (1999).

[89] Gerrard L, Rodgers L, Cui W: Differentiation of human embryonic stem cells to neural lineages in adherent culture by blocking bone morphogenetic protein signaling, *Stem Cells* **23**:1234–1241 (2005).

[90] Tropepe V, Hitoshi S, Sirard C, Mak TW, Rossant J, van der Kooy D: Direct neural fate specification from embryonic stem cells: A primitive mammalian neural stem cell stage acquired through a default mechanism, *Neuron* **30**:65–78 (2001).

[91] Bain G, Kitchens D, Yao M, Huettner JE, Gottlieb DI: Embryonic stem cells express neuronal properties in vitro, *Dev Biol* **168**:342–357 (1995).

[92] Li M, Pevny L, Lovell-Badge R, Smith A: Generation of purified neural precursors from embryonic stem cells by lineage selection, *Curr Biol* **8**:971–974 (1998).

[93] Schuldiner M, Eiges R, Eden A, Yanuka O, Itskovitz-Eldor J, Goldstein RS, Benvenisty N: Induced neuronal differentiation of human embryonic stem cells, *Brain Res* **913**:201–205 (2001).

[94] Barberi T, Klivenyi P, Calingasan NY, Lee H, Kawamata H, Loonam K, Perrier AL, Bruses J, Rubio ME, Topf N et al: Neural subtype specification of fertilization and nuclear transfer embryonic stem cells and application in parkinsonian mice, *Nat Biotechnol* **21**:1200–1207 (2003).

[95] Okabe S, Forsberg-Nilsson K, Spiro AC, Segal M, McKay RD: Development of neuronal precursor cells and functional postmitotic neurons from embryonic stem cells in vitro, *Mech Dev* **59**:89–102 (1996).

[96] Kawasaki H, Suemori H, Mizuseki K, Watanabe K, Urano F, Ichinose H, Haruta M, Takahashi M, Yoshikawa K, Nishikawa S et al: Generation of dopaminergic neurons and pigmented epithelia from primate ES cells by stromal cell-derived inducing activity, *Proc Natl Acad Sci USA* **99**:1580–1585 (2002).

[97] Simon HH, Bhatt L, Gherbassi D, Sgado P, Alberi L: Midbrain dopaminergic neurons: Determination of their developmental fate by transcription factors, *Ann NY Acad Sci* **991**:36–47 (2003).

[98] Ye W, Shimamura K, Rubenstein JL, Hynes MA, Rosenthal A: FGF and Shh signals control dopaminergic and serotonergic cell fate in the anterior neural plate, *Cell* **93**:755–766 (1998).

[99] Billon N, Jolicoeur C, Ying QL, Smith A, Raff M: Normal timing of oligodendrocyte development from genetically engineered, lineage-selectable mouse ES cells, *J Cell Sci* **115**:3657–3665 (2002).

[100] Kang SM, Cho MS, Seo H, Yoon CJ, Oh SK, Choi YM, Kim DW: Efficient induction of oligodendrocytes from human embryonic stem cells, *Stem Cells* **25**:419–424 (2007).

[101] Ying QL, Smith AG: Defined conditions for neural commitment and differentiation, *Methods Enzymol* **365**:327–341 (2003).

[102] Conti L, Pollard SM, Gorba T, Reitano E, Toselli M, Biella G, Sun Y, Sanzone S, Ying QL, Cattaneo E et al: Niche-independent symmetrical self-renewal of a mammalian tissue stem cell, *PLoS Biol* **3**:e283(2005).

[103] Lee SH, Lumelsky N, Studer L, Auerbach JM, McKay RD: Efficient generation of midbrain and hindbrain neurons from mouse embryonic stem cells, *Nat Biotechnol* **18**:675–679 (2000).

[104] Pevny LH, Lovell-Badge R: Sox genes find their feet, *Curr Opin Genet Dev* **7**:338–344 (1997).

[105] Wood HB, Episkopou V: Comparative expression of the mouse Sox1, Sox2 and Sox3 genes from pre-gastrulation to early somite stages, *Mech Dev* **86**:197–201 (1999).

[106] Boheler KR, Czyz J, Tweedie D, Yang HT, Anisimov SV, Wobus AM: Differentiation of pluripotent embryonic stem cells into cardiomyocytes, *Circ Res* **91**:189–201 (2002).

[107] Dani C, Smith AG, Dessolin S, Leroy P, Staccini L, Villageois P, Darimont C, Ailhaud G: Differentiation of embryonic stem cells into adipocytes in vitro, *J Cell Sci* **110 (Pt11)**:1279–1285 (1997).

[108] Bourne S, Polak JM, Hughes SP, Buttery LD: Osteogenic differentiation of mouse embryonic stem cells: Differential gene expression analysis by cDNA microarray and purification of osteoblasts by cadherin-11 magnetically activated cell sorting, *Tissue Eng* **10**:796–806 (2004).

[109] Rohwedel J, Maltsev V, Bober E, Arnold HH, Hescheler J, Wobus AM: Muscle cell differentiation of embryonic stem cells reflects myogenesis in vivo: Developmentally regulated expression of myogenic determination genes and functional expression of ionic currents, *Dev Biol* **164**:87–101 (1994).

[110] Hole N, Graham GJ, Menzel U, Ansell JD: A limited temporal window for the derivation of multilineage repopulating hematopoietic progenitors during embryonal stem cell differentiation in vitro, *Blood* **88**:1266–1276 (1996).

[111] Palacios R, Golunski E, Samaridis J: In vitro generation of hematopoietic stem cells from an embryonic stem cell line, *Proc Natl Acad Sci USA* **92**:7530–7534 (1995).

[112] Chadwick K, Wang L, Li L, Menendez P, Murdoch B, Rouleau A, Bhatia M: Cytokines and BMP-4 promote hematopoietic differentiation of human embryonic stem cells, *Blood* **102**:906–915 (2003).

[113] Menendez P, Wang L, Chadwick K, Li L, Bhatia M: Retroviral transduction of hematopoietic cells differentiated from human embryonic stem cell-derived CD45 (neg)PFV hemogenic precursors, *Mol Ther* **10**:1109–1120 (2004).

[114] Wang L, Li L, Menendez P, Cerdan C, Bhatia M: Human embryonic stem cells maintained in the absence of mouse embryonic fibroblasts or conditioned media are capable of hematopoietic development, *Blood* **105**:4598–4603 (2005).

[115] Kaufman DS, Hanson ET, Lewis RL, Auerbach R, Thomson JA: Hematopoietic colony-forming cells derived from human embryonic stem cells, *Proc Natl Acad Sci USA* **98**:10716–10721 (2001).

[116] Colman A: Making new beta cells from stem cells, *Semin Cell Dev Biol* **15**:337–345 (2004).

[117] Stoffel M, Vallier L, Pedersen RA: Navigating the pathway from embryonic stem cells to beta cells, *Semin Cell Dev Biol* **15**:327–336 (2004).

[118] Hamazaki T, Iiboshi Y, Oka M, Papst PJ, Meacham AM, Zon LI, Terada N: Hepatic maturation in differentiating embryonic stem cells in vitro, *FEBS Lett* **497**:15–19 (2001).

[119] Jones EA, Tosh D, Wilson DI, Lindsay S, Forrester LM: Hepatic differentiation of murine embryonic stem cells, *Exp Cell Res* **272**:15–22 (2002).

[120] Yamada T, Yoshikawa M, Kanda S, Kato Y, Nakajima Y, Ishizaka S, Tsunoda Y: In vitro differentiation of embryonic stem cells into hepatocyte-like cells identified by cellular uptake of indocyanine green, *Stem Cells* **20**:146–154 (2002).

[121] Lin RY, Kubo A, Keller GM, Davies TF: Committing embryonic stem cells to differentiate into thyrocyte-like cells in vitro, *Endocrinology* **144**:2644–2649 (2003).

[122] Ali NN, Edgar AJ, Samadikuchaksaraei A, Timson CM, Romanska HM, Polak JM, Bishop AE: Derivation of type II alveolar epithelial cells from murine embryonic stem cells, *Tissue Eng* **8**:541–550 (2002).

[123] Yamada T, Yoshikawa M, Takaki M, Torihashi S, Kato Y, Nakajima Y, Ishizaka S, Tsunoda Y: In vitro functional gut-like organ formation from mouse embryonic stem cells, *Stem Cells* **20**:41–49 (2002).

[124] Lumelsky N, Blondel O, Laeng P, Velasco I, Ravin R, McKay R: Differentiation of embryonic stem cells to insulin-secreting structures similar to pancreatic islets, *Science* **292**:1389–1394 (2001).

[125] Blyszczuk P, Czyz J, Kania G, Wagner M, Roll U, St-Onge L, Wobus AM: Expression of Pax4 in embryonic stem cells promotes differentiation of nestin-positive progenitor and insulin-producing cells, *Proc Natl Acad Sci USA* **100**:998–1003 (2003).

[126] Miyazaki S, Yamato E, Miyazaki J: Regulated expression of pdx-1 promotes in vitro differentiation of insulin-producing cells from embryonic stem cells, *Diabetes* **53**:1030–1037 (2004).

[127] Hori Y, Rulifson IC, Tsai BC, Heit JJ, Cahoy JD, Kim SK: Growth inhibitors promote differentiation of insulin-producing tissue from embryonic stem cells, *Proc Natl Acad Sci USA* **99**:16105–16110 (2002).

[128] Murtaugh LC, Melton DA: Genes, signals, and lineages in pancreas development, *Annu Rev Cell Dev Biol* **19**:71–89 (2003).

[129] Kahan BW, Jacobson LM, Hullett DA, Ochoada JM, Oberley TD, Lang KM, Odorico JS: Pancreatic precursors and differentiated islet cell types from murine embryonic stem cells: An in vitro model to study islet differentiation, *Diabetes* **52**:2016–2024 (2003).

[130] Micallef SJ, Janes ME, Knezevic K, Davis RP, Elefanty AG, Stanley EG: Retinoic acid induces Pdx1-positive endoderm in differentiating mouse embryonic stem cells, *Diabetes* **54**:301–305 (2005).

[131] Assady S, Maor G, Amit M, Itskovitz-Eldor J, Skorecki KL, Tzukerman M: Insulin production by human embryonic stem cells, *Diabetes* **50**:1691–1697 (2001).

[132] Segev H, Fishman B, Ziskind A, Shulman M, Itskovitz-Eldor J: Differentiation of human embryonic stem cells into insulin-producing clusters, *Stem Cells* **22**:265–274 (2004).

2

ADULT STEM CELLS

LIEF FENNO AND CHAD A. COWAN
Stowers Medical Institute, Center for Regenerative Medicine and Technology, Cardiovascular Research Center, Boston, Massachusetts

Physicians of the early twentieth century noted leukemia as a rare malady common to radiologists and those involved with radioactive research [1]. The mechanism by which radiation causes the hematopoietic system to become cancerous and produce so many leukocytes as to have a disease termed "white blood" in Greek, would not be a central focus of scientific inquiry until the early hours of August 6, 1945 [2]. The destruction of Hiroshima, and days later Nagasaki, marked the first use of atomic bombs in warfare and subsequently the first exposure of a large group of humans to ionizing radiation. Many of those who did not die from the initial blast later died from hematopoietic disease because of the presence of either too many white blood cells [1] (leukemia) or too few white blood cells [2] (infection). The discovery soon after that lead shielding of the spleen during irradiation [3] or transplantation of bone marrow into irradiated mice [4] allows survival of what would be otherwise lethal doses of radiation, spurred James Till and Ernest McCulloch to further experiment with bone marrow transplantation and irradiation, eventually leading them to discover that a single cell was able to repopulate those blood cells killed by radiation damage. This cell is known today as the *hematopoietic stem cell*, and its discovery marked the founding of the field of stem cell biology [5].

Today, science has been able to elucidate the maintenance of tissues through the characterization of multiple types of adult stem cells, including surprises such as the neural stem cell and also show that a stem cell population is not responsible for the maintenance of some tissues, such as the insulin-producing-β-cell of the pancreas [6]. This chapter introduces the adult stem cell, differences between adult stem cells and

Chemical and Functional Genomic Approaches to Stem Cell Biology and Regenerative Medicine
Edited by Sheng Ding

embryonic stem cells, and types of adult stem cells, as well as their maintenance, discovery, disregulation, and therapeutic applications.

2.1 ADULT STEM CELLS

The discovery by Till and McCulloch that bone marrow was able to rescue a lethally irradiated mouse was not in itself groundbreaking, but rather that they observed what others had not. Through an ingenious cell labeling technique, the two demonstrated that one cell was able to divide into many different cell lineages, in this case, cell types of the hematopoietic system [7]. Additionally, by removing colonies formed of injected cells that had grown in the spleen of rescued mice and injecting them into a second irradiated mouse, a technique known as *serial transplantation*, they showed that adult stem cells are able to self-renew [8], laying the foundation for the three fundamental properties of adult stem cells: an ability to differentiate into multiple cell types, an ability to self-renew, and a persistence for the entire life of the animal [9]. These three requirements are fundamental aspects of stem cell biology. The blood of adult humans completely renews itself every 120 days [10], skin every 2–3 days, and the intestinal lining every 3–5 days, all by stem cell proliferation. Furthermore, cellular populations destroyed through normal apoptotic processes, external tissue damage, and invading pathogens must also be replenished, a process that often involves adult stem cells.

Adult stem cells are similar to embryonic stem cells in their ability to terminally differentiate and to self-renew, but differ in their potency. An adult stem cell is multipotent and completely restricted to a specific tissue lineage, whereas an embryonic stem cell is pluripotent and able to differentiate into many different tissue lineages, including adult stem cells (Figure 2.1). Also in contrast to embryonic stem cells, adult stem cells are present for the adult lifespan of the organism; embryonic stem cells are present only early in gestation [3].

Similar to embryonic stem cells, the path from adult stem cell to terminally differentiated somatic cell involves intermediate steps. Cells that fall between adult stem cells and a specific cell type are known as *progenitor cells*, which are more limited than adult stem cells in their proliferative capacity, yet are still able to become one or more cell types [9]. When an adult stem cell or progenitor cell divides, it can do so in two ways, either symmetrically or asymetrically. If a cell undergoes symmetric division, both of the daughter cells will be of the same cell type; in the case of adult stem cells these either will both be adult stem cells [11] (self-renewal) or will each be a cell of more limited proliferative capacity (progenitor or terminally differentiated cells). If a cell undergoes an asymmetric division, the daughter cells are different from one another: one remains an adult stem cell, and the other is a cell type with more limited proliferative capacity. The path from adult stem cell to differentiated cell can be regarded as a hierarchy, with the adult stem cell at the top and the different terminally differentiated cell lineages at the bottom, with all steps in between consisting of more and more limited intermediate progenitor cells. Finally, adult stem cells can be divided into two (**a**) Tissue-specific Adult Stem Cell- Multipotent

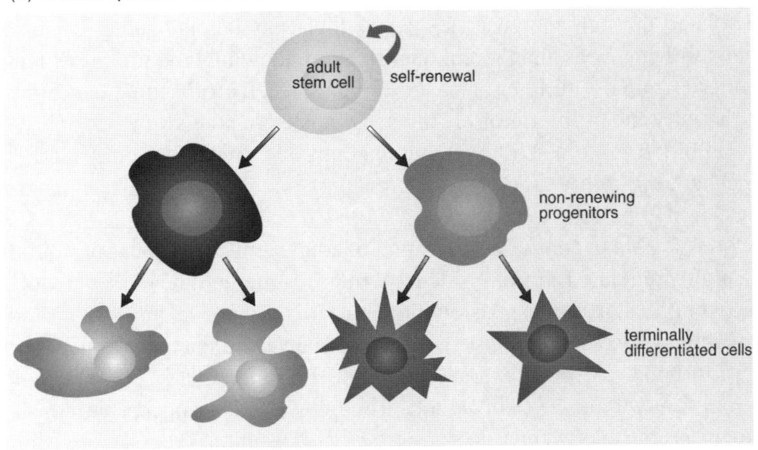

(**b**) Embryonic Stem Cell-Pluripotent

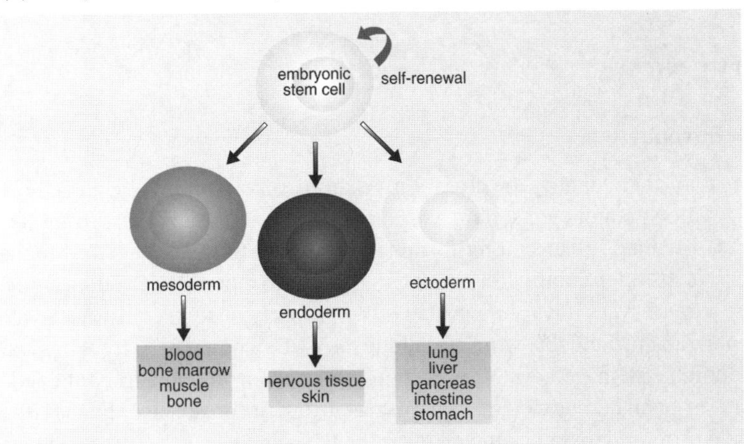

FIGURE 2.1 Adult stem cell lineages. (a) Adult stem cells are a self-renewing, long-lived population of cells able to give rise to multiple types of terminally differentiated cells utilizing a lineage-restricted progenitor intermediate. The progenitor intermediate has a more restricted proliferative capacity than the adult stem cell and is unable to self-renew. (b) This is in contrast to embryonic stem cells, which have the capacity to produce any tissue lineage in the body (including adult stem cells). (See insert for color representation.)

major groups, somatic stem cells (SSCs) and germline stem cells (GSCs) [12]. Organogenesis arises through the proliferation and differentiation of the SSC population, while GSC sustains reproductive activity.

2.1.1 Regulation and Disregulation of Adult Stem Cells

The ability of adult stem cells to be long-lived and self-renew is a behavior similar to that of cancer cells. Disregulation of hematopoietic stem cells is sometimes responsible for leukemias [13]. Over the life of a stem cell mutations may accumulate that will not be ablated through apoptosis, some of which may be oncogenic. Because adult stem cells self-renew as an inherent qualification, they may require fewer mutations than either terminally differentiated or progenitor cells to become cancerous [13]. Mitotic activity and proliferation are inversely related to potency in the adult stem cell system (in blood [13], in skin [14]), but uncontrolled division of progenitor populations that have the ability to become different types of terminally differentiated cells is an unquestioned parallel to cancer.

The body is able to control proliferation of adult stem cell populations through the use of a highly regulated area of residence termed the *niche*. When Till and McCulloch transferred bone marrow suspension into an irradiated mouse, they noticed colonies forming in the spleen of the mouse, made up of many different types of hematopoietic cells [5]. In bone marrow, the location of the HSC niche, HSCs do not proliferate uncontrolled, as was observed in the spleen colonies [5]. The concept and construction of the niche is able to prevent cancerous growth and promote necessary division under normal circumstances.

2.2 THE NICHE

2.2.1 Introduction

Seeking to further quantify the proliferative ability of the recently identified HSC, Till and McCulloch conducted serial transplant experiments using cells from the spleen colonies, so-called spleen colony-forming units (CFU-S) [15]. What they found was a progressive loss of the ability for donor CFU-S cells to repopulate irradiated mice. This became a mystery, as it was known that a key property of stem cells was the ability to self-renew, an attribute that would preclude the CFU-S population from being the true hematopoietic stem cell, yet, the colonies formed in the spleen contained clonal populations of all hematopoietic lineages, requiring the presence of HSC [16]. Robert Schofield proposed that the bone marrow, not the spleen, was the true residence of the HSC. He postulated that bone marrow harbored a special population of cells able to control HSC homeostasis by regulating stemness and division [16]. The niche hypothesis resolved the issue of CFU-S repopulation efficacy by explaining that those HSCs that took up residence in the spleen were not residing in the niche, and therefore were not self-renewing. This caused the irradiated mouse on the receiving end of the transplant to receive fewer and fewer HSCs [16]. Schofield's niche hypothesis has proved to hold up to the test of time, and today scientists are just starting to appreciate the full complexity of this simple idea.

2.2.2 Overview

Although different adult stem cell populations are maintained and regulated in unique ways within their individual niches, there exist in most cases variations on common themes. Regulation of proliferation is tightly and locally controlled by supporting cells

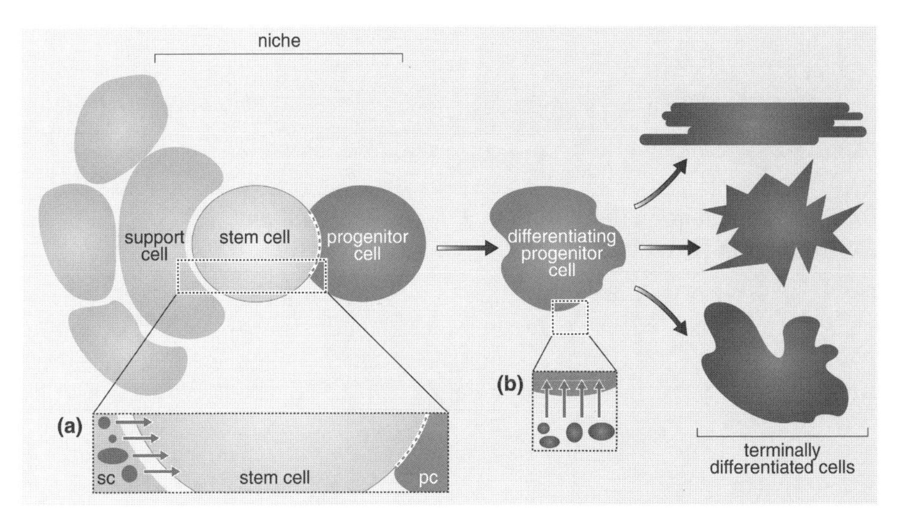

FIGURE 2.2 Adult stem cell niche. Adult stem cells are maintained in a specialized environment known as the *niche* that is characterized by the presence of stem cells and support cells. Paracrine signaling between the support cell and the stem cell is required to prevent differentiation. Asymmetric division of the stem cell produces daughter cells with different proliferative capacities . (a) The daughter stem cell remaining in contact with the support cells will retain its stemness while (b) the daughter progenitor cell not receiving cellular signals from the support cell will begin to differentiate, according to environmental cues, and eventually produce terminally differentiated cells. (See insert for color representation.)

within the niche [14,17]. The niche is generally composed of adult stem cells and supporting cells (*nurse cells*) [18]. It is the nurse cells that are responsible for maintaining homeostasis within the stem population, secreting paracrine cytokines and factors in response to endocrine signals (in blood [17]). Physical contact between nurse cells and adult stem cells is mediated in multiple cases by cadherins and β-catenin, creating an anchorage for stem cells (Figure 2.2) [19–21].

The importance of the regulatory role played by the nurse cells is apparent in experiments attempting to determine the location of stem cells. The transplantation of suspected adult stem cell niches for hair into the immunoprivileged kidney cap of athymic mice causes uncontrolled proliferation [14]. By acting as a molecular gatekeeper, nurse cells ensure that adult stem cell populations are maintained at levels needed to ensure long-term stability of various tissues through the self-renewal pathway [17]. Alternatively, nurse cells are able to respond to tissue damage by stimulating growth and proliferation of stem cells in order to replace cells lost to tissue damage or apoptosis.

Adult stem cells and their lineage-associated progenitors allowed to proliferate in the absence of the strict niche environment can wreak havoc on the organism. The mutation of nurse cells to either continually stimulate adult stem cells to divide or to stop inhibiting division is postulated as a possible oncogenic pathway [22]. This is not surprising, as Till and McCulloch demonstrated the ability of small numbers of adult hematopoietic stem cells to proliferate uncontrolled into clonal populations large enough to be seen without the aid of a microscope [5]. The ability to survive in the absence of the niche has been proposed as the key difference between adult stem cells and cancer stem cells [12].

2.2.3 Regulation of Differentiation and Self-Renewal and the Drosophila GSC Niche

The niche may regulate the fate of stem cells in two different ways. In the lineage mechanism, cellular contact between the nurse and stem populations is required to retain stemness [20]. During division, the mitotic spindle is oriented so that one of the two daughter cells face away from the nurse cell population and is not able to receive signals from the nurse cells to self-renew as a stem cell and thereby differentiates. In the population mechanism, other factors determine the outcome of mitosis [11,23].

While niches regulating mammalian adult stem cell populations are not fully characterized or are the subject of controversy, the germ stem cell niche of the invertebrate drosophila is well understood. In drosophila, the germ stem cell (GSC) niche is well illustrated to operate through a lineage mechanism. Direct contact between the GSC and the hub cells of the drosophila testes niche are required to maintain the stemness of GSC. The contact between the hub nurse cell and the adult stem cell allows for cytokines bone morphogenetic protein (BMP) and unpaired (Upd) protein of hub origin to activate the JAK-STAT signal transduction pathway within the GSC, maintaining stem cell behavior [20,24–26]. On division, the mitotic spindle is arranged perpendicular to the associated hub cell, such that one of the dividing GSC daughter cells will be in contact with a hub to renew the GSC population, and the other will become a primary spermatogonium—a sperm progenitor cell [27]. Orientation of the mitotic spindle is effected by localized concentrations of mammalian tumor suppressor adenomatous polyposis coli (APC) homologs interacting with drosophila cadherin and catenin [20]. Mutations within Apc2, one of the drosophila homologs, can cause misorientation of the mitotic spindle during division and result in both daughter cells remaining in contact with the hub. This disregulation of the niche will, in turn, expand the GSC population [20].

Also in contact with the hub cells are a selection of somatic progenitor cells known as *cyst progenitor cells* [27]. These progenitors, like the drosophila testes GSC, divide asymmetrically such that one of the daughter cells is in contact with the hub and the other is not. Unlike the GSC of the drosophila testes, the cyst progenitor cell is in contact with the hub only through a small cytoplasmic extension. The divisions of the cyst progenitor cells and the GSC are temporally related such that two cyst cells, the product of cyst progenitor division, envelope one primary spermatogonium. These three cells will then proceed to develop into 64 mature sperm through four mitotic and two meiotic divisions of the primary spermatogonium. Throughout the differentiation process of the spermatogonium, the cyst cells do not divide further [27]. Like many adult progenitor populations [28], the differentiating and dividing progeny of the GSC move away from the niche [27] to avoid tumorigenesis [28].

The lineage mechanism is further illustrated in the drosophila ovary. GSC are regularly lost [11] at a low rate to differentiation, requiring the constant replacement of this population. Contact with the Cap cells, which act in a similar manner to the Hub cells in the drosophila testes [11], allows a resident GSC to divide symmetrically into two GSCs when the maintenance of homeostasis requires an increase in GSC population. Drosophila ovary GSC also divides asymmetrically in a fashion similar to drosophila testes GSC.

2.2.4 Location and Identification

Identifying an adult stem cell niche is a technically difficult task [23]. Before a niche may be identified, the resident stem cell and its associated lineages must be known. By replacing resident stem cells with stem cells marked with a reporter gene, such as green fluorescent protein or lacZ, an assay may be implemented that follows marked stem cell progeny. After identifying areas that allow marked stem cells to proliferate at the same rate and potency as expected in a homeostatic environment, experiments to identify protein expression patterns of neighboring cells and subsequent results of interfering with cytokine expression may reveal niche compositions [23,29].

In the case of drosophila testes and ovum, the introduction of mutations in combination with reporter genes was able to give researchers the information necessary to identify the respective niches. By mutating *apc2* within drosophila testes [20], Yamashita and colleagues were able to identify unusual division patterns within the known germ stem cell population that increased GSC numbers. The investigators were able to quantify the disregulation of division after the mutation by examining cellular localization of two proteins: α-tubulin, which is expressed throughout the cell, and γ-tubulin, which is expressed in the centrosome, a key component of division. When mutated, centrosome protein *centrosomin* was also observed to cause misorientation of the mitotic spindle such that the plane of division was not perpendicular to the associated hub cell [20].

The identification of the niche, and more importantly the "how" of its regulation of associated stem populations, is a task intertwined with many therapeutic applications. The disregulation of stem cell proliferation may result in a cancerous phenotype, and their expansion is important for recovery after therapies that destroy rapidly dividing tissues, burn treatments, and neurodegenerative diseases such as Parkinson's disease.

2.2.5 Therapeutic Considerations

Understanding the molecular mechanisms by which the niche controls stem cell self-renewal and differentiation is a critical step in both *in vivo* and *in vitro* manipulation of stem cell populations. Understanding of how stem cells are maintained within the niche has allowed for medical applications by mobilizing stem cells in blood for transplant [30] to improving brain recovery after stroke [31]. The niche is the ultimate regulator of adult stem cells and further understanding of the molecular mechanisms involved in homeostasis will allow researchers to recapitulate the niche *ex vivo*, improving *in vitro* culture, making it possible to expand or preserve adult stem cell populations.

2.3 HEMATOPOIETIC STEM CELLS

The discovery of clonal colonies of many types of hematopoietic cells in the spleen of irradiated mouse bone marrow transfer recipients launched the field of stem cell biology. Today, inroads into diseases involving blood regulation such as leukemia, bone marrow failure, and side effects from disease treatment such as radiotherapy and chemotherapy for cancers are being better managed owing to discoveries in the field of hematopoietic stem cell research. The continuing use of bone marrow transplants in the management of these and other disorders has and will continue to benefit from stem cell research.

The HSC is the only cell able to reconstitute the entire hematopoietic system including erythrocytes (red blood cells), lymphocytes (T, B, and natural killer cells), granulocytes; monocytes, and platelets [2]. The characterization of the HSC using cell surface markers was able to isolate populations [32,33] capable of becoming cells of a specific lineage; however, these do not fit the proliferative capacity of an adult stem cell and are thus progenitors.

To date, the hematopoietic stem cell is the most extensively characterized adult stem cell population. Bone marrow transplants are commonplace in therapeutic applications, and bone marrow banks exist in the United States, Australia, China, and other countries. It is the bone marrow that is banked and not HSC themselves, as the bone marrow has been identified as the HSC niche [5].

2.3.1 Identification

Although it has been known that the HSC resides within bone marrow since the original observations by Till and McCulloch that serially transplanted cells originating within the bone marrow are able to give rise to clonal colonies of many different cell types (Figure 2.3), it wasn't until a quarter of a century later that the HSC was purified [33]. Recognizing that terminally differentiated B cells of the hematopoietic lineage express a cell surface marker not present on B-cell progenitors, Gerald Spangrude and colleagues in the Weissman lab were able to sort B-cell progenitors from B cells using monoclonal antibodies to the differentially expressed surface marker and fluorescence-activated cell sorting [33]. Extending this negative selection process, Spangrude was able to define hematopoietic stem cells as those cells that express low levels of differentiation marker Thy1, do not express T-cell markers CD4 or CD8, B-cell marker B220, myelomonocytic marker Mac1, or granulocyte marker Gr1 [33]. This arrangement of cell surface antigens is commonly referred to as *lineage minus* (lin minus). Experimenting with a set of monoclonal antibodies raised against T-cell progenitors it was discovered that stem cell antigen1 (Sca1)-positive cells, a minor subset of the original negatively sorted population, give a nearly one-to-one ratio of

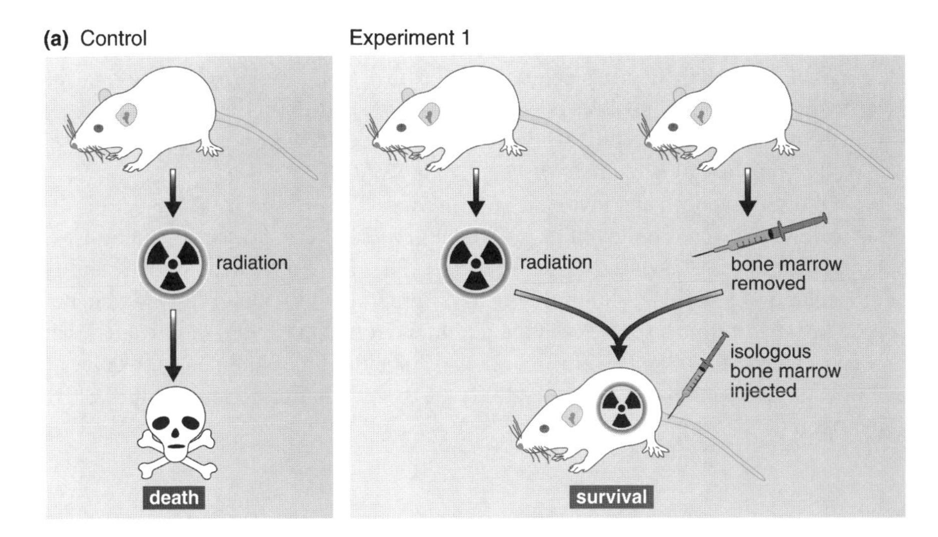

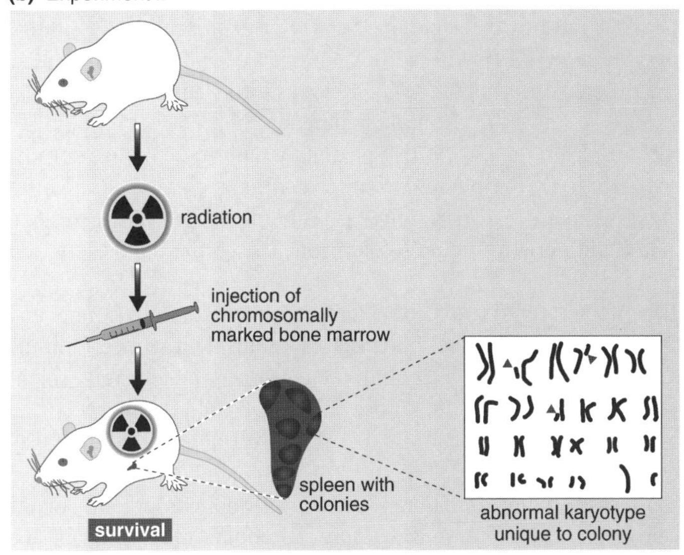

FIGURE 2.3 Discovery of the hematopoietic stem cell. Discovery of adult stem cells resulted from the experimental observation that bone marrow transplants in mice are able to reconstitute blood lineages that had been destroyed by lethal radiation treatment. (a) Mice are unable to survive certain levels of radiation exposure. Transplantations of bone marrow from a mouse that was not exposed to radiation to mice receiving lethal irradiation. Large colonies observed in the spleens of mice received bone marrow suspension transplantations were found to consist of multiple hematopoietic lineages. (b) These lineages were later found to be clonal and thus derived from a single progenitor or "stem" cell using sublethal radiation treatments that caused visible chromosomal abnormalities. A single abnormality was found in multiple cell types, indicating that they originated from a single parent cell. (See insert for color representation.)

injected cells postulated to lodge in the spleen to macroscopic colonies formed, using the original Till–McCulloch spleen colony-forming assay [5,33]. Furthermore, the Thy1 lo, lin minus, Sca1-positive population was able to differentiate into all lineages within the hematopoietic family, while Sca1-negative cells were not. Later this subset was further purified to exclude c-kit-negative cells [34].

While these experiments involved only murine HSC, it was not long before the same techniques were employed to isolate human HSC. The proposed human HSC population is Thy1 lo, lin minus, CD34-positive [32]. The experiment used the ability of an isolated population of human cells to differentiate into donor-derived hematopoietic lineages within a severe combined immunodeficient (SCID) mouse as the bar necessary to constitute an HSC [32]. Today, the standard measurement of whether a cell population contains HSC is the ability of the population to reconstitute a live hematopoietic system [35].

2.3.2 Niche

The HSC is known to reside primarily within a niche situated in the bone marrow [17], but also circulates transiently throughout the cardiovascular system [36]. This transient movement has been shown by the parabiosis of mice expressing a constitutive green fluorescent reporter gene (GFP) with a littermate that does not [37] as well as through the parabiosis of mice differing in the CD45 locus [36]. This surgical joining of the two cardiovascular systems results in chimeric hematopoietic systems; the GFP-expressing mouse harbors HSCs that are wild-type and the wild-type mouse harbors HSCs that are GFP-positive [37]. The parabiosed mice continue to harbor short-lived hematopoietic lineages of partner origin long after being surgically separated, indicating a stable HSC chimerism, including self-renewal of the adult stem cell population [36].

Parabiosis experiments answered a question that had interested physicians and scientists alike for many years: How does bone marrow transplant work? With the knowledge that HSC transiently circulate through the cardiovascular system, scientists knew the mechanism by which a small number of HSC were able to repopulate an entire organism, they next wondered how it was that HSC were able to find their way into the bone marrow and why they stayed there.

Within the bone marrow niche, the bone-forming osteoblasts are the major effectors of HSC proliferation [17]. A subset of osteoblasts residing in the cancellous/trabecular area [21] characterized as spindle-shaped, *N*-cadherin-positive, and CD45-negative (SNO) form *N*-cadherin and β-catenin junctions between themselves and HSCs [21]. The cadherin–catenin interface is thought to provide a physical adherence between the SNO and HSC. The total number of osteoblast nurse cells is thought to be influenced by bone morphogenetic protein (BMP) [21]. The production of HSC-activating factors by osteoblasts may be the result of the binding of parathyroid hormone (PTH, an endocrine cytokine) or its relative, PTH-related protein (PTHrP) to its osteoblast receptor, PTH/PTHrP receptor (PPR). Binding initiates a signal cascade of the Notch1 pathway, eventually causing HSC replication [38,39]. Along with Notch, Wnt and Sonic Hedgehog (Shh) have been implicated as important HSC self-renewal pathways, activated through either direct cell contact or by secreted factors [40].

2.3.3 Therapeutic Applications

Therapeutic treatments for cancer routinely involve radiation and chemical therapies that destroy endogenous HSC populations. Patients must then rely on closely matched bone marrow donors for repopulation of their hematopoietic system. This introduction of foreign-derived blood cells sometimes results in graft-versus-host disease (GVHD) [41], whereas T cells that are present in the donor marrow population recognize the organs of the patient as foreign invaders and mount a massive, systemic attack, killing the patient. There is a period after the bone marrow transplant during which the patient lacks immunity and also a risk of bone marrow rejection by the body. All of these problems could conceivably be avoided if the bone marrow of the patient were saved before the cancer therapies, but cancerous cells often contaminate bone marrow and blood [2]. Repopulating the patient with this population would nullify any therapeutic effects of radiation and chemical therapies by returning cancerous cells to the body.

By recovering a pure HSC population prior to immunoablative therapy, it is possible to reconstitute a patient's hematopoietic system without need of a donor and without relapse. This procedure uses cell surface markers to positively sort HSC, which removes both cancerous and T-cell populations from the donor pool [2,30]. Using these cells to repopulate a patient's hematopoietic system removes any potential for immune rejection, including GVHD, as there is no need for donor cells [30].

Routine organ transplantation may also benefit from a similar manipulation of the immune system. Organ rejection is one of the major concerns of transplantations and is now managed using immunosuppressants, exposing the body to pathogens of opportunity. A novel method of managing organ transplantation involves using HSC transplantation [2]. Transplantation of HSC triggers the body's innate T-cell-negative selection process, removing immune agents prone to cause organ rejection [42]. Using a purified HSC population is thought to reduce immune rejection by inducing a selection process in the thymus that selects against both donor-derived T cells and host T cells [42], similar to the way that fetal development of the immune system eliminates T cells recognizing the self as foreign [43]. The HSC used in this procedure is not major histocompatability (MHC) immunomatched with the patient, but the HSC population must originate from the donor of the organ [42]. While this method allows for self-recognition of donor organs MHC-matched to the HSC, transplants of organs not from either the donor or recipient do not benefit from this procedure [42].

2.3.4 Conclusion

From humble beginnings with Till and McCulloch's observations in the mouse spleen to today's life-saving applications in hematopoietic reconstitution after otherwise fatal therapies, hematopoietic stem cells have transformed the research and medical communities and laid the groundwork for the field of adult stem cell biology. HSC continue to offer hope to those unable to find matched organs for transplantation or matched bone marrow for postcancer therapy.

Hematopoietic stem cells remain the best-characterized adult stem cell in humans and other mammals. Further studies into regulation of population expansion in the osteoblast niche will assist scientists in expanding rare HSC for therapeutic uses *ex vivo*, a technically difficult task still under intense research [2]. Already, the use of circulating HSC purified from blood is possible using agents designed to mobilize HSC from the marrow niche, but the potential long-term effects of this procedure are unknown [30]. The ability to expand blood-derived HSC *ex vivo* may nullify the need to expand circulating HSC concentrations. The future of adult hematopoietic stem cells and their therapeutic potential is extremely promising.

2.4 NEURAL STEM CELLS

It is almost intuitive that a complex system for renewal of the hematopoietic system must exist owing to the continual generation of blood components, especially erythrocytes (nucleus-free red blood cells) and T cells, which must be produced in large numbers to fight infection. The brain, however, is traditionally [44] regarded as a non-renewable organ: tissue damage to the central nervous system (CNS) is seldom repaired, resulting in paralysis, neurodegenerative disease, and blindness, among other conditions.

The challenge of the nonproliferation dogma gave researchers reason to attempt localization and characterization of the neural stem cells (NSCs) in a hunt to find cures or therapeutic applications, research that is today beginning to bear fruit. Evidence suggests that neural stem cells are able to respond to CNS damage by proliferation and migration [45,46], but the ultimate reasons that these cells lack the efficacy to repair routine damage remains to be seen.

The neural stem cell, more than any other adult stem cell, remains mysterious. It is known to reside in only two places in the human brain [19], but can be cultured and expanded successfully *ex vivo* into the three major cells types of the central nervous system: neurons, astrocytes, and oligodendrocytes [47].

2.4.1 Identification—A Case of Proliferation Markers

Evidence of postnatal neurogenesis first surfaced in 1965 [44], showing that neurons of young rats incorporated tritiated thymidine, a radioactive analog of the DNA base integrated into genomic DNA during mitosis. In 1982, bird researchers (ornithologists) Steven Goldman and Fernando Nottebohm used tritiated thymadine in an experiment designed to examine the possibility that male hormones caused the brains of female songbirds to enlarge to the size of a male's brain. However, both the canary given testosterone and the canary given a negative vehicle control incorporated the thymadine marker, suggesting that neurogenesis was a normal part of adulthood in the bird [48].

It was not until 1998 that Peter Eriksson and colleagues, studying postmortem brain sections from terminally ill cancer patients, discovered strong evidence of a human adult neural precursor [49]. Cancer patients being intravenously administered a different DNA base analog, BrdU, for diagnostic purposes related to their cancer treatment regimen [49] were found to have multiple lineages of CNS neurons that stained positive for neuronal markers and for BrdU. As BrdU is incorporated into DNA during mitosis, these neurons were assumed to have arisen after the administration of the analog, ultimately originating from a neuronal precursor.

In 1999 several groups used a combination of drugs targeting mitotically active cells and tritiated thymadine to identify dividing CNS cells and their progeny [49–51]. This research identified astrocytes of the subventricular zone (SVZ) as able to survive the antimitotic treatment and give rise to multiple cell lineages, including migratory neurons destined for the olfactory bulb. These cells were identified by their incorporation of tritiated thymadine, given systemically 12 h after the destruction of the dividing progeny of the stem population [50].

Adult neurogenesis also occurs within the subgranular zone (SGZ) of the hippocampus as discovered by Bettina Seri and colleagues using a similar technique to describe the mitotic activity of an astrocyte population [52]. Proliferation in this brain region occurs at a lower rate than that of the SVZ, and progeny neurons are not migratory [19,47]. The potency of the NSC to differentiate into all neuronal subtypes has yet to be fully established, partially because of the extensive nature of the human brain [53]. However, of the two characterized proliferation centers of the adult brain, evidence exists that NSC from the SGZ transplanted into the SVZ are able to differentiate into olfactory neurons that both migrate and express proteins not transcribed by SGZ neurons [54].

2.4.2 Niche–SVZ

During embryonic neural development, neurons and glia develop via division and differentiation of cells in the ventricular zone (VZ), a layer of the neural tube [55]. In the mammalian adult, a region of the VZ, the subventricular zone (SVZ) remains a site of neural proliferation, generating neurons that migrate to the olfactory bulb (OB) via the rostral migratory stream [19,47,50]. The SVZ germinal area is composed of three cell types, conveniently labeled A cells (migratory neuroblasts), B cells (SVZ astrocytes), and C cells (progenitor or transit-amplifying) [19,56]. These three cell types are walled on one side by ependymal cells and are also contacted by the blood vessel–originating basal lamina [19], a layer of extracellular matrix (ECM) in this case so complex as to be termed "fractone"—a reference to fractals [57].

The B-cell astrocytes are the self-renewing stem cell population within the SVZ niche [50]. This population divides to produce a C-cell progenitor, which then rapidly divides to produce A-cell neuroblasts [19]. Cell surface markers can identify these three cell types. Self-renewing B cells express glial fibrillary acidic protein (GFAP), a glial marker, while C cells and A cells are GFAP-negative and Dlx2-positive. A cells can be distinguished from C cells by the presence of the polysialic acid form of neural cell adhesion molecule (PSA-NCAM) [19]. It is thought that ependymally produced Noggin, in conjunction with other morphogens and signally molecules, including BMPs, Sonic Hedgehog (Shh), and Notch, among others, may be involved in the regulation of the SVZ niche, although ultimate roles for these molecules have not been elucidated [56]. Additionally, the basal lamina layer may influence homeostasis of the SVZ niche through the concentration and regulation of signaling molecules [19,57].

2.4.3 Niche–SGZ

The location and proliferation of subgranular zone (SGZ) cells is closely associated with vascular proliferation and angiogenesis, suggesting a possible influence by vascular signaling factors [58]. The SGZ niche consists of three cell types as well, including the self-renewing stem cell population of astrocytes (SGZ-A), progenitor D cells, and granule neurons [52]. The niche is closely associated with blood vessels and is otherwise composed of granule neurons [56]. The D cells are so named as to distinguish them from C cells of the SVZ niche, as D cells are not as mitotically active and are smaller [19,52].

Similar to the SVZ niche, the astrocyte stem population divides to produce intermediate progenitor cells, in this case, the D cell. These may be distinguished by the expression of GFAP by the astrocytes and the lack of expression by the D cells [52]. The progeny of D cells are functional granule cells, which integrate into the surrounding dentate gyrus [56]. Vascular endothelial growth factor (VEGF) and Shh play roles in the regulation of the SGZ niche, with intraventricular infusion of VEGF causing increased mitotic activity in both the SGZ and SVZ [59].

2.4.4 Therapeutic Considerations

Although many aspects of NSC biology remain mysterious, their value as a therapeutic tool has not been overlooked, and advances have been made regarding the potential of this stem population to repair CNS injury. The human brain is a complex and compartmentalized system that relies on the ability of more than 100 different types of neurons to work together in harmony. Injuries of the central nervous system consist either of localized injury to a certain type of neuron, as in Parkinson's disease (PD) or spinal cord injury (SCI), or the disorder can be systemic, as in the damage caused by a stroke or by Alzheimer's disease (AD) [60]. The rate of division of NSC is regulated not only by many environmental factors, such as stress, but also by insults such as stroke and seizure [61]. Observations of a differential rate of proliferation coupled with attempts to culture NSC *ex vivo* has spurred research into the molecular mechanisms regulating homeostasis of the NSC niche [61].

Applications of NSC to localized disorders such as Parkinson's disease and spinal cord injuries may occur in one of two ways. The first involves the targeted introduction of NSC to the site of injury, in PD, the substantia nigra *pars compacta*, in SCI, the site of spinal insult. A spinal insult does not induce endogenous NSC populations to differentiate into neurons at the insult site. Targeted injection of NSC-derived cells to the site of injury causes the naive cells to take on a neuronal phenotype, and these cells are long-lived [62].

In systemic injuries, such as the ischemia resulting from a stroke, some evidence of endogenous neurogenesis and chemokine-mediated neural migration to the site of injury exists in animal models [46]. Also being investigated is the potential of exogenous NSC-mediated repair. Unlike localized events, there is not a focal point of injury that would be ideal for NSC injection and, as such, intravenous or intracerebroventricular injection is a possible route of administration [60]. In experimental models of multiple sclerosis, injected NSC were observed to express cell adhesion molecules that may assist in their intravasation of the blood–brain barrier [63], which normally ensures that the brain remains privileged by blocking the entrance of most bloodborne cells [64]. *In vivo*, massive recovery of certain regions of the brain by endogenous NSC after ischemic insult has been observed in mice. This regeneration was augmented by infusion of epidermal growth factor (EGF) [31], a mitogen known to cause expansion of NSC *in vitro* [65].

Although much data have been collected showing some functional recovery and engraftment of activated or transplanted NSC, controversy exists over which is the best NSC candidate for transplantation [61]. NSC differentiate mainly into glial cells when transplanted into sites of spinal injury; however, when neural progenitors — lineage-restricted precursors derived from NSC *in vitro* — are transplanted, they differentiate mainly into neurons [62]. Evidence that environmental influences are the main cause of lineage fate in the NSC [54,62] combined with the wide range of neurological diseases complicates potential therapeutic applications.

2.4.5 Conclusion

The neural stem cell today remains a mysterious entity. NSCs are able to become all three resident cellular lineages of the central nervous system, but their capacity to become all different neuronal subtypes remains to be shown. NSCs can be activated in response to tissue damage within the brain, yet neurodegenerative diseases by definition are diseases in which neurons are lost and not replaced.

The field of neural stem cell biology is one of the newest, most exciting, and most complex, owing to the nature of the nervous system. Although much remains to be understood, research is progressing in therapeutic applications to diseases such as Alzheimer's and Parkinson's, and spinal cord injuries.

2.5 EPITHELIAL STEM CELLS

The inner and outer parts of the body are covered and compartmentalized by a special cell type called the *epithelium*. Epithelium is found on the outer part of the cornea and the inner wall of the stomach. Generally, the epithelium functions as a barrier cell, ensuring that things on one side do not traverse to the other. In two cases where this segregation is critical, that of the skin and the intestine, an epithelial stem cell has been found to be involved in the homeostasis of the epithelium, ensuring that the body is protected from pathogens both inside and out.

Both skin and intestinal epithelial stem cells were discovered using label-retaining experiments [66,67]. While label-retaining techniques are powerful for identifying potential progenitor populations, they are unable to give molecular information and are thus limited in describing adult stem cells. Today, protocols for laser capture and more bountiful and powerful antibody staining along with full genomic transcriptional profiling are facilitating the description of cell populations, and the application of novel technology to skin stem cells is already producing results [68].

The discovery of self-renewing adult stem cell populations within the gastrointestinal tract and the epidermis, two areas routinely exposed to carcinogens and often the site of oncogenesis, may pave the way to therapeutic treatments for cancer. However, the inability to purify epithelial stem cells by cell surface markers hampers their uses in medical application. The innovation and rapid pace of adult stem cell technology is helping to overcome this barrier, in a strong and steady march toward the end of cancer.

2.5.1 Skin Stem Cells

Skin is the outermost and largest organ in the body. The skin consists of a layer of epithelial cells that makes up what is traditionally regarded as skin (the epidermis), hair follicles, and sebaceous glands. The squamous layer that acts as the physical barrier between the outside world and an organism is continuously shed and replenished throughout the lifetime of the organism. Individual hairs also undergo regular cycling, including, as many pet owners know, a shedding phase. The sebaceous gland is responsible for producing a waxy substance known as *sebum*, the function of which is unknown, although implicated as a necessary ingredient of acne pathogenesis [69].

Although some question exists [70] as to their actual function, a population of epithelial adult stem cells able to produce all three of these structures has been identified. *Keratinocytes*, as they are sometimes called, exist within a niche located in the bulge region of the hair follicle [71]. In the less hairy *Homo sapien*, it has been postulated that a potential second stem cell exists in the basal layer [72] to regularly produce epithelial cells for day-to-day homeostasis of the epidermis and that the bulge population assists in repairing wounds [73]. The potential of this second population to produce hair follicles and sebaceous glands has yet to be shown.

Characterization [71] of the stem population followed the initial discovery of label-retaining cells within the bulge region [67]. Further research led to cell surface marker proteins, allowing the purification of adult skin stem cells. Understanding the molecular mechanisms of skin stem cell self-renewal and propagation may lead to a further understanding of carcinogenesis of skin cancers, possible local drug targets, and may also assist in the repair of the epidermis after severe burns.

2.5.2 Identification

It is established that the body—be it one made for land, sea, or air—is able to repair external wounds. Serial passaging of human skin biopsies found that single epidermal cells are able to give rise to entire sheets of epidermis, opening the idea that a single cell may underlie epidermal creation and homeostasis, including the repair of injury [74]. The extension of this idea to include hair follicles as well gave rise to the idea of the skin stem cell. The first evidence that skin and hair may be formed from a single progenitor was observed in an experiment that stripped a mouse of its epidermis and found that cells originating from the hair follicle shaft assisted in skin repair [75]. Following up on this observation years later using tritiated thymadine label-retaining experiments, it was found that the likely stem cell of the hair shaft resides not in the hair matrix, which gives rise to the physical hair, but in the bulge, located at the lower extremity of the permanent partition of the hair follicle [76]. The theory put forth by Cotaserlis and colleagues was that bulge cells were the stem cell of the hair follicle, and the matrix cells were a limited progenitor as supported by observations that matrix cells do not retain label for extended periods of time [76].

This division of hair shaft and skin, which form a seamless sheet, into two separate stem cell components was challenged in 2000 by Taylor and colleagues, working out of the same laboratory as Cotaserlis [67]. To examine the possibility of a bulge cell contribution to the epithelia, the researchers used a double-labeling scheme whereby only slowly cycling cells that had divided shortly before the examination point would contain both labels. After giving the label pulse, they applied a small, penetrating wound to the back of a mouse. Their findings showed that doubly labeled cells did contribute to the epithelia that grew to heal the inflicted puncture wound, showing that cells of a bulge origin were involved in homeostasis of all three epithelial lineages of the skin.

A year later [14] Oshima and colleagues created chimeric hair follicles consisting of bulges that contained a genetic marker and hair follicle shafts that did not. After transplantation of this follicle in the immunoprivledged kidney cap of athymic mice, they observed cells containing the genetic marker of the transplanted bulge region in sebaceous gland, hair follicle, and epithelia, confirming the ability of the bulge region for generation of these three tissues. They also found that removal of the bulge region severely impaired hair growth. Together, these two findings suggested that the bulge was necessary for proper follicle growth. A further experiment by the same group [14] transplanted genetically marked individual bulges onto the backs of immunodeficient mice. These transplants gave rise to multiple hair follicles, showing that the cells of the bulge region not only contribute to multiple lineages but also are self-renewing.

2.5.3 Niche

The bulge region of the hair follicle is not in itself a cell type but rather a morphological landmark within the hair follicle. The epithelial adult stem cells that reside within the bulge are maintained by an unidentified cell type, possibly a population of mesenchymal cells adjacent to the bulge [12]. The cyclic fashion of hair growth is also implicated in control of stem cell mitosis, coordinating stem cell proliferation through the Wnt pathway in a model known as the *bulge activation hypothesis* [12]. Injury is also a key regulator of stem cell proliferation, as wounds to the epidermis are repaired with cells originating from the bulge region [75].

The epithelial stem cell niche of the bulge contains three cell types: the epithelial stem cells themselves, supporting cells, and a highly proliferative progenitor cell known as a *transit-amplifying* (TA) cell [76]. On exit of the bulge, TA cells will either migrate up to exit the follicle and contribute to the epidermis as epithelial cells or will migrate down into the shaft to contribute to the growing hair follicle [12]. Although these cell types are known to exist, the interactions between these cells driving homeostasis of the stem cell niche and signals stimulating proliferation have yet to be thoroughly explored.

While the genetic profile of the niche cells has yet to be elucidated, that of the epithelial stem cells themselves has been. Early molecular markers of skin stem cells such as cytokeratin-15 (K15) [77] have been improved on by full genomic expression profiling of the bulge and surrounding cells, which allows scientists to find genes specifically upregulated in the stem cell. Using a fluorescently retained marker, Tumbar and colleagues were able to purify slowly dividing skin stem cells from the bulge region and from these found genes expressed in the stem population [71]. They compared the profile of these cells to that of hematopoietic, human embryonic, and neural stem cells and found a high incidence of similar expression between stem cell types relative to other cells found within the follicle shaft [71]. These data have allowed scientists a window into the receptors expressed on the surface of the skin stem cell, insight that includes possible morphogenetic pathways. Working backward from what pathways may affect stem cell behavior may allows scientists to locate cells secreting the ligand that binds to an expressed receptor, eventually finding the cells that supply factors necessary for stem cell homeostasis.

2.5.4 Therapeutic Considerations

A direct clinical use of adult skin stem cells is *ex vivo* epidermal culture. Skin grafting has therapeutic applications in many arenas. The *in vitro* culturing of patient-derived adult skin stem cells has been used routinely in some clinics for large area skin replacement. Many cases have been planned far enough in advance to allow for large-scale *ex vivo* culturing of the stem cells (such as tattoo removal) [78]. The delicacy of large sheets of stem cell–derived epithelia has prevented the creation of large sheet transplants; instead transplants are a patchwork of smaller grafts. This conundrum has been partially solved using transplantable growth matrices. Transplantable growth matrices are much stronger than thin epidermal sheets and ensure that the delicate epithelia does not tear during culture manipulations [79]. Graftable matrices for autologous skin transplants were pioneered on human victims of wide-ranging third-degree burns (covering >50% of the body) [79]. *Ex vivo* adult stem cell skin culture, as pioneered in serial passaging experiments of Green and Rheinwald, is essential for these injuries, as there is not enough healthy donor-derived skin available for a simple transplant and in third-degree burns the skin itself, including stem cells, has been destroyed.

Another major medical and social application of epithelial stem cells is alopecia, commonly known as *baldness*. A subtype of alopecia is thought to be caused by an autoimmune reaction specifically targeting cells within the bulge region (the stem population), thereby permanently destroying the ability to produce new hair within affected follicles [80]. As the discovery of the skin stem cell itself [14] utilized a technique creating new hairs from grafted bulge regions, the use of these cells to treat baldness is a realistic therapy in the near future.

2.5.5 Conclusion

The epithelial skin stem cell residing in the bulge region of the hair follicle is critical to the maintenance of the body's first line of defense against the outside world. It contributes to the maintenance and repair of the epidermis, produces the hair follicle shaft, and can be extensively cultured *ex vivo* for therapeutic applications. Although the skin stem cell is now routinely manipulated and is involved in even cosmetic medical procedures, the cellular processes underlying homeostasis of the niche and the signals directing division and differentiation of the stem cell and the transit-amplifying progenitor remain unknown. Although the skin stem cell is unique in its ability to produce the hair follicle, sebaceous gland, and epidermis, it is one of a small, but potentially growing [78], family of epithelial stem cells, including an epithelial layer that performs the same function as skin, but is instead located on the innermost wall of the body: the intestinal stem cell.

2.6 INTESTINAL STEM CELLS

The intestine is segregated from the body by a highly proliferative epithelial layer with a specific twofold mission: to keep toxins and pathogens out of the body and to bring nutrients in. The barrier between the intestinal lumen and the body is only one cell thick [81] and is rapidly replaced, with old nutrient-absorbing villi supplanted by new ones every few days [82].

Producing more than 10 billion new cells per day within the human intestine [83] are the resident intestinal stem cells (ISCs) [84]. Researchers Leblond and Stevens first hinted in 1948 at the idea that cells making up the villi of the intestine originate from a population in the crypt area at its base, when investigating the notion that massive proliferation in the crypt was not matched by massive cell death in the crypt [85]. In order to mark cells in the crypt to explore their eventual fate, they used the mitotic arresting agent colchecine, which creates a visible mitotic figure. Their experiments led to the conclusion that cells in mitosis migrate away from the crypt and eventually exit through an *extrusion zone* at the apex of the villus.

ISCs reside in a unique niche referred to as the crypt, several of which are located at the base of each intestinal villi [66]. The villi are arranged in such a way as to expose a large surface area to the intestinal lumen, creating finger-like projections. The epithelium making up the villi progressively differentiate as they make their way toward the tip, creating a differentiation gradient [66,82]. ISCs can also migrate downward, to the base of the crypt [66].

The intestinal stem cell produces four morphologically distinguishable cell types, including Paneth cells (after Austrian physiologist Josef Paneth, 1857–1890), which reside at the base of the crypt below the stem population, enteroendocrine cells, goblet cells, and columnar cells [66]. The crypt is also able to divide through a process referred to as *fission* to produce two daughter crypts. This routine niche multiplication may help ensure that the thin line between intestinal lumen and tissue remains sealed [84].

2.6.1 Identification

With preliminary evidence in hand [85], Leblond pushed on toward a definite intestinal stem cell, this time with colleague B. E. Walker. Using a label-retaining assay incorporating adenine and thymadine containing heavy carbon-14, the two researchers confirmed the earlier observation that cells from the crypt migrate up the villi [86]. Knowing that cells of a crypt origin were shed through the villi, Leblond and Cheng in 1974 provided the direct evidence of a specific intestinal stem cell with the ability to become all major cell types of the intestine [66], and to localize it to a crypt base position directly above the Paneth cell [87].

The approach used by the two scientists to identify the stem cell differed from that of most label-retaining experiments in one crucial way—instead of simply marking dividing cells, they intended to kill them and take advantage of an interesting property of what was termed the *crypt-base columnar cell*. These cells phagocytose dead cells in their vicinity [66]. With the knowledge that crypt-base columnar cells are located in the area of the crypt found earlier to be the source of massive proliferation, they intended to destroy some of them and to have these dead, label-containing cells engulfed by neighboring cells.

After injecting a mouse with a sublethal dose of tritiated thymadine, some of the cells incorporating the base analog died, and the neighboring living cells that had phagocytosed the dead columnar neighbor were labeled (as the dead cell contained the label). Leblond and Cheng followed the lineages of cells containing these label-retaining phagosomes as they migrated and progressively differentiated into the different lineages of the intestinal lumen. This led them to postulate that crypt-base columnar cells were the intestinal stem cell, and also to identify an intermediate lineage-restricted progenitor [66] that was later confirmed using a genomic labeling technique [82].

During the same period, other research also showed that, similar to hematopoietic stem cells, the intestinal stem cell is able to repopulate the intestine following sublethal irradiation. In this case a rescue assay was not performed, but rather the ability of surviving crypts to regenerate the intestinal epithelium was observed [88].

2.6.2 Niche

The intestinal stem cell niche is interesting in that the epithelium of the intestinal wall is only one cell thick. The location of the actual adult stem cell is near the base of the crypt, a rounded area several of which make up the base of each villus. The base of the crypt is made up of Paneth cells, while the intestinal stem cell resides approximately four or five cells upward, continuing from there with a progressively differentiating cells, including lineage restricted progenitors [66,89].

Regulation of ISC proliferation is promoted through the Wnt pathway, whose downstream targets include cell proliferation genes regulated in conjunction with β-catenin [84]. Mesenchymal cells adjacent to the ISC region of the crypt express BMP, which suppresses β-catenin activity. It's thought that the expression of a fourth protein, the BMP antagonist Noggin, is required in conjunction with Wnt to promote self-renewal within the intestinal crypt [90].

BMP-secreting mesenchymal cells adjacent to the ISC are known to be involved in the regulation of ISC proliferation. BMP receptor (BMPr) mutants have been shown to have irregularly expanded stem cell populations mimicking juvenile polyposis syndrome [90]. Unfortunately, a lack of definite molecular markers has prevented researchers from deciding with certainty the locations of ISC; however, some preliminary markers such as Mushashi-1, a protein expressed in neural stem cells and upregulated in cells occupying positions 4 and 5 in the crypt, are starting to emerge [84]. Differential regulation of lineage-restricted progenitors has also been shown to be regulated by specific proteins, including glucagon-like peptide 2 (Glp2), which increases proliferation of enterocytes through a pathway involving intestinal resident enteric neurons [91].

2.6.3 The Immortal Strand Theory

The proliferative potential of adult stem cells, combined with their long lifespan, make it likely that they will incur genomic mutations during mitosis [92]. Intestinal stem cells are thought to complete mitosis approximately 5000 times over the course of the human lifespan [93]. Why adult stem cells do not all become cancerous over time as a result of mutation remains an open question. Evidence does exist that adult stem cells may be the source of some cancers by being transformed into so-called cancer stem cells [22,93,94]. One hypothesis put forth to explain the lack of oncogenesis in adult stem cell populations suggests that nonrandom asymmetric DNA segregation during mitosis may be responsible [92].

The *immortal strand* theory, put forth by Cairns and Potten in 1975, theorizes that adult stem cells retain the template DNA strand during mitosis, ensuring that replicative mutation within the genome is unable to occur. This is most easily explained by thinking of the adult stem cell as being the daughter cell that retains the oldest DNA strand. The idea is irrelevant in a first round of mitosis, but taking it another round of division to the granddaughter stage illustrates the point; if one were to label one of the two genomic strands of DNA the immortal strand, the labeled strand would be segregated into one of two daughter cells during division along with a newly synthesized complement. When this cell divides, the complement synthesized during the first round of mitosis will segregate with its new partner and the immortal strand will yet again have a new complement. By nonrandomly segregating the complement of the immortal strand into the non–stem cell during division, it ensures that no cumulative damage can occur to the genome of the adult stem cell as a result a DNA replication error .

Mammalian evidence for the immortal strand hypothesis was first discovered in the mouse intestine, using the intestinal stem cell [89]. Intestinal crypts, containing all major types of intestinal epithelia, are able to themselves divide and produce a sister crypt, creating an inherent mechanism to preserve the intestinal lining. By treating either young mice or mice that had been sublethally irradiated with tritiated thymadine, all newly synthesized DNA would be labeled. Any cells that retained the radioactive label for extended periods of time are called label-retaining cells (LRCs). By combining this labeling technique with a second labeling technique applied later, researchers were able to double label ISC DNA during mitosis and follow DNA segregation. Their results agreed with Cairns and Potten's theory of an immortal strand by showing that the DNA labeled with tritiated thymadine remained in the cell presumed to be the stem cell, while the DNA labeled with the second technique segregated to cells that differentiated [89].

2.6.4 Therapeutic Considerations

Cancers of the gastrointestinal tract are one of the most common carcinogenic events in the human body, and carcinogenesis is well understood in the majority of cases [84]. As the intestinal stem cell is the only long-lived cell within the intestinal epithelium, it has been examined extensively for possible common oncogenic pathways [84]. Today, the path from adult stem cell to cancer is best characterized molecularly within intestinal stem cells.

Mutations of the adenomatous polyposis coli (APC) tumor suppressor gene were originally implicated in familial adenomatous polyposis (FAP), a rare, inherited, colorectal carcinoma [95]. In the ISC, APC functions in concert with two other proteins to form a β-catenin destruction complex [95]. As β-catenin is in the ISC self-renewal pathway, this destruction complex prevents stem cell self-renewal [95]. Activation of Wnt receptors stops the destruction complex, allowing β-catenin to accumulate in the nucleus and effect transcription in concert with T-cell factor (TCF). Together, β-catenin and TCF impose a self-renewal instead of differentiation fate on ISC [96].

Most spontaneous gastrointestinal cancers involve an inactivation of the APC/β-catenin destruction complex early in progression [95]. Increased β-catenin levels due to functional APC or destruction complex loss lead to a selective advantage of the mutated stem cell, causing expansion, and eventual progression to adenoma, and finally, carcinoma. While this is not a one-hit hypothesis, nearly all gastrointestinal cancers include a functional loss of APC [95].

A molecular map from intestinal stem cell to intestinal cancer provides researchers and pharmaceutical companies with potential drug targets. Conservation of different mutations between different types of cancer suggests a critical mutation, and, in the case of the intestinal stem cell, these mutations are well characterized.

2.6.5 Conclusion

The existence of an epithelial stem cell is indispensable for maintaining barriers between the human body and organs and the outside world. The action of an epithelial stem cell ensures that a tight cellular barrier is maintained and repaired in the case of injury. Cancers of the skin and intestine are two of the most common maladies and the mapping of the oncogenic path in gastrointestinal cancer from self-renewing intestinal stem cell to transformed cancer stem cell has provided researchers with targets and understanding of the mechanisms of oncogenesis. Even with all that is known about these two adult stem cell types, much remains to be learned. Molecular markers of intestinal stem cells are beginning to be found, and techniques used to identify other adult stem cell populations may be useful in the future for purification and further characterization. What is definitely known about them is that they'll be there, dividing and self-renewing to keep our bodies clean and dry.

2.7 CANCER STEM CELLS

Cellular growth is tightly maintained within the body through numerous autocrine, paracrine, and endocrine signaling pathways. When a cell fails to respond to signals promoting quiescence or acquires the ability to grow in the absence of exogenous growth signals, it may proliferate uncontrolled, a hallmark of cancerous behavior [97]. These two properties make up half of the four cellular changes classically proposed to be necessary in oncogenesis; the others are the ability to evade apoptosis and limitless replicative potential [97]. These factors together make cancers an attractive target for therapies looking to ablate quickly dividing cells, including radiotherapy and some chemotherapies. However, cancer patients undergoing these therapies often relapse, leading researchers to pursue further elucidation of oncogenic pathways.

Owing to the similarities between cancerous cells and adult stem cell populations, stem cells and progenitors have been examined as possible sources of oncogenesis. Further, evidence pointing to the existence of a rare cancer stem cell has been observed in some cancers, including acute myeloid leukemia [22,98].

2.7.1 Stem Cells and Cancer: Different Sides of the Same Coin?

By definition, adult stem cells are able to undergo self-renewal and extensive proliferation. Similarly, cancers are defined by uncontrolled growth and can metastasize to produce many tumors throughout the body. Although unable to evade apoptosis in a cancerous sense, adult stem cells are long-lived and also harbor a limitless replicative ability, two of the four changes classically thought to be required for a cell to become cancerously transformed.

More recently, the idea of a cancer stem cell, sometimes referred to as a *tumor-initiating cell*, has been proposed to be responsible for oncogenesis in many systems, including patients with acute myeloid leukemia (AML) [98]. Peripheral blood xenographs from humans afflicted with AML into severe combined immunodeficient (SCID) mice were able to give rise to AML within the recipient mice. Cells from these mice can then form colonies when plated *in vitro*, indicating a self-renewing ability. Colony-forming cells were found to be present at only 1 in 250,000 cells within the peripheral blood of human AML patients. Characterization of cell surface markers found that cells from peripheral AML patient blood expressing markers of an undifferentiated state were able to form colonies in SCID mice while others were not [98]. Together, these observations promote the idea of a cancer stem cell (CSC)

(**a**) Normal Adult Stem Cell

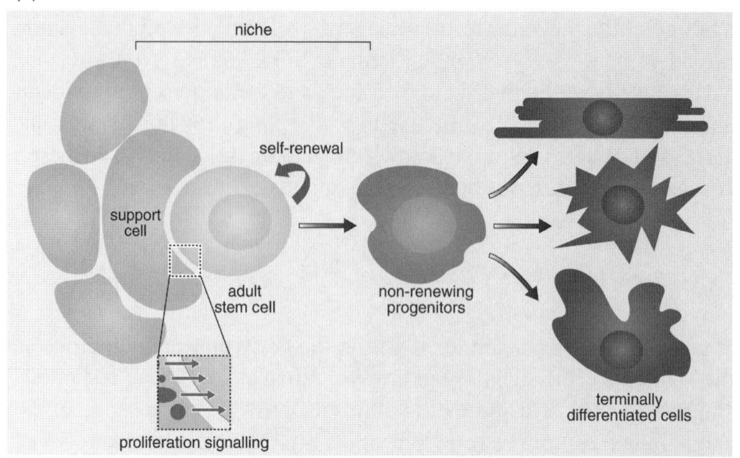

(**b**) Cancer Model I

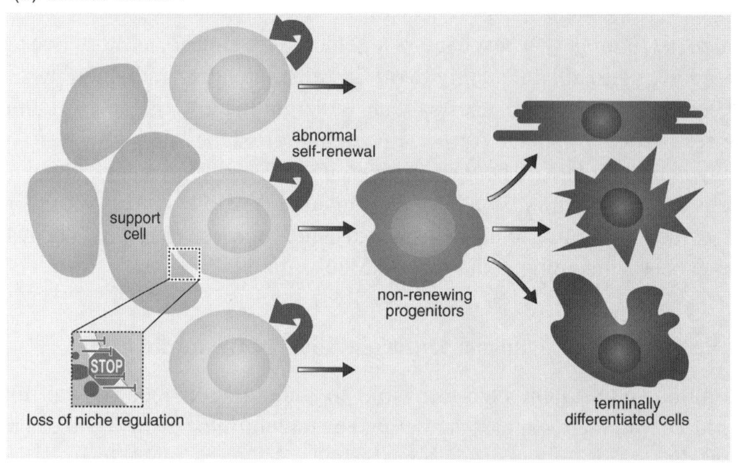

(**c**) Cancer Model II

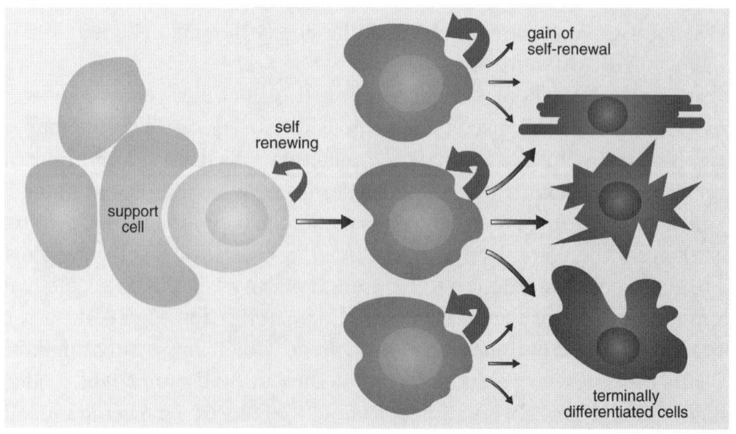

within at least some AML patients. Along with the hematopoietic system, tumor-initiating populations have been found in breast and brain cancers [99].

2.7.2 The Cancer Stem Cell Model

Although quite neatly drawn up as a stepwise pathway to cancer, the classical idea that terminally differentiated cells accumulate sufficient mutations to transform has been criticized as unlikely [100]. Observations that (1) terminally differentiated cells most likely do not live long enough to collect multiple specific mutations [100], (2) adult stem cell populations already contain some of the characteristics of cancer cells, and (3) cancer cells self-renewal (including the same self-renewal pathways) have driven the creation of a cancer model originating from adult stem cells (Figure 2.4) [13].

The cancer stem cell model postulates two possible oncogenic targets: the adult stem cell population and the limited progenitor population [13]. The transformation from stem cell to cancer stem cell relies on the manipulation of niche-mediated regulation. The niche is responsible for maintaining stem cells in an undifferentiated state and signaling division and quiescence. If a stem cell were to ignore signals from the niche and begin self-directed self-renewal, it would have become oncogenic. As described earlier, this disregulation seems to occur in colorectal cancers, as functional loss of the APC tumor suppressor is found in almost all cancers of the gastrointestinal tract and its function is known to directly involve regulation of self-renewal [95]. Alternatively, cancerous mutations may affect lineage-restricted, but massively dividing, progenitor cells. Experimental manipulations targeting mutations to promoters uniquely expressed in progenitor populations has shown that AML may arise from a strictly progenitor origin, raising the possibility of a progenitor cancer pathway [101].

2.7.3 Therapeutic Considerations

A stem cell–based cancer model has two immediate implications for the field of cancer treatment. A lack of rapid or frequent division by cancerous stem cells would render them immune to therapies targeting dividing cells. While the total tumor volume may

FIGURE 2.4 Cancer stem cell. Two major theories of the cancer stem cell arise from the common theme of aberrant self-renewal. (a) Normally, adult stem cell proliferation is tightly regulated by paracrine signals received from the niche. Stem cells divide asymmetrically to produce a stem cell (self-renewal) and a progenitor cell. The progenitor then divides to become multiple types of terminally differentiated cells, but does not self-renew. (b) In the first cancer model, adult stem cells are thought to begin a program of unregulated self-renewal. This requires either the loss of function of the signaling pathway between the stem cell and the niche, the loss of function of certain cell-cycle-inhibiting proteins within the stem cell, or the gain of function of certain cell-cycle-promoting proteins within the stem cell to promote division in the absence of proper signal. (c) In the second theory, the lineage-restricted progenitor daughter cell acquires novel self-renewal ability. Both theories cause uncontrolled proliferation of a cell that is able to create multiple cell lineages. (See insert for color representation.)

decrease to what appears to be nothing, the cancer stem cell model predicts a rare population of slowly cycling cells that would not be affected and that may be responsible for relapse [100]. The ATP-binding cassette (ABC) transporter family of proteins commonly expressed in stem cells is responsible for pumping toxic compounds out of the cell [99]. This second concern has implications for chemotherapy, which relies on the selective uptake of cytotoxic compounds to kill cancerous cells.

Further characterization of major cancer stem cell pathways and molecular markers, such as those for AML, colorectal cancer, and others will allow researchers to target pathways that are specific to a possible stem cell source of cancer. Targeted treatment of cancer stem cell pathways may be able to spare noncancerous cells in the future, but softer therapies will only follow a solid foundation of basic research.

2.7.4 Conclusion

Traditional views of cancer have recently been challenged with the observation of phenotypic and molecular parallels between cancer cells and adult stem cells. This has led to the creation of a cancer stem cell model postulating that mutations in stem cells may lead to oncogenic transformation. The cancer stem cell theory has major implications for cancer therapy, as treating tumors with agents targeting dividing cells would not affect slowly cycling stem populations. The characterization of cancer stem cell–specific pathways could lead to better and more effective treatments that spare noncancerous cells.

2.8 SUMMARY

Since concrete evidence of adult stem cells was uncovered through the bone marrow transplantation experiments by Till and McCulloch, the field of adult stem cell biology has exploded to encompass therapeutic uses, niche characterization, and *in vitro* function of multiple adult stem cell lineages while simultaneously providing evidence that some tissues lack adult stem cells. The key properties of adult stem cells, namely, being long-lived and self-renewing and able to differentiate into multiple terminally differentiated cell lineages have created a vocabulary from which experimental designs testing stemness of a population have arisen. The search to find adult stem cell populations that can be described by this set of terms has developed technologically using cell surface markers to purify and characterize populations, base analogs to track cell proliferation, and fluorescent transgenics to examine transplanted tissues. However, even with the advent of new technology, the gold standard by which adult stem cells are defined remains that used by Till and McCulloch in their original radiation experiments, the ability of a cell to repeatedly repopulate all lineages of a depleted tissue.

The concept of the adult stem cell has been thoroughly explored, but knowledge of stem cell regulation remains shallow. In some instances adult stem cell regulation is understood at a molecular level, but a deeper understanding of these mechanisms will allow members of the scientific and medical communities to manipulate cellular control of stem cells for therapeutic applications. Already, some applications of stem cell manipulation are being used *in vitro* for patient-specific applications, but further characterization may allow for *in vivo* therapies.

Understanding how adult organisms maintain tissues and repair injuries merely scratches the surface of stem cell biology. Problems within adult stem cells or progenitor cells may result in cancerous transformation, highlighting the fragile balance between division and quiescence. From the incidence of leukemia in atomic bomb survivors to the characterization of transformations within the intestinal crypt, cancer is only one of many arenas that adult stem cells have been used as a tool to study an unknown biological event. Future research will make adult stem cells the basis of a new generation of medical therapy.

REFERENCES

[1] Brill AB, Tomonaga M, Heyssel RM: Leukemia in man following exposure to ionizing radiation. A summary of the findings in Hiroshima and Nagasaki, and a comparison with other human experience, *Ann Intern Med* **56**:590–609 (1962).

[2] Weissman IL: Stem cells: Biology, transplantation, and political ethics, *Proc Am Phil Soc* **150** (1):121–147 (2004).

[3] Jacobson LOM, Marks EK, Gaston EO, Robson M, Zirkle RE: The role of the spleen in radiation injury, *Proc Soc Exp Biol Med* **1949** (70):740–742 (1949).

[4] Lorenz E et al: Modification of irradiation injury in mice and guinea pigs by bone marrow injections, *J Natl Cancer Inst* **12** (1):197–201 (1951).

[5] Till JE, McCulloch EA: A direct measurement of the radiation sensitivity of normal mouse bone marrow cells, *Radiat Res* **14**:213–222 (1961).

[6] Rajagopal J et al: Insulin staining of ES cell progeny from insulin uptake, *Science* **299** (5605):363 (2003).

[7] Becker AJ, McCulloch EA, Till JE: Cytological demonstration of the clonal nature of spleen colonies derived from transplanted mouse marrow cells, *Nature* **197**:452–454 (1963).

[8] Siminovitch L, McCulloch EA, Till JE: The distribution of colony-forming cells among spleen colonies, *J Cell Physiol* **62**:327–336 (1963).

[9] Lanza RP, *Handbook of Stem Cells*, Elsevier Academic, Burlington, MA, 2004, pp. xxv–xxxi.

[10] Shermin DR: Studies on the formation of heme and on the average life time of the human red cell, *Fed Proc* **5** (Pt II):153(1946).

[11] Xie T, Spradling AC: A niche maintaining germ line stem cells in the Drosophila ovary, *Science* **290** (5490):328–330 (2000).

[12] Li L, Xie T: Stem cell niche: Structure and function, *Annu Rev Cell Dev Biol* **21**:605–631 (2005).

[13] Reya TM, Morrison SJ, Clarke MF, Weissman IL: Stem cells, cancer, and cancer stem cells, *Nature* **414**:106–111 (2001).

[14] Oshima HR, Rochat A, Kedzia C, Kobayashi K, Barrandon Y: Morphogenesis and renewal of hair follicles from adult multipotent stem cells, *Cell* **104**:233–245 (2001).

[15] Siminovitch L, Till JE, McCulloch EA: Decline in colony-forming ability of marrow cells subjected to serial transplantation into irradiated mice, *J Cell Physiol* **64**:23–31 (1964).

[16] Schofield R: The relationship between the spleen colony-forming cell and the haemopoietic stem cell, *Blood Cells* **4** (1–2):7–25 (1978).

[17] Calvi L, Adams GB, Weibrecht KW, Weber JM, Olson DP, Knight MC, Martin RP, Schipani E, Divieti P, Bringhurst FR, Milner LA, Kronenberg HM, Scadden DT: Osteoblastic cells regulate the haematopoietic stem cell niche, *Nature* **425**:841–846 (2003).

[18] Moore KA, Lemischka IR: Stem cells and their niches, *Science* **311** (5769):1880–1885 (2006).

[19] Doetsch F: A niche for adult neural stem cells, *Curr Opin Genet Dev* **13** (5):543–550 (2003).

[20] Yamashita YM, Jones DL, Fuller MT: Orientation of asymmetric stem cell division by the APC tumor suppressor and centrosome, *Science* **301** (5639):1547–1550 (2003).

[21] Zhang J et al: Identification of the haematopoietic stem cell niche and control of the niche size, *Nature* **425** (6960):836–841 (2003).

[22] Li L, Neaves WB: Normal stem cells and cancer stem cells: the niche matters, *Cancer Res* **66** (9):4553–4557 (2006).

[23] Spradling A, Drummond-Barbosa D, Kai T: Stem cells find their niche, *Nature* **414** (6859):98–104 (2001).

[24] Kiger AA et al: Stem cell self-renewal specified by JAK-STAT activation in response to a support cell cue, *Science* **294** (5551):2542–2545 (2001).

[25] Kawase E et al: Gbb/Bmp signaling is essential for maintaining germline stem cells and for repressing bam transcription in the *Drosophila* testis, *Development* **131** (6):1365–1375 (2004).

[26] Tulina N, Matunis E: Control of stem cell self-renewal in *Drosophila* spermatogenesis by JAK-STAT signaling, *Science* **294** (5551):2546–2549 (2001).

[27] Hardy RW et al: The germinal proliferation center in the testis of *Drosophila melanogaster*, *J Ultrastruct Res* **69** (2):180–190 (1979).

[28] Ohlstein B et al: The stem cell niche: Theme and variations, *Curr Opin Cell Biol* **16** (6):693–699 (2004).

[29] Turner DL, Cepko CL: A common progenitor for neurons and glia persists in rat retina late in development, *Nature* **328** (6126):131–136 (1987).

[30] Herve P: Donor-derived hematopoietic stem cells in organ transplantation: Technical aspects and hurdles yet to be cleared, *Transplantation* **75** (Suppl 9):55S–57S (2003).

[31] Nakatomi H et al: Regeneration of hippocampal pyramidal neurons after ischemic brain injury by recruitment of endogenous neural progenitors, *Cell* **110** (4):429–441 (2002).

[32] Baum CM et al: Isolation of a candidate human hematopoietic stem-cell population, *Proc Natl Acad Sci USA* **89** (7):2804–2808 (1992).

[33] Spangrude GJ, Heimfeld S, Weissman IL: Purification and characterization of mouse hematopoietic stem cells, *Science* **241** (4861):58–62 (1988).

[34] Okada S et al: *In vivo* and *in vitro* stem cell function of c-kit- and Sca-1-positive murine hematopoietic cells, *Blood* **80** (12):3044–3050 (1992).

[35] van Os R, Kamminga LM, de Haan G: Stem cell assays: Something old, something new, something borrowed, *Stem Cells* **22** (7):1181–1190 (2004).

[36] Wright DE et al: Physiological migration of hematopoietic stem and progenitor cells, *Science* **294** (5548):1933–1936 (2001).

[37] Wagers AJS, Sherwood RI, Christensen JL, Weissman IL: Little evidence for developmental plasticity of adult hematopoietic stem cells, *Science* **297**:2256–2259 (2002).

[38] Taichman RS, Emerson SG: Human osteoblasts support hematopoiesis through the production of granulocyte colony-stimulating factor, *J Exp Med* **179** (5):1677–1682 (1994).

[39] Taichman RS, Reilly MJ, Emerson SG: Human osteoblasts support human hematopoietic progenitor cells *in vitro* bone marrow cultures, *Blood* **87** (2):518–524 (1996).

[40] Calvi LM: Osteoblastic activation in the hematopoietic stem cell niche, *Ann NY Acad Sci* **1068**:477–488 (2006).

[41] Blazar BR, Murphy WJ: Bone marrow transplantation and approaches to avoid graft-versus-host disease (GVHD), *Phil Trans Roy Soc Lond B Biol Sci* **360** (1461):1747–1767 (2005).

[42] Shizuru JA et al: Purified hematopoietic stem cell grafts induce tolerance to alloantigens and can mediate positive and negative T cell selection, *Proc Natl Acad Sci USA* **97** (17):9555–9560 (2000).

[43] Billingham RE, Brent L, Medawar PB: Actively acquired tolerance of foreign cells, *Nature* **172** (4379):603–606 (1953).

[44] Altman J, Das GD: Autoradiographic and histological evidence of postnatal hippocampal neurogenesis in rats, *J Compar Neurol* **124** (3):319–335 (1965).

[45] Martino G, Pluchino S: The therapeutic potential of neural stem cells, *Nat Rev Neurosci* **7** (5):395–406 (2006).

[46] Thored P et al: Persistent production of neurons from adult brain stem cells during recovery after stroke, *Stem Cells* **24** (3):739–747 (2006).

[47] Galli R et al: Neural stem cells: An overview, *Circ Res* **92** (6):598–608 (2003).

[48] Goldman SA, Nottebohm F: Neuronal production, migration, and differentiation in a vocal control nucleus of the adult female canary brain, *Proc Natl Acad Sci USA* **80** (8):2390–2394 (1983).

[49] Eriksson PS et al: Neurogenesis in the adult human hippocampus, *Nat Med* **4** (11):1313–1317 (1998).

[50] Doetsch F et al: Subventricular zone astrocytes are neural stem cells in the adult mammalian brain, *Cell* **97** (6):703–716 (1999).

[51] Johansson CB et al: Identification of a neural stem cell in the adult mammalian central nervous system, *Cell* **96** (1):25–34 (1999).

[52] Seri B et al: Astrocytes give rise to new neurons in the adult mammalian hippocampus, *J Neurosci* **21** (18):7153–7160 (2001).

[53] Gage FH: Mammalian neural stem cells, *Science* **287** (5457):1433–1438 (2000).

[54] Suhonen JO et al: Differentiation of adult hippocampus-derived progenitors into olfactory neurons *in vivo*, *Nature* **383** (6601):624–627 (1996).

[55] Purves D, Williams SM: *Neuroscience*, 2nd ed., Sinauer Associates, Sunderland, MA, 2001, pp. xviii, 681 [16, 3, 25].

[56] Alvarez-Buylla A, Lim DA: For the long run: Maintaining germinal niches in the adult brain, *Neuron* **41** (5):683–686 (2004).

[57] Mercier F, Kitasako JT, Hatton GI: Anatomy of the brain neurogenic zones revisited: Fractones and the fibroblast/macrophage network, *J Compar Neurol* **451** (2):170–188 (2002).

[58] Palmer TD, Willhoite AR, Gage FH: Vascular niche for adult hippocampal neurogenesis, *J Compar Neurol* **425** (4):479–494 (2000).

[59] Jin K et al: Vascular endothelial growth factor (VEGF) stimulates neurogenesis *in vitro* and *in vivo*, *Proc Natl Acad Sci USA* **99** (18):11946–11950 (2002).

[60] Pluchino S et al: Neural stem cells and their use as therapeutic tool in neurological disorders, *Brain Res Brain Res Rev* **48** (2):211–219 (2005).

[61] Zhu J, Wu X, Zhang HL: Adult neural stem cell therapy: expansion *in vitro*, tracking *in vivo* and clinical transplantation, *Curr Drug Targ* **6** (1):97–110 (2005).

[62] Han SS et al: Grafted lineage-restricted precursors differentiate exclusively into neurons in the adult spinal cord, *Exp Neurol* **177** (2):360–375 (2002).

[63] Pluchino S et al: Injection of adult neurospheres induces recovery in a chronic model of multiple sclerosis, *Nature* **422** (6933):688–694 (2003).

[64] Ballabh P, Braun A, Nedergaard M: The blood-brain barrier: An overview: structure, regulation, and clinical implications, *Neurobiol Dis* **16** (1):1–13 (2004).

[65] Svendsen CN, Caldwell MA, Ostenfeld T: Human neural stem cells: Isolation, expansion and transplantation, *Brain Pathol* **9** (3):499–513 (1999).

[66] Cheng H, Leblond CP: Origin, differentiation and renewal of the four main epithelial cell types in the mouse small intestine. V. Unitarian theory of the origin of the four epithelial cell types, *Am J Anat* **141** (4):537–561 (1974).

[67] Taylor G et al: Involvement of follicular stem cells in forming not only the follicle but also the epidermis, *Cell* **102** (4):451–461 (2000).

[68] Ohyama M et al: Characterization and isolation of stem cell-enriched human hair follicle bulge cells, *J Clin Invest* **116** (1):249–260 (2006).

[69] Thiboutot D: Regulation of human sebaceous glands, *J Invest Dermatol* **123** (1):1–12 (2004).

[70] Cotsarelis G: Epithelial stem cells: A folliculocentric view, *J Invest Dermatol* **126** (7):1459–1468 (2006).

[71] Tumbar T et al: Defining the epithelial stem cell niche in skin, *Science* **303** (5656):359–363 (2004).

[72] Bickenbach JR, Mackenzie IC: Identification localization of label-retaining cells in hamster epithelia, *J Invest Dermatol* **82** (6):618–622 (1984).

[73] Fuchs E, Raghavan S: Getting under the skin of epidermal morphogenesis, *Nat Rev Genet* **3** (3):199–209 (2002).

[74] Rheinwald JG, Green H: Serial cultivation of strains of human epidermal keratinocytes: the formation of keratinizing colonies from single cells, *Cell* **6** (3):331–343 (1975).

[75] Argyris T: Kinetics of epidermal production during epidermal regeneration following abrasion in mice, *Am J Pathol* **83** (2):329–340 (1976).

[76] Cotsarelis G, Sun TT, Lavker RM: Label-retaining cells reside in the bulge area of pilosebaceous unit: Implications for follicular stem cells, hair cycle, and skin carcinogenesis, *Cell* **61** (7):1329–1337 (1990).

[77] Lyle S et al: The C8/144B monoclonal antibody recognizes cytokeratin 15 and defines the location of human hair follicle stem cells, *J Cell Sci* **111** (Pt 21):3179–3188 (1998).

[78] Inoue H et al: Application for regenerative medicine of epithelial cell culture-vistas of cultured epithelium, *Congenit Anom (Kyoto)* **46** (3):129–134 (2006).

[79] Ronfard V et al: Long-term regeneration of human epidermis on third degree burns transplanted with autologous cultured epithelium grown on a fibrin matrix, *Transplantation* **70** (11):1588–1598 (2000).

[80] Mobini N, Tam S, Kamino H: Possible role of the bulge region in the pathogenesis of inflammatory scarring alopecia: Lichen planopilaris as the prototype, *J Cutan Pathol* **32** (10):675–679 (2005).

[81] Lodish HF: *Molecular Cell Biology*, 5th ed., Freeman, New York, 2003, pp. xxxiii, 973 [79].

[82] Bjerknes M, Cheng H: Clonal analysis of mouse intestinal epithelial progenitors, *Gastroenterology* **116** (1):7–14 (1999).

[83] Holmberg J et al: EphB receptors coordinate migration and proliferation in the intestinal stem cell niche, *Cell* **125** (6):1151–1163 (2006).

[84] Leedham SJ et al: Intestinal stem cells, *J Cell Mol Med* **9** (1):11–24 (2005).

[85] Leblond CP, Stevens CE: The constant renewal of the intestinal epithelium in the albino rat, *Anat Rec* **100** (3):357–377 (1948).

[86] Walker BE, Leblond CP: Sites of nucleic acid synthesis in the mouse visualized by radioautography after administration of C14-labelled adenine and thymidine, *Exp Cell Res* **14** (3):510–531 (1958).

[87] Cheng H: Origin differentiation and renewal of the four main epithelial cell types in the mouse small intestine, *IV. Paneth cells. Am J Anat* **141** (4):521–535 (1974).

[88] Withers HR, Elkind MM: Microcolony survival assay for cells of mouse intestinal mucosa exposed to radiation, *Int J Radiat Biol Relat Stud Phys Chem Med* **17** (3): 261–267 (1970).

[89] Potten CS, Owen G, Booth D: Intestinal stem cells protect their genome by selective segregation of template DNA strands, *J Cell Sci* **115** (Pt 11):2381–2388 (2002).

[90] He XC et al: BMP signaling inhibits intestinal stem cell self-renewal through suppression of Wnt-beta-catenin signaling, *Nat Genet* **36** (10):1117–1121 (2004).

[91] Mills JC, Gordon JI: The intestinal stem cell niche: There grows the neighbourhood, *Proc Natl Acad Sci USA* **98** (22):12334–12336 (2001).

[92] Cairns J: Mutation selection and the natural history of cancer, *Nature* **255** (5505):197–200 (1975).

[93] Schier S, Wright NA: Stem cell relationships and the origin of gastrointestinal cancer, *Oncology* **69** (Suppl 1):9–13 (2005).

[94] Vescovi AL, Galli R, Reynolds BA: Brain tumour stem cells, *Nat Rev Cancer* **6** (6):425–436 (2006).

[95] Fodde R: The APC gene in colorectal cancer, *Eur J Cancer* **38** (7):867–871 (2002).

[96] van de Wetering M et al: The beta-catenin/TCF-4 complex imposes a crypt progenitor phenotype on colorectal cancer cells, *Cell* **111** (2):241–250 (2002).

[97] Hanahan D, Weinberg RA: The hallmarks of cancer, *Cell* **100** (1):57–70 (2000).

[98] Lapidot T et al: A cell initiating human acute myeloid leukaemia after transplantation into SCID mice, *Nature* **367** (6464):645–648 (1994).

[99] Jordan CT, Guzman ML, Noble M: Cancer stem cells, *N Engl J Med* **355** (12):1253–1261 (2006).

[100] Al-Hajj M et al: Therapeutic implications of cancer stem cells, *Curr Opin Genet Dev* **14** (1):43–47 (2004).

[101] Traver D et al: Mice defective in two apoptosis pathways in the myeloid lineage develop acute myeloblastic leukemia, *Immunity* **9** (1):47–57 (1998).

3

GENOMEWIDE EXPRESSION ANALYSIS TECHNOLOGIES

JOHN R. WALKER
Group Leader, RNA Dynamics, Genomics Institute of the Novartis Research Foundation, San Diego, California

Global gene expression profiling using microarrays has become a common practice in today's biomedical laboratories. Because their use has superseded earlier-developed whole-genome gene expression technologies, microarrays will be the focus of this chapter. As microarray technologies and analysis methods have matured, so have the various applications. Microarrays have been used to discover genes involved in basic biological processes, to identify and validate targets for potential pharmacotherapy, to prioritize compounds in drug development using efficacy and toxicity gene expression signatures in vitro and *in vivo*, to diagnose disease, and to determine likely response to drug therapy.

Gene expression microarrays promise, and have already delivered, to provide better markers for stem cell isolation and characterization that will no doubt lead to a clearer understanding of how stem cells function in normal and disease conditions. This will, in turn, accelerate the ability to manipulate these cells to treat disease.

3.1 HISTORY OF MICROARRAYS

If a microarray is defined as an ordered matrix of DNA sequences, then the first microarrays were Southern blots [1]. Similarly, the process of screening colonies

Chemical and Functional Genomic Approaches to Stem Cell Biology and Regenerative Medicine
Edited by Sheng Ding

or plaques from cDNA libraries using a DNA or RNA probe of a single sequence can be considered an early microarray application. But as a method to measure differential gene expression between different complex samples, differential colony hybridization, a modification of the colony hybridization technique [2], can be regarded as the first microarray application. In this application, radiolabeled cDNAs from two different populations are used to screen replicate plates of a cDNA library from a single population (Figure 3.1a). Signal differences between cDNA populations would indicate a potential differentially expressed gene. When it was necessary to organize putative positive plasmids or phage from colony or plaque screens, clones were sometimes purified, amplified, and then arrayed on filters and hybridized with the same probe used for the colony or plaque screen (Figure 3.1b). In a sense, these were the first arrays of ordered clones, although their identities were not known until after they were chosen for sequencing.

A modification of differential colony or plaque screening involved a subtractive hybridization step. An enrichment of one population over another took place when a single-stranded cDNA population from one sample was hybridized to an excess of an mRNA population from another sample. Ideally, sequences in common between the populations would hybridize and be removed. Unhybridized mRNA sequences would then be degraded, and an enriched cDNA population would remain. This cDNA could then be hybridized to colonies or plaques from a cDNA library to identify which genes were present in the subtracted population [3].

Further refinements of this primitive microarray technology took place when a novel method was used to amplify an mRNA population. Eberwine developed a method to produce amplified antisense RNA (cRNA) from a small heterogeneous sample (Figure 3.1c) [4]. This technique allowed for a linear amplification of an mRNA population and broadened the applications of differential screens to smaller amounts of material, from single cells to small tissue samples [5,6].

Simultaneously, other advances in molecular biology led to easier ways to order sequences on surfaces. Advances in DNA sequencing technologies led to expansion of cDNA clone collections, so many sequences of interest could be obtained and arrayed. Methods to synthesize oligonucleotides were improved, which led to the application of oligonucleotides to glass slides [7]. Soon, groups at Stanford University were applying this technology to answer broad biological questions in model organisms, and sharing methods and advice so that other groups could also spot high-density arrays [8]. In a short time, commercial spotted and synthesized arrays became available (Figure 3.1d) [9]. Further advancements in oligonucleotide spotting and synthesizing methods have led to consistent production of high-quality commercial arrays. The completed sequencing of the mouse and human genomes, and improvements in scanner technologies, has allowed for whole-genome representation on single arrays. Finally, data extraction, analysis, and mining of microarray data have advanced significantly since the mid-1990s. All of these improvements have led to the almost ubiquitous use of microarrays for gene expression in today's biomedical laboratories.

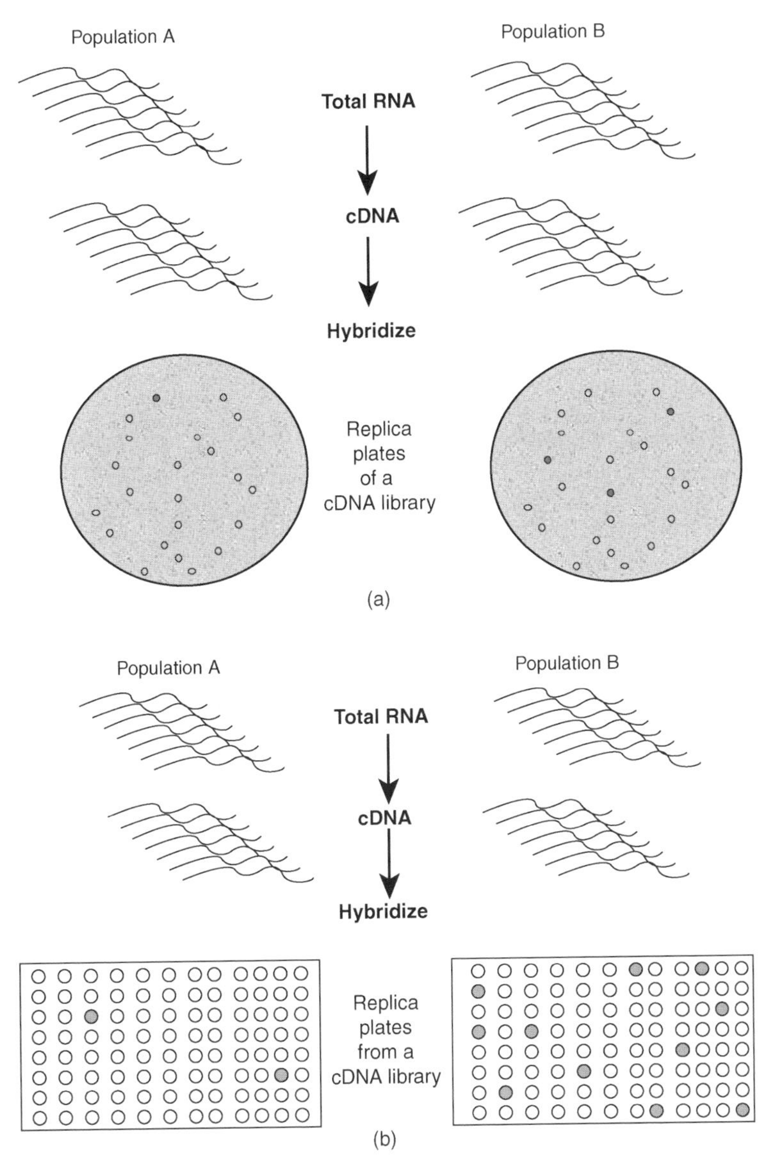

FIGURE 3.1a,b Differential gene expression technology progression. (a) Differential colony hybridization screen of the 1970s. cDNA libraries were screened by replica plating of colonies or plaques onto nylon or nitrocellulose filters. Radiolabeled cDNA was synthesized from total RNA of different populations and hybridized to the replica plates. A difference in radioactive signal between replicate colonies or plaques represented a potential differentially expressed gene. (b) A modification of the differential colony hybridization screen, but replicate purified cDNA colonies, plaques, or cloned cDNAs were stored and plated in 96-well format. The replica plates were screened with radio labeled cDNAs of different populations. The advantage of this technique is that the clones were purified and ordered, and often the identity of the clones was known.

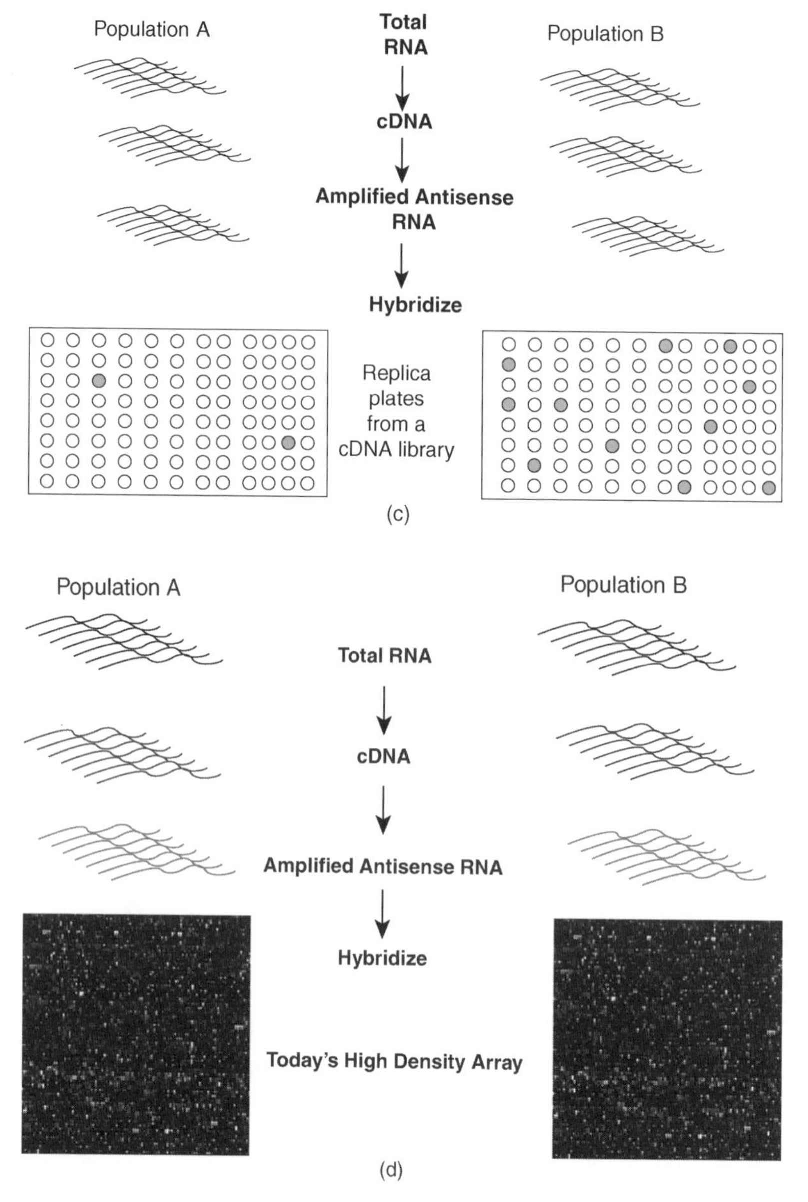

FIGURE 3.1c,d (c) Use of amplified antisense RNA to screen duplicate arrayed 96-well plates. The advantage of this technique is the extension of these applications to rare tissues or cell types because the material could be amplified linearly. (d) Standard microarray procedure of today. Compared to (c), arrays are higher-density, array manufacturing is more consistent, and material to be hybridized to the array does not need to be radioactive. (See insert for color representation of part (d).)

3.2 COMMERCIAL MICROARRAY VENDORS

Today there are many microarray vendors promoting slightly different technologies. Affymetrix is the current market leader and produces "whole-transcriptome" arrays for several different species (http://www.Affymetrix.com). Their technology relies on photolithography to synthesize 25-mer oligonucleotides directly onto a high-density array matrix [10]. Every transcript is currently interrogated with 22 different probes, 11 of which are a perfect match to a different portion of mRNA; the other 11 are identical to the perfect-match oligonucleotides except for a single base pair mismatch in the middle of the 25-mer. This redundant design requires an additional processing step to determine the signal for each mRNA. Affymetrix, like many of the other vendors mentioned below, also requires a specialized scanner and other associated hardware to process the arrays.

Agilent (www.agilent.com) is a major microarray manufacturer that relies on a patented inkjet printer that promises to minimize spot artifacts due to its "no contact" printing process. Agilent uses 60-mer oligonucleotides that they claim offer better hybridization properties than do shorter ones simply because portions of these long oligonucleotides are at a greater distance from the array surface [11].

Illumina (www.illumina.com) offers a bead-based microarray technology, in which 3-μm beads are coated with 50-mer oligonucleotides concatenated with shorter "address" oligonucleotides. Beads with different oligonucleotide sequences are mixed together and added to a microarray that contains the ends of fiberoptic bundles that have been etched to form tiny wells. The address oligonucleotide is then used to determine the position of the 50-mers on the array using a decoding process. There are on average 30 identical 50-mer probes per mRNA on each array. The unique design of beads and microwells allows Illumina to assemble arrays in very high densities [12]. Since the address oligonucleotide is attached to the bead and the probe oligonucleotide is away from the bead, this arrangement would theoretically achieve better hybridization than would short probes directly attached to arrays. Because the manufacturing process is simple, an advantage of this technology over its competitors is price.

ABI was a fairly late entry into the microarray market. They rely on chemiluminescent detection and 60-mer oligonucleotides synthesized onto a nylon support. The chemiluminescent detection approach claims to provide backgrounds lower than those achievable with fluorescent detection (https://www2.appliedbiosystems.com).

CodeLink from Amersham uses a noncontact, proprietary piezoelectric dispensing method to apply 30-mer oligonucleotides to a three-dimensional (3D) surface [13]. This process allows for validation of full-length synthesis of the oligonucleotides. In addition, because of the 3D nature of the array, they claim that hybridization kinetics are better, which should result in a lower background (http://www4.amershambiosciences.com/aptrix/upp01077.nsf/Content/codelink_bioarray_system).

Other microarray vendors provide greater flexibility in design of custom arrays. Nimblegen (http://www.nimblegen.com/) uses a proprietary oligonucleotide synthesis technology that allows for custom high-density array designs at a very low cost [14]. Because this allows for greater flexibility in custom designs, the technology is also widely used for other microarray applications such as chromatin immunoprecipitation (ChIP on chip) and array comparative genome hybridization (aCGH). Also because they use longer probes for expression analysis, Nimblegen claims better probe behavior than for shorter oligonucleotides. Like Affymetrix and Illumina, they have redundant probes for every transcript. One unique feature of this company is that they operate as a service, so hybridizations are performed by Nimblegen. However, a partnership with Affymetrix allows them to provide their custom arrays in Affymetrix array cartridges, so hybridizations can be done on any Affymetrix system.

The probe design process that CombiMatrix (`http://www.combimatrix.com`) uses also promises design flexibility [15]. They provide an array synthesizer, so customers are able to synthesize their own arrays. An additional advantage of their system is that their catalog arrays can be read on any standard microarray slide scanner so that no additional capital equipment needs to be purchased. Arrays are synthesized and not spotted, promising high array quality. Because density is not as high as with other technologies, this system is ideal for use in model organisms for which whole-genome catalog arrays are not available. A list of microarray providers as of this writing is available in a recent *Nature* review [16].

3.3 MICROARRAY SERVICE PROVIDERS

If an academic institution or biotechnology company does not have a microarray core facility, it is still possible to perform microarray experiments without a significant capital investment. Service providers have formed which support technologies of the major microarray vendors. Academic core facilities have also made a business out of performing experiments for "outside" laboratories. Service providers can supply everything for a microarray experiment from experimental design to analysis. All that a customer needs to supply is the samples. Because of these services, the technology has become much more readily available to all who can afford it. Another advantage to using the services of these providers is that they are already trained in design, execution, and analysis of these experiments. Therefore, a submitter can immediately obtain consistent high-quality data. Service providers can be found by visiting the Websites of current microarray vendors or core labs of major academic centers. However, if an organization is planning to run more than 1000 samples across microarrays per year for several years, it makes sense for them to consider setting up their own array core facility.

3.4 SAMPLE AMPLIFICATION AND LABELING TECHNOLOGIES

Detection on a microarray requires that mRNA or its derivative be labeled, and in some cases, amplified. There are myriad options for this process, and the techniques can have a major impact on the results. Most sample-labeling techniques incorporate a fluorescent dye into the nucleic acid that is hybridized to the array. The method of incorporation and the type of fluorescent dye can vary. The fluorescent dye, or a modified nucleotide that will later be tagged with a fluorescent dye, can be incorporated during mRNA conversion to cDNA (direct labeling) or later (indirect labeling).

Some protocols require that nucleic acids be amplified in order to achieve adequate detection sensitivity. There are many different methods to do this, but the most common is the Eberwine amplification scheme [4]. In this protocol, cDNA is made from mRNA using a T7-RNA polymerase promoter concatenated with an oligo-dT primer. After first- and second-strand cDNA synthesis from mRNA, cRNA is amplified by transcription of the T7-cDNA using T7-RNA polymerase. A modified nucleotide is incorporated at this step, which can later be tagged with a fluorescent dye. This material can be hybridized to a microarray (Figure 3.1c). If necessary, two rounds of amplification can be used to produce labeled amplified material from picogram amounts of total RNA, in some instances.

Other less laborious protocols can be used to amplify and label nucleic acids from very small quantities of total RNA. For example, NuGen uses a proprietary amplification scheme using a hybrid RNA-DNA primer to make labeled antisense cDNA that can be hybridized to arrays [17]. There are some indications that this technique might allow for a greater signal-to-noise ratio by reducing cross-hybridization [18].

All of these amplification technologies can introduce protocol-specific biases that render cross-technique comparisons difficult. In addition, two rounds versus one round of amplification of the same material will often result in detection of sequences specific to one protocol or another. At present, researchers should choose techniques that are compatible with their microarray vendor of choice and use the same protocol for all samples in a given experiment. Future improvements in these methods should be geared toward protocols that are simpler and compatible with automation, introduce little amplification bias, and can be used with small and/or degraded samples.

Some protocols are now recommended for partially degraded RNA samples, such as those from formalin-fixed and paraffin-embedded tissues. Many of these samples have been archived from valuable disease populations over the decades. Because these samples may not have been collected with RNA integrity in mind, the samples may be old, and the formalin fixation might interfere with sample amplification and labeling. Several companies, including Epicentre (`http://www.epibio.com`), Genisphere (`http://www.genisphere.com`), and Illumina (`http://www.illumina.com`), claim success in using these types of samples for microarray analysis.

3.5 MICROARRAY ANALYSIS TOOLS

Because microarray gene expression datasets yield much information, development of tools to manage this information has also had to keep pace with the development of the technology. Today, an Affymetrix dataset using 50 whole-genome arrays that can be created in just a few days generates over 60 million data points. Software packages now exist to capture relevant sample information and track samples, exclude outlier and

TABLE 3.1 Typical Steps in Analyzing Gene Expression Data[a]

1. *Image analysis*—align grid to image, eliminate outlier pixels and features
2. *Preprocessing*—summarize probes, subtract background, normalize overall intersity between arrays
3. *Differential expression analysis*—statistical analysis, clustering
4. *Pathway analysis*—determine whether differentially expressed genes are functionally related

[a]Regardless of platform technology, gene expression data is analyzed in a series of ordered steps. *Image analysis* occurs after the microarray is scanned. At this stage, obvious image artifacts are flagged and/or eliminated from further analysis. A grid is aligned to the image to allow for location of each DNA sequence on the array. If a spotted microarray was used, optimal regions of each spot can be determined for analysis. Obvious outlier pixels (a subset of each spot) or spots are flagged and/or eliminated. *Preprocessing* occurs before any differential gene expression measurements are made. In this step, array background needs to be determined and subtracted, normalization must be done so that differential expression is not driven by overall differences in array intensity, and some type of probe summarization needs to be done. If all probes for a given gene are the same, then nonoutlier probes can be averaged. For Affymetrix technology where multiple probes per mRNA are not identical, more complicated summarization algorithms need to be applied (see Ref. [35]). *Differential expression* measurements depend on the design of the experiment, but can involve classical statistical methods as well as clustering. The latter method groups transcripts and samples with similar expression values so that potentially relevant patterns can emerge. Heatmaps, which involve experimenter-specified transcript and/or sample ordering, can also be used to display relevant gene expression patterns. *Pathway analysis* involves layering lists of differentially expressed genes onto known moleular interactions so that some biological relevance can be drawn from the experiment.

artifactual data points from a scan image, statistically analyze and provide data visualization, and summarize relevant biological findings. There are various open-source and commercial software options to aid in each of these stages in the workflow of a microarray experiment (Table 3.1). Many of these options are listed on the Websites of microarray core facilities. The Bioconductor project [19] (`http://www.bioconductor.org`) assembles microarray analysis packages created and freely shared by scientists. Commercial software compatible with a particular microarray platform is listed on the array manufacturers' Websites. A summary of the various steps in which analysis software is needed will be outlined here. Since new software vendors are continually being introduced, the focus here will be on the general procedures and not the vendors. The reader is referred to a complete recent list of microarray-related software vendors and open-source providers in a review by Eisenstein [16].

Once a microarray experiment has been planned, it is useful to collect as much experimental information up front as this information will not only provide for a more informed analysis but will also prepare the experimenter for the eventual submission of the data to a public database on publication of a manuscript. The type of information that might be tracked includes details about the samples such as tissue type, amount of starting material, age, gender, genotype, drug treatments, and developmental stages. Information about the experiment should also be tracked, including labeling and amplification method, microarray type, array washing protocols, and scanner properties. Experiment performance should be tracked, including amplification yields and array background and signal-to-noise measurements. Although the possible types of experimental parameters that can be captured are virtually endless, a laboratory needs to balance the possible information gained with potential lack of experimenter compliance in requiring a large amount of information.

Keeping track of this information is not difficult if a laboratory is performing only a handful of microarray experiments per year. However, research institutes, biotechnology companies, and pharmaceutical companies have found it useful to share data across the organization to allow for later data mining. To do this, some type of laboratory information management system (LIMS) needs to be put into place. Many of these LIMS databases are custom-built or provided by the microarray manufacturer, but some are commercially available.

Once an experiment has been completed, the first data captured are the array image. Software, often provided by the manufacturer, is then used to align a grid to the image so that the location of features can be determined. If the array features were spotted, then analysis needs to be done to determine the best possible portion of the spot to measure and to eliminate results due to spotting artifacts. For all array types, background subtraction must be performed and outlier features need to be eliminated. At some point, multiple image pixels per feature need to be averaged. Redundant or identical probes need to be averaged in some manner. Finally, if multiple arrays are to be analyzed together, normalization is necessary to enable comparison of arrays with differing overall intensities. At this stage, arrays subject to obvious tissue preparation, sample processing, or array-processing error can often be found and eliminated.

The next step in data analysis is to measure mRNA-level differences between relevant experimental groupings. Replicate samples need to be identified, and outlier samples can be flagged. If the number of replicates per grouping is sufficient, then classical statistical tests such as analysis of variance (ANOVA) can be used to calculate the significance of the differences between the groups. Because of the number of simultaneous measurements in microarray experiments, the results are subject to multiple testing errors. For example, if one is measuring the difference in expression between two experimental groups on a 10,000-probe array and a Student's *t*-test is used for analysis, a *p*-value cutoff of 0.01 by chance alone will yield 100 putative differentially expressed genes. So, if 200 genes with this *p* value are found in this experiment, approximately half are likely false positives. A strict correction for the number of measurements (Bonferroni) often is much too stringent so that *p* values of less than 1×10^{-7} will be required to state that a given gene is differentially expressed. Therefore, alternative methods besides a strict multiple testing correction should be considered. Some more recent methods that were developed specifically for microarray analysis can estimate a "false discovery rate" so that some level of confidence in the findings can be estimated [20–22]

Because of the relatively high cost of microarray experiments, experimental designs seldom incorporate enough replicates for traditional statistical analyses. In order to assess differential expression, researchers often simply rely on determining fold-change differences between conditions, in other words, magnitude of change. This can be adequate when one is simply attempting to catalog the most differentially expressed genes, and then will verify those changes with another technology such as quantitative RT-PCR (QPCR). But because of the multiple testing problems described above, confidence in the findings cannot be assessed. Also, magnitude of change in mRNA levels doesn't necessarily correlate with biological significance, as groups of genes in the same biological pathway might be differentially expressed at a low level and add up to a biologically meaningful event [23]. To improve the chances of finding relevant changes in experiments with few replicates, two more measures besides fold change should be considered. The first is the intrareplicate expression level relative to the between-group expression level. In other words, how reliable is the group measurement? If the replicates do not give similar measurements, then that gene should not be considered for differential expression. The second measure is the minimum expression threshold, so that at least background expression levels must be reached in at least one experimental grouping.

Another method to assess differential expression and to display data, and a very popular one in microarray studies, is to cluster the data. The advantage of this technique is that we can begin to see patterns in the expression data, which can be difficult given the number of simultaneously changing mRNAs and possible different experimental conditions. However, clustering can make patterns out of random data. Since there are a large number of clustering algorithms available, the best advice is to choose at least two different ones and compare results. It is also a good idea to compare your results to randomized data to ensure that the patterns you see with your data are distinct. The article by D'Haeseleer is a useful primer on clustering analysis [24].

Regardless of which method is used to find differentially expressed genes, it is still useful to display the changes so that patterns may emerge. One way to do this is with heatmaps, where one orders all experimental groups in one dimension and mRNAs in the other dimension. The third dimension is expression level and is most likely displayed with a color scheme. Upregulated, across conditions, genes are displayed in one color and downregulated genes in another. The magnitude of the change can be indicated by the intensity of the color. In order to display genes in this manner, expression levels need to be normalized so that the magnitude of the change, not the absolute expression level, is highlighted.

Once differentially expressed genes are found, then experimenters need to put the data into a biological context. The completed sequence of various genomes has aided the acceleration of knowledge of gene function. That information is being assembled in databases so that large webs of gene product interactions can now be described. Thanks to projects such as Gene Ontology (GO) and the *Kyoto Encyclopedia of Genes and Genomes* (KEGG), which assemble this information, both commercial and open-source pathway-building software packages have been created [25,26]. So now, one can fairly rapidly determine whether their gene expression experiment has led to changes in biological pathways that could explain the underlying biology behind the experiment. Often it turns out that no obvious biological pathways seem to be affected. This could be due to faults in the experimental design or analysis, limitations in the technology, because the biological question cannot be assessed with gene expression alone, or simply that the genes affected are not yet adequately described. Integration of pathway databases with other emerging genomics datasets, like single-nucleotide polymorphisms (SNPs) and methylation collections, will in the future lead to an even greater understanding of gene function. The hope is that more information, when managed properly, will aid in interpretation of gene expression experiments.

3.6 GROWING PAINS

As described above, microarray designs, sample preparation techniques, and analysis methods vary widely. More recently, there has been quite a bit of concern about the reproducibility of microarray experiments. For example, comparing the same samples across microarray platforms yielded very little concordance [27]. But one encouraging set of studies showed that data collected using standardized methods from experienced laboratories can be reproducible across platforms and laboratories, at least for a subset of differentially expressed genes [28–30]. In addition, the Tan group study was reanalyzed and it was found that the analysis methods and errors in the experimental procedures, and not platform inconsistencies, led to the low concordance in the datasets [31]. Because of these concerns and the Food and Drug Administration's (FDA's) consideration of the use of genomic datasets in the drug approval process, there have been efforts to compare gene expression platforms and associated sample and data-processing methods (`http://www.fda.gov/cde/genomics`). The Microarray Quality Controls Consortium (MAQC) at the FDA is charged with establishing a large dataset of microarray data across several platforms from different sites and making tightly controlled comparisons across these platforms [31]. Similarly, the External RNA Controls Consortium (ERCC) has been developing a set of controls that can be used across platforms to assess the efficiency of different stages of sample and array processing [32]. The results from these consortia concluded that cross-platform comparisons were actually quite good if proper comparisons are made [33]. Finally, the Microarray Gene Expression Data (MGED) Society has made progress in requiring the submission of at least a minimal amount of experimental information before submitting data to journals [34]. These efforts have led major journals to also require submission of all microarray data referenced in a manuscript to a public database on submission of the manuscript. This transparency should result in greater success in replication of experiments and, as described below, allows for sharing of microarray data that can lead to additional discoveries from the same dataset.

In parallel with the microarray platforms themselves, microarray analysis methods have matured since 1997 or so. Many algorithms have been developed to aid in all steps in the analysis process. For example, more than 50 algorithms have been developed simply to condense probe data from Affymetrix arrays so that transcript intensity values can be derived [35]. Because these methods often give widely varying results, it can be confusing to determine which one to choose, and it is probably safest to use at least two of these methods for analysis [36]. Allison and colleagues listed basic recommendations to dispel analysis myths and provided clarity to five components of microarray analysis: design, preprocessing, inference, classification, and validation [37]. The authors point out that many of these supposedly new methods are simply modifications of methods used by statisticians for decades, and they suggest that algorithm developers spend more effort in describing and analyzing the merits of current algorithms than trying to invent new ones.

Because microarrays attempt to quantify gene expression across many genes simultaneously, detection is not optimal for every nucleic acid sequence. The resulting loss in sensitivity means that many messenger RNAs are simply not detectable with microarrays. In addition, compared to QPCR, technical replicates are more variable, and there is a loss in estimating relative changes in expression level across microarray samples in an experiment [38]. Further improvements in microarray designs, sample amplification and labeling, array-scanning technologies, and analysis algorithms in the future will give more comparable results to QPCR.

Many technologies have benefited from automation. As any microarray core lab manager knows, variability can be introduced at all levels of sample and array processing, and the experimenter alone can be a major factor. Therefore, automation of sample and array processing will reduce experimenter artifacts. Regardless of array platform type, sample preparation steps from RNA isolation to array hybridization can benefit from automated liquid-handling robots that have been used for years in other areas of the biotechnology industry. As far as array processing steps, array manufacturers like Affymetrix and Illumina are attempting to introduce systems to simultaneously hybridize, wash, and scan 96 whole-genome expression arrays simultaneously.

There are still several areas of improvement for the technology, independent of analysis methods and measurement limitations. The most obvious problem at the current time is monetary expense of these experiments. Although the cost per messenger RNA is lower for a whole-genome microarray than for other less multiplexed methods, typical experiments can run in the tens of thousands of dollars for just the arrays and reagents. Moreover, most platforms require purchase of specialized equipment that can cost in the hundreds of thousands of dollars. Until these costs drop significantly, most laboratories will limit the sizes and frequency of microarray experiments.

3.7 APPLICATION OF PUBLIC GENE EXPRESSION DATASETS

Although direct comparison across platforms and laboratories warrants caution, as described above, very useful information can be extracted from other people's experiments performed in other laboratories. There are many publicly available gene expression data sources that can be queried for particular genes and biological processes of interest. There are two general categories of gene expression databases. One category involves the compilation of information about every gene in the genome. Information can consist of basic details such as sequence, biological function, and location in the genome, but the key feature is gene expression levels across tissues. These "tissue gene expression" databases typically contain microarray data performed by a single group using a single microarray type, so cross-sample comparisons are possible. The second type of expression database catalogs complete expression datasets performed in a wide array of laboratories. These "data warehouses" are the preferred data deposit sites of major journals on submission of a manuscript containing microarray gene expression data.

Tissue gene expression data sources like Symatlas (http://symatlas.gnf.org), SOURCE (http://source.stanford.edu), and The Mouse Gene Prediction Database (http://mgpd.med.utoronto.ca/) can be valuable tools to determine gene function [39–41]. These free resources allow researchers to perform simple queries to determine in which tissues their genes of interest are expressed. These resources can also be used in a much broader fashion. For example, Mootha's group (2003) used tissue gene expression to find a previously unknown gene responsible for a mitochondrial disorder [42]. Tissue gene expression databases, along with other databases, have also been used to categorize, at a whole-genome level, genes potentially involved in a particular type of disease category [43]. These datasets can, in addition, help prioritize potentially causative genes in rodent quantitative trait locus (QTL) or ethyl-*N*-nitrosourea (ENU) mutagenesis studies [44,45]. Finally, tissue expression can determine whether a gene product is a realistic target for pharmacotherapy. For example, if one is interested in targets for prostate cancer therapy, an ideal candidate would be one whose expression is activated in cancer yet whose expression is low in other normal tissues besides prostate [46].

Often researchers have a particular gene or set of genes on which they would like information in addition to tissue gene expression. Besides *where* a gene is expressed, it would also be useful to know *when* it is expressed. For example, it could be beneficial to determine what happens to expression of a particular gene when another gene is knocked down or overexpressed in mice or in cell culture, when animals or cells are treated with a particular drug or given a specific stimulus, or in a particular disease condition. Genewise queries across all samples are not yet possible in large publicly available datasets, probably because it would take a huge effort to organize the data and set up the analyses. But two large data repositories have been making it easier to query experiments that other groups have performed, and to download that data for further analysis. Both Array Express from EMBL (http://www.ebi.ac.uk/arrayexpress), and Gene Expression Omnibus (GEO; http://www.ncbi.nlm.nih.gov/geo) contain tens of thousands of microarrays from hundreds of published microarray experiments [47,48]. Before considering performing a microarray experiment, it could be beneficial to first see if a similar experiment has been performed and submitted to one of these data repositories.

Some data repositories have been created specifically for stem cell microarray experiments. These data collections are intended to put relevant gene expression datasets in one place so that many laboratories can mine data that have already been collected. Two of those deserve mentioning here. StemBase from the Ottawa Health Research Institute (http://www.ottawagenomecenter.ca/projects/stembase) is a collection of about 200 microarrays with specialized analysis tools [49]. The Stem Cell Database (http://stemcell.princeton.edu/v1/index.html) contains information on genes enriched in hematopoietic stem cells [50]. The hope is that these specialized stem cell datasets will aid in discovery of genes important in the unique functions of stem cells.

TABLE 3.2 Potential Microarray Applications

Identify genes potentially relevant to biological system of interest
Validate therapeutic targets of disease
Optimize candidate therapeutics
Identify targets of therapeutics, when in doubt
Clinical applications: diagnostics and prognostics

[a]The most common application has been to find genes that might participate in a biological process by examining their expression patterns across a set of conditions such as a disease state or disease model. The technology has also been used widely in other areas of the drug discovery process. A *therapeutic target* refers to a gene product that could be used as a target for pharmacological intervention. *Validation* refers to the process of examining the role of a particular gene product in a disease state or disease model. A *candidate therapeutic* is a pharmacological agent in consideration for eventual administration to humans. Often small molecule screens are developed against a disease process without a particular target in mind. Gene expression profiling can be used to determine the targets of candidate compounds.

3.8 THE WIDE ARRAY OF MICROARRAY APPLICATIONS

The first global microarray experiments broadly described, at a transcriptional level, genes that were likely to be involved in a biological process of interest [8,51]. Application of gene expression microarray technology has ventured beyond a global analysis of gene expression in a biological system in order to broadly describe that system. It is now used at all levels of the drug discovery process, including disease target identification, disease target validation, drug candidate optimization, confirmation of targets of drug candidates, and for diagnostic purposes in clinical studies (Table 3.2).

Gene expression microarrays have successfully identified potential therapeutic targets for disease across many areas of biology. In one example, genes not previously known to be involved in multiple sclerosis (MS) were found by microarray to be differentially expressed between disease and control samples. Manipulation of the levels of a subset of these transcripts in a mouse model of MS reduced the disease symptoms [52]. In another example, expression profiling of tumor samples from various organs identified EphB2 as a cell surface target in colorectal cancers. Using that information, a therapeutic antibody was developed against EphB2 and shrunk tumors in mouse models [53].

Expression profiling using microarrays has been successfully applied in the target validation process. Target validation can be seen as the process in which a protein that has been determined as a target for drug therapy is validated by overexpression or knockdown in another *in vivo* or *in vitro* system to determine whether disease characteristics are maintained. Disease characteristics can be either phenotypes or gene expression profiles consistent with the disease. As an example of the latter approach, expression profiling was used to identify target genes of a previously identified candidate disease gene [54]. This approach uncovered a transcriptional network that played a key role in the disease. Thus, expression profiling can serve as a phenotypic marker in target validation as well as reveal additional genes for possible therapeutic intervention.

Another step in the drug discovery process where microarrays can be used is the process of lead compound optimization, where candidate compounds can be prioritized by showing a combination of beneficial gene expression profiles in target organs as well as minimal toxicity profiles in other vital organs. This approach might be aided by comparing expression profiles to a competitor's marketed drug for which the candidate compounds are intended to compete. This is still an unvalidated approach.

Global gene expression analysis has been shown to be useful in identifying the mechanism of action of potential therapeutics in biological systems. Some small-molecule functional assays used in drug development are not designed against a specific target. An example of this would be cell viability or proliferation as an endpoint in screens for anticancer compounds. Once effective candidate compounds are found, it is imperative to discover the target protein or pathway to provide a greater understanding of the mechanism of action. In 2004 Wu and colleagues used gene expression to find the target pathway of a small molecule that induces osteogenesis in mesenchymal progenitor cells. The authors found by global gene expression that the Hedgehog signaling pathway was stimulated by this compound, and they were able to narrow down the site of action to a specific point in the pathway [55]. Even in targeted small-molecule screens, microarray profiling can be used to confirm specificity of the compounds against the intended target. Bedalov and coworkers [56], compared expression profiles of a candidate compound in yeast against a yeast deletion mutant of the intended target of the small molecule. The expression profile of the small molecule overlapped the expression profile of the deletion mutant of the intended target better than did a collection of deletion mutants from the same protein family that were not believed to be targets of the compound [56]. These approaches demonstrate that global gene expression profiling can be used to identify specific protein targets of biologically active small molecules.

Global gene expression profiling has found many diagnostic and prognostic applications since 1997 or so, although the reliability of some of these early studies is in question [57]. Nevertheless, the technology has been shown to effectively find subgroups within disease populations as well as predict disease outcome. In addition, it has the promise of being able to predict whether a patient will respond to a particular pharmacotherapy. Beer and others [58] used expression patterns of lung adenocarcinomas to find an optimal subset of genes that was able to stratify patients into high- and lower-risk categories. This approach could possibly be used to identify alternative therapeutic strategies for the high-risk groups. In a kidney transplantation study, Flechner's group [59] used expression profiling to predict rejection status using kidney biopsies. Importantly, they were also able to use peripheral blood lymphocytes for the same purpose, leading to the possibility of a minimally invasive approach to predict organ rejection status in patients. Finally, some studies have used gene expression measures in tumors or disease-relevant cell lines to predict the likelihood of response to a particular therapy. Some examples can be found in acute myeloid leukemia [60], colon carcinoma [61], gastric cancer [62], and non-small cell lung cancer [63], among others.

Microarrays promise to provide much new information about stem cells that allow us to more easily manipulate them for therapeutic use and understand their roles in cancer. Ramalho-Santos compared expression profiles between hematopoeitic, neural, and embryonic stem cells [64]. Genes that were commonly expressed in all populations might define the ability of stem cells to self-renew and generate differentiated progeny. Forsberg and coworkers used microarrays to identify novel regulators of self-renewal and proliferation in hematopoietic stem cells [65]. One caveat to expression profiling in defining the roles of various genes in stem cell properties is that the populations used in profiling experiments might not be completely pure. Hopefully cell surface markers can be found from these experiments that may provide for isolation and identification of more pure populations.

3.9 NEW AND COMPLEMENTARY MICROARRAY TECHNOLOGIES

The measurement of steady state mRNA levels using microarrays was the first use of a whole-genome array technology. Many additional complementary technologies are being developed that can detect genetic and epigenetic changes (SNP, array comparative genomic hybridization, and methylation analyses), determine which areas of the genome are transcriptionally active under a given condition (chromatin immunoprecipitation on arrays or ChIP on chip), measure non-protein-coding RNA expression (noncoding RNAs and microRNAs), and measure expression of individual exons and thus find splice variants (exon arrays). Combined application of these tools has already identified novel genes involved in stem cell self-renewal [66].

Identification of single-nucleotide polymorphisms (SNPs) among individuals has accelerated due to the development of array platforms for SNP analysis [67]. A lower-throughput SNP assay together with gene expression analysis in recombinant inbred mouse strains led to the implication of multiple genes in various characteristics of hematopoeitic stem cells [68]. It is only a matter of time until these new array-based SNP technologies will be used to characterize genes involved in human stem cell function. Josephson's, group, used a combination of gene expression and SNP microarray technologies, along with other methods, to characterize human embryonic stem cell lines [69].

Methylation analysis is also being performed on high-density arrays. Since embryonic stem cells are derived during an active period of epigenetic reprogramming, it would make sense that DNA methylation status controls many properties of these cells [70]. Bibikova and colleagues used Illumina arrays to compare methylation patterns between human ES cells and other cell types [71]. This approach identified many regions likely to control ES cell properties, and could serve as a "fingerprint" for human ES cells.

MicroRNAs are an abundant class of noncoding RNA species that have been shown to be important in development and cancer [72]. It is also believed that they will play an important role in stem cell function [73]. Commercial microarrays are now available to examine expression of all known microRNAs simultaneously. Some current vendors include Ambion (`www.ambion.com`), Invitrogen (`www.invitrogen.com`), and Exiqon (`http://www.exiqon.com`). Annotation of known microRNAs is being compiled at the Sanger Institute [74].

Alternative mRNA splicing is also likely to play an important role in stem cell function, as it has in cancer [75]. Genomewide studies of alternative splicing under certain conditions have only begun to be performed. Pritsker's group computationally predicted alternative splice sites on a genomewide scale and confirmed the existence of a high percentage of those predictions [76]. Shoemaker and coworkers examined alternative splicing on a genomewide level using microarrays [77]. Affymetrix now manufactures a full-transcriptome exon array, bringing the possibility of global splice-variant analyses to any laboratory (www.affymetrix.com).

Tracking changes in DNA copy number can be efficiently performed on microarrays. The technique, named aCGH for array comparative genomic hybridization, has been used mainly for cancer studies [78]. Its application in stem cell biology has not been fully realized beyond tracking stability of stem cell lines [79].

ChIP on chip approaches have been used to globally define areas of the genome that are under active transcription in a given experimental condition, as well as identify targets of a particular transcription factor [80]. Several different microarray manufacturers market arrays for such approaches, and there are some service providers who will perform these experiments on a fee-for-service basis (http://www.nimblegen.com; http://www.genpathway.com). A genomewide ChIP on chip study of stem cells under certain conditions would allow us to focus on particular chromosomal regions under active transcription [66,81].

There are now many microarray technologies that can examine, at a genomewide level, genetic, epigenetic, and transcriptional characteristics between cell and tissue types. These evolving technologies will be used in parallel with gene expression analysis to better explain how RNA dynamics controls cell function. We are beginning to learn our lessons from the first genomewide array technology and will apply those lessons to these new technologies. Bringing these various technologies together in a controlled and intelligent manner will allow us to better manipulate stem cells for biomedical purposes.

REFERENCES

[1] Southern EM: Detection of specific sequences among DNA fragments separated by gel electrophoresis, *J Mol Biol* **98** (3):503–517 (1975).

[2] Grunstein M, Hogness DS: Colony hybridization: A method for the isolation of cloned DNAs that contain a specific gene, *Proc Natl Acad Sci USA* **72** (10):3961–3965 (1975).

[3] Travis GH, Naus CG, Morrison JH, Bloom FE, Sutcliffe JG: Subtractive cloning of complementary DNAs and analysis of messenger RNAs with regional heterogeneous distributions in primate cortex, *Neuropharmacology* **26** (7B):845–854 (1987).

[4] Van Gelder RN, von Zastrow ME, Yool A, Dement WC, Barchas JD, Eberwine JH: Amplified RNA synthesized from limited quantities of heterogeneous cDNA, *Proc Natl Acad Sci USA* **87** (5):1663–1667 (1990).

[5] Eberwine J, Yeh H, Miyashiro K, Cao Y, Nair S, Finnell R, Zettel M, Coleman P: Analysis of gene expression in single live neurons, *Proc Natl Acad Sci USA* **89** (7):3010–3014 (1992).

[6] Walker JR, Sevarino KA: Regulation of cytochrome c oxidase subunit mRNA and enzyme activity in rat brain reward regions during withdrawal from chronic cocaine, *J Neurochem* **64** (2):497–502. (1995).

[7] Schena M, Shalon D, Davis RW, Brown PO: Quantitative monitoring of gene expression patterns with a complementary DNA microarray, *Science* **270** (5235):467–470 (1995).

[8] DeRisi JL, Iyer VR, Brown PO: Exploring the metabolic and genetic control of gene expression on a genomic scale, *Science* **278** (5338):680–686 (1997).

[9] Lockhart DJ, Dong H, Byrne MC, Follettie MT, Gallo MV, Chee MS, Mittmann M, Wang C, Kobayashi M, Horton H, Brown EL: Expression monitoring by hybridization to high-density oligonucleotide arrays, *Nat Biotechnol* **14** (13):1675–1680 (1996).

[10] Fodor SP, Rava RP, Huang XC, Pease AC, Holmes CP, Adams CL: Multiplexed biochemical assays with biological chips, *Nature* **364** (6437):555–556 (1993).

[11] Hughes TR, Mao M, Jones AR, Burchard J, Marton MJ, Shannon KW, Lefkowitz SM, Ziman M, Schelter JM, Meyer MR, Kobayashi S, Davis C, Dai H, He YD, Stephaniants SB, Cavet G, Walker WL, West A, Coffey E, Shoemaker DD, Stoughton R, Blanchard AP, Friend SH, Linsley PS: Expression profiling using microarrays fabricated by an ink-jet oligonucleotide synthesizer, *Nat Biotechnol* **19** (4):342–347 (2001).

[12] Kuhn K, Baker SC, Chudin E, Lieu MH, Oeser S, Bennett H, Rigault P, Barker D, McDaniel TK, Chee MS: A novel, high-performance random array platform for quantitative gene expression profiling, *Genome Res* **14** (11):2347–2356 (2004).

[13] Ramakrishnan R, Dorris D, Lublinsky A, Nguyen A, Domanus M, Prokhorova A, Gieser L, Touma E, Lockner R, Tata M, Zhu X, Patterson M, Shippy R, Sendera TJ, Mazumder A: An assessment of Motorola CodeLink microarray performance for gene expression profiling applications, *Nucl Acids Res* ,**30** (7):e30 (2002).

[14] Nuwaysir EF, Huang W, Albert TJ, Singh J, Nuwaysir K, Pitas A, Richmond T, Gorski T, Berg JP, Ballin J, McCormick M, Norton J, Pollock T, Sumwalt T, Butcher L, Porter D, Molla M, Hall C, Blattner F, Sussman MR, Wallace RL, Cerrina F, Green RD: Gene expression analysis using oligonucleotide arrays produced by maskless photolithography, *Genome Res* **12** (11):1749–1755 (2002).

[15] Liu RH, Dill K, Fuji HS, McShea A: Integrated microfluidic biochips for DNA microarray analysis, *Expert Rev Mol Diagn* **6** (2):253–261 (2006).

[16] Eisenstein M: Microarrays: Quality control, *Nature* **442** (7106):1067–1070 (2006).

[17] Dafforn A, Chen P, Deng G, Herrler M, Iglehart D, Koritala S, Lato S, Pillarisetty S, Purohit R, Wang M, Wang S, Kurn N: Linear mRNA amplification from as little as 5 ng total RNA for global gene expression analysis, *Biotechniques* **37** (5):854–857 (2004).

[18] Eklund AC, Turner LR, Chen P, Jensen RV, Defeo G, Kopf-Sill AR, Szallasi Z: Replacing cRNA targets with cDNA reduces microarray cross-hybridization, *Nat Biotechnol* **24** (9):1071–1073. (2006).

[19] Dudoit S, Gentleman RC, Quackenbush J: Open source software for the analysis of microarray data, *Biotechniques* , Suppl 45–51. (2003).

[20] Pounds S, Morris SW: Estimating the occurrence of false positives and false negatives in microarray studies by approximating and partitioning the empirical distribution of p-values, *Bioinformatics* **19** (10):1236–1242 (2003).

[21] Datta S: Empirical Bayes screening of many p-values with applications to microarray studies, *Bioinformatics* **21** (9):1987–1994 (2005).

[22] Tusher VG, Tibshirani R, Chu G: Significance analysis of microarrays applied to the ionizing radiation response, *Proc Natl Acad Sci USA* **98** (9):5116–5121 (2001).

[23] Hakak Y, Walker JR, Li C, Wong WH, Davis KL, Buxbaum JD, Haroutunian V, Fienberg AA: Genome-wide expression analysis reveals dysregulation of myelination-related genes in chronic schizophrenia, *Proc Natl Acad Sci USA* **98** (8):4746–4751 (2001).

[24] D'Haeseleer P: How does gene expression clustering work? *Nat Biotechnol* **23** (12):1499–1501 (2005).

[25] Ashburner M, Ball CA, Blake JA, Botstein D, Butler H, Cherry JM, Davis AP, Dolinski K, Dwight SS, Eppig JT, Harris MA, Hill DP, Issel-Tarver L, Kasarskis A, Lewis S, Matese JC, Richardson JE, Ringwald M, Rubin GM, Sherlock G: Gene ontology: Tool for the unification of biology. The Gene Ontology Consortium, *Nat Genet* **25** (1):25–29 (2000).

[26] Goto S, Bono H, Ogata H, Fujibuchi W, Nishioka T, Sato K, Kanehisa M: Organizing and computing metabolic pathway data in terms of binary relations, *Pac Symp Biocomput* :175–186 (1997).

[27] Tan PK, Downey TJ, Spitznagel EL, Jr, Xu P, Fu D, Dimitrov DS, Lempicki RA, Raaka BM, Cam MC: Evaluation of gene expression measurements from commercial microarray platforms, *Nucl Acids Res* **31** (19):5676–5684 (2003).

[28] Bammler T, Beyer RP, Bhattacharya S, Boorman GA, Boyles A, Bradford BU, Bumgarner RE, Bushel PR, Chaturvedi K, Choi D, Cunningham ML, Deng S, Dressman HK, Fannin RD, Farin FM, Freedman JH, Fry RC, Harper A, Humble MC, Hurban P, Kavanagh TJ, Kaufmann WK et al: Standardizing global gene expression analysis between laboratories and across platforms, *Nat Meth* **2** (5):351–356 (2005).

[29] Irizarry RA, Warren D, Spencer F, Kim IF, Biswal S, Frank BC, Gabrielson E, Garcia JG, Geoghegan J, Germino G, Griffin C, Hilmer SC, Hoffman E, Jedlicka AE, Kawasaki E, Martinez-Murillo F, Morsberger L, Lee H, Petersen D, Quackenbush J, Scott A, Wilson M, Yang Y, Ye SQ, Yu W: Multiple-laboratory comparison of microarray platforms, *Nat Meth* **2** (5):345–350 (2005).

[30] Larkin JE, Frank BC, Gavras H, Sultana R, Quackenbush J: Independence and reproducibility across microarray platforms, *Nat Meth* **2** (5):337–344 (2005).

[31] Shi L, Tong W, Fang H, Scherf U, Han J, Puri RK, Frueh FW, Goodsaid FM, Guo L, Su Z, Han T, Fuscoe JC, Xu ZA, Patterson TA, Hong H, Xie Q, Perkins RG, Chen JJ, Casciano DA: Cross-platform comparability of microarray technology: Intra-platform consistency and appropriate data analysis procedures are essential, *BMC Bioinformatics* **6** (Suppl 2): S12 (2005).

[32] Baker SC, Bauer SR, Beyer RP, Brenton JD, Bromley B, Burrill J, Causton H, Conley MP, Elespuru R, Fero M, Foy C, Fuscoe J, Gao X, Gerhold DL, Gilles P, Goodsaid F, Guo X, Hackett J, Hockett RD, Ikonomi P, Irizarry RA, Kawasaki ES, Kaysser-Kranich T, Kerr K, Kiser et al: The External RNA Controls Consortium: A progress report, *Nat Meth* **2** (10):731–734 (2005).

[33] Shi L, Reid LH, Jones WD, Shippy R, Warrington JA, Baker SC, Collins PJ, de Longueville F, Kawasaki ES, Lee KY, Luo Y, Sun YA, Willey JC, Setterquist RA, Fischer GM, Tong W, Dragan YP, Dix DJ, Frueh FW, Goodsaid FM, Herman D, Jensen RV, Johnson CD et al: The MicroArray Quality Control (MAQC) project shows inter- and intraplatform reproducibility of gene expression measurements. *Nat Biotechnol* **24** (9):1151–1161. (2006).

[34] Ball CA, Brazma A: MGED standards: Work in progress, *Omics* **10** (2):138–144 (2006).

[35] Irizarry RA, Wu Z, Jaffee HA: Comparison of Affymetrix GeneChip expression measures, *Bioinformatics* **22** (7):789–794 (2006).

[36] Ahmed SH, Lutjens R, van der Stap LD, Lekic D, Romano-Spica V, Morales M, Koob GF, Repunte-Canonigo V, Sanna PP: Gene expression evidence for remodeling of lateral hypothalamic circuitry in cocaine addiction, *Proc Natl Acad Sci USA* **102** (32):11533–11538 (2005).

[37] Allison DB, Cui X, Page GP, Sabripour M: Microarray data analysis: From disarray to consolidation and consensus, *Nat Rev Genet* **7** (1):55–65. (2006).

[38] Wang Y, Barbacioru C, Hyland F, Xiao W, Hunkapiller KL, Blake J, Chan F, Gonzalez C, Zhang L, Samaha RR: Large scale real-time PCR validation on gene expression measurements from two commercial long-oligonucleotide microarrays, *BMC Genomics* **7**:59. (2006).

[39] Su AI, Wiltshire T, Batalov S, Lapp H, Ching KA, Block D, Zhang J, Soden R, Hayakawa M, Kreiman G, Cooke MP, Walker JR, Hogenesch JB: A gene atlas of the mouse and human protein-encoding transcriptomes, *Proc Natl Acad Sci USA* **101** (16):6062–6067 (2004).

[40] Diehn M, Sherlock G, Binkley G, Jin H, Matese JC, Hernandez-Boussard T, Rees CA, Cherry JM, Botstein D, Brown PO, Alizadeh AA: SOURCE: A unified genomic resource of functional annotations, ontologies, and gene expression data, *Nucl Acids Res* **31** (1):219–223 (2003).

[41] Zhang W, Morris QD, Chang R, Shai O, Bakowski MA, Mitsakakis N, Mohammad N, Robinson MD, Zirngibl R, Somogyi E, Laurin N, Eftekharpour E, Sat E, Grigull J, Pan Q, Peng WT, Krogan N, Greenblatt J, Fehlings M, van der Kooy D, Aubin J, Bruneau BG, Rossant J et al: The functional landscape of mouse gene expression, *J Biol* **3** (5):21. (2004).

[42] Mootha VK, Lepage P, Miller K, Bunkenborg J, Reich M, Hjerrild M, Delmonte T, Villeneuve A, Sladek R, Xu F, Mitchell GA, Morin C, Mann M, Hudson TJ, Robinson B, Rioux JD, Lander ES: Identification of a gene causing human cytochrome c oxidase deficiency by integrative genomics, *Proc Natl Acad Sci USA* **100** (2):605–610(2003).

[43] Calvo S, Jain M, Xie X, Sheth SA, Chang B, Goldberger OA, Spinazzola A, Zeviani M, Carr SA, Mootha VK: Systematic identification of human mitochondrial disease genes through integrative genomics, *Nat Genet* **38** (5):576–582 (2006).

[44] Brown AC, Olver WI, Donnelly CJ, May ME, Naggert JK, Shaffer DJ, Roopenian DC: Searching QTL by gene expression: Analysis of diabesity, *BMC Genet* **6** (1):12. (2005).

[45] Wen BG, Pletcher MT, Warashina M, Choe SH, Ziaee N, Wiltshire T, Sauer K, Cooke MP: Inositol (1,4,5) trisphosphate 3 kinase B controls positive selection of T cells and modulates Erk activity, *Proc Natl Acad Sci USA* **101** (15):5604–5609 (2004).

[46] Welsh JB, Sapinoso LM, Kern SG, Brown DA, Liu T, Bauskin AR, Ward RL, Hawkins NJ, Quinn DI, Russell PJ, Sutherland RL, Breit SN, Moskaluk CA, Frierson HF, Jr, Hampton GM: Large-scale delineation of secreted protein biomarkers overexpressed in cancer tissue and serum, *Proc Natl Acad Sci USA* **100** (6):3410–3415 (2003).

[47] Brazma A, Parkinson H, Sarkans U, Shojatalab M, Vilo J, Abeygunawardena N, Holloway E, Kapushesky M, Kemmeren P, Lara GG, Oezcimen A, Rocca-Serra P, Sansone SA: ArrayExpress—a public repository for microarray gene expression data at the EBI, *Nucl Acids Res* **31** (1):68–71. (2003).

[48] Edgar R, Domrachev M, Lash AE: Gene Expression Omnibus: NCBI gene expression and hybridization array data repository, *Nucl Acids Res* **30** (1):207–210 (2002).

[49] Perez-Iratxeta C, Palidwor G, Porter CJ, Sanche NA, Huska MR, Suomela BP, Muro EM, Krzyzanowski PM, Hughes E, Campbell PA, Rudnicki MA, Andrade MA: Study of stem cell function using microarray experiments, *FEBS Lett* **579** (8):1795–1801 (2005).

[50] Phillips RL, Ernst RE, Brunk B, Ivanova N, Mahan MA, Deanehan JK, Moore KA, Overton GC, Lemischka IR: The genetic program of hematopoietic stem cells, *Science* **288** (5471):1635–1640 (2000).

[51] Wodicka L, Dong H, Mittmann M, Ho MH, Lockhart DJ: Genome-wide expression monitoring in Saccharomyces cerevisiae, *Nat Biotechnol* **15** (13):1359–1367 (1997).

[52] Lock C, Hermans G, Pedotti R, Brendolan A, Schadt E, Garren H, Langer-Gould A, Strober S, Cannella B, Allard J, Klonowski P, Austin A, Lad N, Kaminski N, Galli SJ, Oksenberg JR, Raine CS, Heller R, Steinman L: Gene-microarray analysis of multiple sclerosis lesions yields new targets validated in autoimmune encephalomyelitis, *Nat Med* **8** (5):500–508 (2002).

[53] Mao W, Luis E, Ross S, Silva J, Tan C, Crowley C, Chui C, Franz G, Senter P, Koeppen H, Polakis P: EphB2 as a therapeutic antibody drug target for the treatment of colorectal cancer, *Cancer Res* **64** (3):781–788 (2004).

[54] Mootha VK, Handschin C, Arlow D, Xie X, St Pierre J, Sihag S, Yang W, Altshuler D, Puigserver P, Patterson N, Willy PJ, Schulman IG, Heyman RA, Lander ES, Spiegelman BM: Erralpha and Gabpa/b specify PGC-1alpha-dependent oxidative phosphorylation gene expression that is altered in diabetic muscle, *Proc Natl Acad Sci USA* **101** (17):6570–6575 (2004).

[55] Wu X, Walker J, Zhang J, Ding S, Schultz PG: Purmorphamine induces osteogenesis by activation of the hedgehog signaling pathway, *Chem Biol* **11** (9):1229–1238 (2004).

[56] Bedalov A, Gatbonton T, Irvine WP, Gottschling DE, Simon JA: Identification of a small molecule inhibitor of Sir2p, *Proc Natl Acad Sci USA* **98** (26):15113–15118 (2001).

[57] Ntzani EE, Ioannidis JP: Predictive ability of DNA microarrays for cancer outcomes and correlates: An empirical assessment, *Lancet* **362** (9394):1439–1444 (2003).

[58] Beer DG, Kardia SL, Huang CC, Giordano TJ, Levin AM, Misek DE, Lin L, Chen G, Gharib TG, Thomas DG, Lizyness ML, Kuick R, Hayasaka S, Taylor JM, Iannettoni MD, Orringer MB, Hanash S: Gene-expression profiles predict survival of patients with lung adenocarcinoma, *Nat Med* **8** (8):816–824 (2002).

[59] Flechner SM, Kurian SM, Head SR, Sharp SM, Whisenant TC, Zhang J, Chismar JD, Horvath S, Mondala T, Gilmartin T, Cook DJ, Kay SA, Walker JR, Salomon DR: Kidney transplant rejection and tissue injury by gene profiling of biopsies and peripheral blood lymphocytes, *Am J Transplant* **4** (9):1475–1489 (2004).

[60] Okutsu J, Tsunoda T, Kaneta Y, Katagiri T, Kitahara O, Zembutsu H, Yanagawa R, Miyawaki S, Kuriyama K, Kubota N, Kimura Y, Kubo K, Yagasaki F, Higa T, Taguchi H, Tobita T, Akiyama H, Takeshita A, Wang YH, Motoji T, Ohno R, Nakamura Y: Prediction of chemosensitivity for patients with acute myeloid leukemia, according to expression levels of 28 genes selected by genome-wide complementary DNA microarray analysis, *Mol Cancer Ther* **1** (12):1035–1042 (2002).

[61] Mariadason JM, Arango D, Shi Q, Wilson AJ, Corner GA, Nicholas C, Aranes MJ, Lesser M, Schwartz EL, Augenlicht LH: Gene expression profiling-based prediction of response of colon carcinoma cells to 5-fluorouracil and camptothecin, *Cancer Res* **63** (24):8791–8812 (2003).

[62] Tanaka T, Tanimoto K, Otani K, Satoh K, Ohtaki M, Yoshida K, Toge T, Yahata H, Tanaka S, Chayama K, Okazaki Y, Hayashizaki Y, Hiyama K, Nishiyama M: Concise prediction models of anticancer efficacy of 8 drugs using expression data from 12 selected genes, *Int J Cancer* **111** (4):617–626 (2004).

[63] Coldren CD, Helfrich BA, Witta SE, Sugita M, Lapadat R, Zeng C, Baron A, Franklin WA, Hirsch FR, Geraci MW, Bunn PA, Jr: Baseline gene expression predicts sensitivity to gefitinib in non-small cell lung cancer cell lines, *Mol Cancer Res* **4** (8):521–528 (2006).

[64] Ramalho-Santos M, Yoon S, Matsuzaki Y, Mulligan RC, Melton DA: "Stemness": Transcriptional profiling of embryonic and adult stem cells, *Science* **298** (5593):597–600. (2002).

[65] Forsberg EC, Prohaska SS, Katzman S, Heffner GC, Stuart JM, Weissman IL: Differential expression of novel potential regulators in hematopoietic stem cells, *PLoS Genet* **1** (3): e28. (2005).

[66] de Haan G, Gerrits A, Bystrykh L: Modern genome-wide genetic approaches to reveal intrinsic properties of stem cells, *Curr Opin Hematol* **13** (4):249–253 (2006).

[67] Ng JK, Liu WT: Miniaturized platforms for the detection of single-nucleotide polymorphisms, *Anal Bioanal Chem* (2006).

[68] Bystrykh L, Weersing E, Dontje B, Sutton S, Pletcher MT, Wiltshire T, Su AI, Vellenga E, Wang J, Manly KF, Lu L, Chesler EJ, Alberts R, Jansen RC, Williams RW, Cooke MP, de Haan G: Uncovering regulatory pathways that affect hematopoietic stem cell function using "genetical genomics", *Nat Genet* **37** (3):225–232 (2005).

[69] Josephson R, Sykes G, Liu Y, Ording C, Xu W, Zeng X, Shin S, Loring J, Maitra A, Rao MS, Auerbach JM: A molecular scheme for improved characterization of human embryonic stem cell lines, *BMC Biol* **4** (1):28. (2006).

[70] Jacob S, Moley KH: Gametes and embryo epigenetic reprogramming affect developmental outcome: Implication for assisted reproductive technologies, *Pediatr Res* **58** (3):437–446 (2005).

[71] Bibikova M, Chudin E, Wu B, Zhou L, Garcia EW, Liu Y, Shin S, Plaia TW, Auerbach JM, Arking DE, Gonzalez R, Crook J, Davidson B, Schulz TC, Robins A, Khanna A, Sartipy P, Hyllner J, Vanguri P, Savant-Bhonsale S, Smith AK, Chakravarti A, Maitra A, Rao M, Barker DL, Loring JF, Fan JB: Human embryonic stem cells have a unique epigenetic signature, *Genome Res* **16** (9):1075–1083 (2006).

[72] Tsuchiya S, Okuno Y, Tsujimoto G: MicroRNA: Biogenetic and functional mechanisms and involvements in cell differentiation and cancer, *J Pharmacol Sci* **101** (4):267–270 (2006).

[73] Zhang B, Pan X, Anderson TA: MicroRNA: A new player in stem cells, *J Cell Physiol* **209** (2):266–269 (2006).

[74] Griffiths-Jones S: The microRNA Registry, *Nucl Acids Res* **32** (Database Issue):D109–D111 (2004).

[75] Venables JP: Unbalanced alternative splicing and its significance in cancer, *Bioessays* **28** (4):378–386 (2006).

[76] Pritsker M, Doniger TT, Kramer LC, Westcot SE, Lemischka IR: Diversification of stem cell molecular repertoire by alternative splicing, *Proc Natl Acad Sci USA* **102** (40):14290–14295 (2005).

[77] Shoemaker DD, Schadt EE, Armour CD, He YD, Garrett-Engele P, McDonagh PD, Loerch PM, Leonardson A, Lum PY, Cavet G, Wu LF, Altschuler SJ, Edwards S, King J,

Tsang JS, Schimmack G, Schelter JM, Koch J, Ziman M, Marton MJ, Li B, Cundiff P, Ward T, Castle J et al: Experimental annotation of the human genome using microarray technology. *Nature* **409** (6822):922–927 (2001).

[78] Mosse YP, Greshock J, Weber BL, Maris JM: Measurement and relevance of neuroblastoma DNA copy number changes in the post-genome era, *Cancer Lett* **228** (1–2):83–90 (2005).

[79] Caisander G, Park H, Frej K, Lindqvist J, Bergh C, Lundin K, Hanson C: Chromosomal integrity maintained in five human embryonic stem cell lines after prolonged *in vitro* culture, *Chromos Res* **14** (2):131–137 (2006).

[80] Kim TH, Ren B: Genome-wide analysis of protein-DNA interactions, *Annu Rev Genomics Hum Genet* (2006).

[81] Kim TH, Barrera LO, Zheng M, Qu C, Singer MA, Richmond TA, Wu Y, Green RD, Ren B: A high-resolution map of active promoters in the human genome, *Nature* **436** (7052):876–880 (2005).

4

GENOMIC cDNA AND RNAi FUNCTIONAL PROFILING AND ITS POTENTIAL APPLICATION TO THE STUDY OF MAMMALIAN STEM CELLS

JIA ZHANG, MYLEEN MEDINA, GENEVIEVE WELCH, DEANNA SHUMATE, ANTHONY MARELLI, AND ANTHONY P. ORTH

Genomics Institute of the Novartis Research Foundation, San Diego, California

Since their discovery in the 1960s, stem cells have been the focus of intense research interest. More recent recognition that stem cell fate determinations may underlie human pathologies and that stem cell differentiation *ex vivo* might offer therapeutic hope for disorders like Parkinson's has only increased this interest. Because of the fragile nature that plagues primary human cells in an *ex vivo* state, a limited supply of stem cells for research use, and the lack of specific tools for their study, a thorough understanding of the cellular programs directing their self-renewal and differentiation has yet to be achieved. One technique that has not been widely applied is genomic functional profiling, a technique capable of testing the contributions of many genes individually in high-throughput fashion for their impact on key cellular processes. Similar methods exploited in traditional low-throughput fashion have demonstrated the importance of the transcription factors Nanog and Oct 4 in maintaining embryonic stem cells in an undifferentiated, pluripotent state [1]. Differentiation, in contrast, involves changes in the expression patterns of hundreds of genes, implying programs of great complexity that remain to be explored [2,3]. High-throughput technologies such as genomic functional profiling are ideally suited to the rapid, unbiased identification of key players in processes such as these. The purpose of this chapter

Chemical and Functional Genomic Approaches to Stem Cell Biology and Regenerative Medicine
Edited by Sheng Ding

is to introduce the concept of genomic functional profiling and what it offers to the study of stem cell function.

4.1 EARLY GENE FUNCTIONALIZATION IN MAMMALIAN CELLS

Until the early 1960s, scientists characterized gene function by determining general nucleotide or amino acid composition; spatiotemporal expression patterns; and metabolic, enzymatic, or structural activities. Elucidation in the 1960s of the genetic code, which lead to development of forward genetic, reverse genetic, and cell culture technologies, greatly accelerated the search for gene function. Over time the functions of thousands of mammalian genes have been characterized *in vitro* or *in vivo* using technologies focusing on one or a few proteins at a time. Functional information derived from such direct experimentation has remained surprisingly restricted, however. The Entrez Gene database developed and maintained by the National Center for Biotechnology Information indexes individual genes to published reports on their functions [4]. Figure 4.1 illustrates the focused nature of this knowledge by representing the distribution of PubMed citations across the human genome. Of 35,566 human genes annotated by Entrez Gene, the most frequently cited 100 account for 12% of all indexed PubMed citations. The least frequently cited half together account for just 3%. Thousands of genes are indexed to one or a few functionally relevant citations and many, to none at all. Moreover, there is little

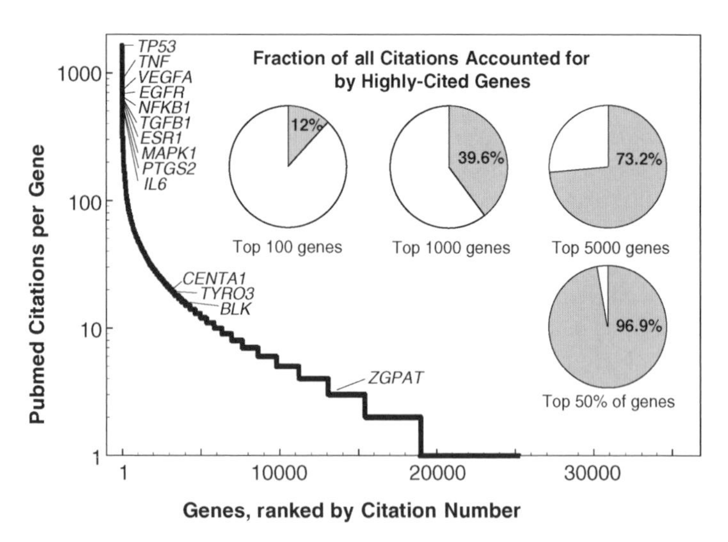

FIGURE 4.1 PubMed citations per human gene according to the Entrez Gene database. The `gene2pubmed.gz` file dated 2/22/07 was obtained from `ftp://ftp.ncbi.nih.gov/gene/DATA/`. This file links PubMed citations reporting well-validated annotation information to canonical genes reported by Entrez Gene. Human genes and linked PubMed citations were parsed by species ID, the number of citations counted for each of the 35,566 obtained human loci, and then genes ranked from most to least frequently cited.

evidence that the degree of citation correlates with the importance of particular genes in key cellular processes [5]; several genomic functional profiling studies have identified putative novel oncogenes lacking either published support or information that would have enabled hypothesis driven prediction of their roles in this highly relevant and intensely studied cellular process.

In the late 1990s, a series of technological innovations spurred the genomics revolution. The availability of the sequences of entire genomes fundamentally altered views of gene function, as traditional gene-centric knowledge achieved through traditional means could be supplemented by whole-genome sequence homology searches rooted in our understanding of the relationship between domains and molecular function. Early genomics tended to view function in terms of genes numbers and protein domains rather than in terms of individual genes themselves. Mining the human genome assemblies using sophisticated gene prediction algorithms and homology-finding hidden Markov models yielded predictive or inferential annotations for 25,000 genes, generally without any experimental data supporting functional prediction. Accompanying this increase in raw sequence data was an increase in the production and availability of complementary cDNA clones used as expressed sequence tags (ESTs) for genome assembly. Libraries of EST clones in formats and vectors permitting overexpression on introduction into mouse and human cells became available from organizations such as ATCC (`www.atcc.org`) and the Mammalian Genome Consortium [6]. Introducing these clones into appropriate cell types became one way to test sequence-derived functional hypotheses.

A second important development in the field of genomics was the widespread adoption of DNA microarrays for measuring global RNA expression. Exposing RNA from cells to a microarray permits relative levels of each RNA to be precisely quantified. Expression patterns for thousands of known or predicted genes have been characterized by comparing RNA levels in different tissues or cell lines at different times [7]. Integrating spatiotemporal RNA expression data, experimentally determined functional knowledge, and domain signature predictions has provided even greater levels of functional information for poorly characterized genes. A scan of the genome using well-trained HMM models from Pfam can easily identify every region capable of encoding a canonical protein kinase and profiling their expression in diverse tissues via microarray analysis has further refined our understanding of putative kinase functions [8,9].

Their considerable advantages notwithstanding, domain prediction and mRNA expression profiling provide information from which function can only be inferred; predicted activities must still be tested and confirmed by direct experimentation. In addition, annotation based strictly on inferred domains or characteristic expression patterns remains powerless to tell us anything about the functions of genes that do not resemble those with known molecular functions or with uninformative expression patterns. Fully 41% of gene products described by the original human genome sequence assemblies were found to be unclassifiable [10]. Evaluation of cDNAs encoding 19,574 unique human proteins using InterProScan found classifiable domains, repeats, or family associations for only 50.1%, leaving the remainder un-annotated [11]. Gene expression profiles determined by microarray analysis may be similarly uninformative owing to very low expression levels, such as occur throughout the GPCR family, or to broad and even tissue expression. Moreover, neither gene expression patterning or knowledge of molecular function necessarily tell us what roles gene products play in key processes of interest to us. All human protein kinases and nuclear hormone receptors have likely been identified, but many remain incompletely uncharacterized in terms of their cellular function, despite our knowledge of their mechanistic capabilities.

4.2 EXPLORING CELLULAR GENE FUNCTION VIA GENOMIC FUNCTIONAL PROFILING

Genomic functional profiling attempts to fill this gap in our knowledge of gene function by subjecting the genome itself to direct functional experimentation. At its most basic, genomic functional profiling systematically overexpresses or depletes genes in cellular models seeking those involved in key processes. By profiling thousands of genes at once without regard to what is known about them in advance, relationships between cellular processes and the genes eliciting them can be characterized in an unbiased fashion. Gene overexpression and depletion strategies are not new; our understanding of molecular function owes much to their application in the context of individual genes. What genomic functional profiling does differently is miniaturize these approaches for use in high-throughput cellular assays, lowering the barrier to testing well-characterized genes for new cellular roles and providing a high-throughput, low-input means of characterizing the roles of genes about which we know little or nothing. CENTA1, a poorly studied GTPase-activating protein regulating neural signaling (see Figure 4.1 for level of functional support in the scientific literature), has been shown by genomic functional profiling to also modulate AP1 signaling leading to anchorage-independent growth, raising the possibility that CENTA1 may also have oncogenic potential [12]. Mammals possess only 2–3 times as many genes as the fruitfly [10,13,14]. One way to achieve greater organismal complexity without dramatically increasing gene number is to use the same molecular activities in diverse cellular contexts. As-yet undetected functions may therefore exist for many well-characterized genes. Detection of these functions and roles will rely on techniques such as genomic functional profiling that can scan through thousands of events for the few true players. Genomic functional profiling offers additional advantages over analyses rooted in inference-based domain mining or studies of RNA dynamics. Expression profiling of pre- and postdifferentiated stem cells has revealed global changes in gene expression [2,3]. Genomic functional profiling assays are capable, in theory, of filtering through this gene set for those that specifically direct differentiation or those that merely implement orders. Assays executed in differentiated cell types have identified tens to hundreds of genes performing rate-limiting steps in diverse cellular processes, and many of these genes have subsequently proved to be functionally related. Genomic functional profiling therefore offers the possibility of unmasking the components of entire cellular pathways or biological networks in a single high-throughput experiment.

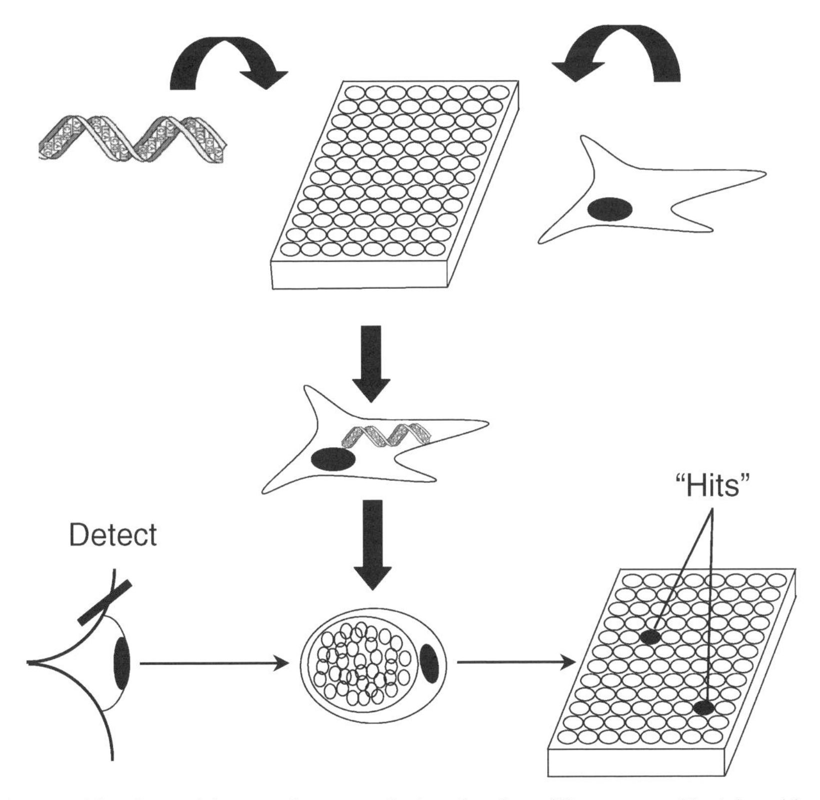

FIGURE 4.2 General layout of a genomic functional profiling screen. Nucleic acids are arrayed in plates, arrayed on glass slides, or pooled into a tube. These nucleic acids form a screening library and usually comprise synthetic siRNAs, encoded shRNAs, or encoded cDNAs. By some means the library is introduced into cells, depicted on the right, producing cells bearing the individual nucleic acids constituting the library, depicted just below the upper 96-well plate. Typical delivery mechanisms include transfection or viral transduction. Those nucleic acids impacting the biological process under investigation generate a detectable change in cellular behavior, depicted by the round, vacuolar cell in the lower center. These primary screen "hits" are usually subjected to considerable follow-up using more traditional, lower-throughput methods to verify the biological relevance to the process under investigation.

Figure 4.2 illustrates the general methodology of a genomic functional profiling experiment. An experiment, frequently referred to as a *screen* or *assay*, comprises three elements: a library of functional nucleic acids, some means of introducing the library into cells, and a measurable and informative assay endpoint linking function and nucleic acid. Overexpression nucleic acid libraries comprise mostly protein-coding cDNA clones. Depletion nucleic acid libraries comprise constructs acting through RNA interference (RNAi) mechanisms, typically either synthetic small interfering RNAs (siRNAs) or encoded short-hairpin RNAs (shRNAs). Libraries are introduced into cells via lipid-mediated transfection or viral delivery. Assay endpoints are varied, reflecting the wide variety of cellular mechanisms to be tested and methods for determining the impact that entities have on them. They range from whole-well assays using cotransfected reporters detecting activation or repression of particular signaling cascades to complex multiparametric and cell-autonomous endpoints detected by high-content imaging. Virtually any cell-based assay developed for low-throughput analyses can be adapted for high-throughput genomic functional profiling. Functional profiling assays can be conducted in arrayed format, where each nucleic acid is interrogated independently or in pooled format, where the assay endpoint incorporates some selection mechanism favoring those nucleic acids eliciting the desired cellular phenotype. Novel hits can generally be divided into three categories: well-annotated genes with previously undetected roles, unannotated genes with predicted domains, and unannotated genes devoid of domain signatures.

4.2.1 Profiling the Function of Overexpressed cDNAs

Initial efforts at overexpression-based functional profiling adapted the reverse-transfection approach to microwell plates, employing cDNAs randomly picked from plated cDNA libraries [15]. Using a library of 10,000 randomly arrayed cDNAs from a CD1 mouse library, Albaryrak and Grimm transfected HEK293T human embryonic kidney cells seeking genes capable of inducing apoptosis, identifying several known and novel proapoptotic factors, including the novel activator SHDC [Cyb(L)], a member of the cytochrome *B* large subunit [16]. Using an adenoviral library of 13,000 randomly selected arrayed cDNAs from a human placental library, Michiels et al. infected mesenchymal stem cells seeking those capable of inducing osteoblast formation among other phenotypes, reidentifying the known osteogenic genes BMP4, BMP7, and FosB as well as several genes hitherto unassociated with this role [17]. Using a library of 145,000 clones semipooled to four clones per well, Matsuda et al. transfected HEK293T cells cotransfected with luciferase reporters sensitive to activation of the NFκB and MAPK signaling pathways, identifying 34 copies of the TNF receptor, 19 of MyD88, 8 of TRADD, but just one copy of the RelA DNA-binding p65 subunit of the NFKB transcription factor complex itself [18].

The composition of cDNA libraries made from cells or tissues tends to be biased toward the most frequently expressed mRNAs. To overcome this bias, most genomic functional profiling efforts employing overexpression have exploited arrays of individual, unique, preselected clones. By purposefully mining large collections of arrayed, expressible, presequenced cDNAs, unique clone arrays can be restricted to desired gene sets, offering far greater gene diversity for any given number of individual clones screened and control over clone diversity. Screening large or rarely expressed genes using cDNA library-based arrays may require hundreds of thousands of clones, and even the very best libraries will contain just a fraction of all genes whose representation cannot be dictated, and be sized-biased against the largest. Of the nearly 7,000,000 human EST clones sequenced to date, only 10 code for a portion of the proangiogenic protein erythropoietin. These 10 clones appear just 11 times among 86,923 clones randomly selected for sequencing from six different cDNA libraries (data compiled from Unigene, http://www.ncbi.nlm.nih.gov/UniGene/cust.cgi?ORT=Hs&CID=2303).

Using unique clone arrays consisting of 7000–20,000 human and/or mouse cDNAs, several groups have functionally profiled the AP1, p53, Wnt, and CREB signaling pathways and probed the mechanism of action of the cyanobacterial cytotoxin apratoxin A. In each of these overexpression-based functional profiling experiments, tens to hundreds of known and novel activators were obtained and confirmed and, in the case of select novel activators, specific activities and mechanisms of action reported. In functionally profiling AP1 signaling with ~20,000 cDNAs transfected independently into 293T cells, the AP1 regulatory activities of cFos, cJun, GADD45, TRAF6, ERB-B2, ATF3, MAP3K11, and others were confirmed [12]. Among the >100 activators, approximately 80 had not been previously associated with a regulatory role in AP1 signaling. Novel AP1 activators of all three varieties were obtained that appear to act at multiple points in the AP1 signaling cascade. The GTPase-activating protein CENTA1 and the uncharacterized protein ZGPAT demonstrated novel oncogenic potential, supplementing what little was known regarding their cellular roles (see Figure 4.1 for functional literature citation number).

Similar results were obtained in pathway-based reporter assays interrogating p53, Wnt, and CREB signaling. In each case known activities were reconfirmed, novel activities identified, and mechanisms of action pursued by more traditional means. In the p53 screen, conducted in p53+/+ HCT116 cells, known p53 activators HIF1A and CDKN2A (p14^{ARF}) were identified as were seven previously undocumented activators and two undocumented repressors [19]. Novel activators HES1, HEY1, TFAP4, and OSR1 were subsequently shown to activate p53 via repression of HDM2. In the Wnt screen, conducted by transfecting a library of ~31,000 unique cDNAs into HEK293 cells cotransfected with a β-catenin/TCF/LEF-responsive TOPflash reporter, the well-documented Wnt activation function of β-catenin and Dishevelled1 DVL1 were reconfirmed [20], the activity of the novel Wnt activator LRRFIP2 revealed, and a biochemical interaction with DVL1 demonstrated after substantial follow-up by traditional means. The activities of two previously unrecognized CREB regulated transcriptional coactivators, CRTC1 and CRTC2 (TORC1 and TORC2), were suggested after functional profiling of CREB reporters by two groups using unique clone arrays of 20,000 and 7000 cDNAs revealed CREB-dependent activation exceeding that of CREB itself [21,22]. Subsequent studies have implicated roles for the CRTC1 and CRTC2 in gluconeogenesis and muscle biogenesis. Overexpression-based profiling has also proved useful in target deconvolution. Functional profiling combined with expression profiling has been employed, for example, to demonstrate that overexpression of the fibroblast growth factor receptor FGFR can complement the cytotoxicity of the potent tumor cell toxin apratoxin A through action on Stat3, a downstream mediator of FGFR signaling [23].

To broaden profiling coverage to as yet un-available clones, Koenig-Hoffman's group screened HEK293 cells for tumor suppressors using a unique clone array containing 17,680 human cDNA clones and an embryo-derived cDNA library array containing ~600,000 clones [24]. Known proapoptotic hits included the small

GTPase ras homologs RHOB and RHOC and the cysteine protease CASP2. Novel proapoptotic hits included the aarF domain containing kinase ADCK1 and ubiquitin-like HERPUD1, neither of which had been hitherto linked to a cellular or molecular function. In an interesting variation on the theme of arrayed well functionalization assays, Fiscella et al. generated a library of 8000 human cDNA clones encoding known and predicted secreted molecules, transfected the library into HEK293T cells generating conditioned media, and then applied the conditioned media to primary T cells and T-cell lines seeking secreted proteins increasing cytokine production [25]. A previously unannotated gene, Integrin Alpha FG-GAP Repeat Containing 1 (ITFG1), was identified and demonstrated to stimulate T-cell secretion of both pro- and anti-inflammatory cytokines and, subsequently, to activate the MAPK pathway leading to *in vivo* induction of interferon-γ. Using the same library and approach, a second group discovered a previously unrecognized role for bone morphogenic protein-9 BMP9 in regulating glucose homeostasis *in vitro* and in an *in vivo* diabetic mouse models [26]. Using a combination of unique clone and random cDNA library arrays comprising over 250,000 clones, Korherr et al. profiled the activity of individual conditioned medias produced similarly on proangiogenic proliferation of human umbilical vein endothelial (HUVEC) cells, implicating a signaling cascade consisting of IKK family member TANK-binding kinase TBK1, the TLR adaptor TICAM1, and the interferon regulator IRF3 in mediating proliferation and angiogenesis in HUVECs [27].

Several overexpression-based profiling assays have employed automated microscopy to detect the cell-autonomous effects of cDNA overexpression. In a proof-of-concept study, Harada et al. profiled the effects of 7364 human and mouse cDNAs on growth and survival of U2OS osteocarcinoma cells using a live-cell, multitimepoint assay measuring individual cell viability via image analysis [28]. Images of representative fields in assay wells were acquired using an automated microscope before and after deprivation of growth factors, growth rates calculated, and subcellular images retained for detailed analysis. As expected, previously identified growth factors such as PCNA and CDK9 permitted growth under serum-starved conditions whereas known proapoptotic factors such as BAD and BAX accelerated death. Two novel genes capable of promoting cellular proliferation and survival of primary cells were also identified and confirmed. Cho et al. used high-content imaging to map the effects of the same library on the insulin-stimulated nuclear import of the diabetes-related transcription factor FOXO1 in U2OS cells, identifying the protein–tyrosine phosphatase PTP-MEG2 as a negative regulator of FOXO1 translocation. Subsequent *in vivo* experiments demonstrated the ability of PTP-MEG2 to regulate the phosphorylation status of the insulin receptor and antagonize insulin signaling [29].

4.2.2 Genomic Functional Profiling Using RNA Interference Libraries

Enforced overexpression of specific proteins has a long history as a research tool in biological discovery. RNAi, although a relatively recent discovery, has rapidly evolved into an equally useful tool for exploring the cellular functions of genes from the perspective of genomic functional profiling. As with overexpression-based studies, initial RNAi-based genomic functional profiling efforts were modest in scope. Using a library of 200 shRNA-expressing plasmids targeting 50 human deubiquitinating proteins, Brummelkamp et al. sought to determine the role that these important proteins play in regulating protein stability leading to PMA-stimulated activation of the NF-κB pathway [30]. Among the targets, only depletion of the ubiquitin-specific processing protease CYLD proved capable of enhancing transcription of PMA stimulation NF-κB activation. Follow-up biochemical experiments revealed that CYLD modifies the ubiquitination status of TRAF2, a key component of the NFκB signaling pathway and a regulator of the IkB kinase complex that normally maintains NFκB in an inactivated cytoplasmic state.

The first large-scale RNAi screens were executed using shRNA libraries. Paddison et al. constructed and screened a library of 28,000 shRNAs targeting 9610 human and 5563 mouse genes by transfection, seeking to identify genes required for proper proteosome function in HEK293 cells [31]. Of the 100 shRNAs impacting turnover of a proteosome reporter, 22 targeted 15 different proteosome subunit proteins. Overall, over 50% of the shRNAs targeting a known proteosome protein in the screen scored as hits, not only validating the utility of this particular library and RNAi-based genomics but also reinforcing the notion that genomic functional profiling can reveal entire sets of related, rate-limiting components of interesting cellular processes. A similar strategy was pursued by Berns et al., who generated a library of 23,742 shRNAs targeting 7914 unique human genes in a retroviral vector configured with molecular barcodes [32]. To identify genes critical for normal p53 function, the library was packaged as a single infectious pool and transduced into primary human fibroblasts that proliferate rapidly on p53 inactivation. In addition to p53 itself, arrest depended on five other genes, including ribosomal protein S6 kinase PRS6KA6 and the HIV-1 TAT interacting protein HTATIP, none of which had been previously associated with mediating p53 function in this context. Similar screens with this library lead to the discovery and functional validation of the novel tumor suppressors PITX1 and REST [33,34]. Applied to MCF7 cells treated with or without a potent cytotoxic inhibitor of the p53 interactor MDM2, this strategy yielded a component of the ATM-CHK-53BP1 pathway, 53BP1, which, when depleted in the presence of the inhibitor, nutlin-3, permitted continued cell survival [35].

RNAi libraries that target genes whose functional products are RNA rather than protein have also been deployed for functional genomic profling. Noncoding RNAs include ribosomal and tRNAs, microRNAs, which trigger endogenous RNAi-mediated suppression of other cellular RNAs, and longer RNAs of unknown function such as the human accelerated region (HAR) RNAs recently hypothesized to be associated with cortical neuron development [36]. Because only minimal knowledge of target sequence is required to design and synthesize active profiling constructs, RNAi-based strategies are ideal for investigating the functions of noncoding RNAs, whose complete sequences are frequently unknown, or, if known, are uncloned. Reasoning that long conserved ncRNAs would have important cellular functions, Willingham et al. screened an shRNA library targeting 512 uncharacterized ncRNAs for those impacting the activity of the nuclear factor of activated T cells NFAT [37]. They identified an ncRNA termed NRON that when repressed significantly increases

NFAT translocation into the nucleus via a physical RNA–protein interaction with importin family member importin-β KPNB1. Libraries that target microRNA noncoding RNAs have also become available. A library of 197 mature microRNAs was assayed for those with oncogenic potential, identifying miRNA-372 and miRNA-373 as potent suppressors of wild-type p53 in human cancers [38].

Two groups have independently expanded shRNA coverage to encompass virtually all known human and mouse protein-coding genes and tested the resulting RNAi libraries in arrayed-well functional profiling screens. One group, whose library is commonly known as the *Elledge–Hannon library,* configured the shRNA with a microRNA loop linking the guide and passenger strands of the self-complementary RNA, achieving enhanced RNAi potency through superior triggering of the cellular RNAi response [39]. The Elledge–Hannon library comprises some 140,000 shRNA constructs targeting 34,711 human and 32,628 mouse known or predicted genes in an MSCV-based retroviral vector coexpressing a puromycin resistance marker. In proof-of-concept arrayed-well experiments seeking to confirm and compare the activities of shRNAs targeting proteosome subunits, 47 shRNAs targeting proteosome components and 562 shRNAs targeting protein kinases were assayed by transfection for their ability to influence proteosome degradation of a PESTdomain containing fluorescent protein [39]. In this assay, most shRNAs influencing protein turnover targeted a known proteosome component, validating the ability of the library to achieve phenotype-appropriate responses. The second group, known as The RNAi Consortium (TRC), constructed a similarly sized shRNA library incorporating a nonmicroRNA loop in a puromycin-marked lentiviral vector [40,41]. This library proposes to target the entire human and mouse protein-encoding transcriptome with four to five shRNAs per target and at the time of its first report consisted of 104,000 shRNAs targeting 22,000 human and mouse genes. To demonstrate the utility of the TRC lentiviral shRNA library, a subset of 4903 unique lentiviral constructs targeting 1028 human genes was packaged in arrayed-well format as infective virus and screened by high-content imaging in HT29 human colon cancer cells for components involved in mitotic regulation [40]. Targeted depletion of genes known to regulate cell cycle progression like CDC2 and CDK1 induced G2/M arrest, as did a number of genes whose cell cycle regulatory roles had not previously been appreciated. A third group has constructed and validated a smaller library featuring a doxycycline-regulated RNA pol II promoter driving conditional shRNA expression [42], a feature likely to be incorporated in future shRNA libraries.

Advantages of virally encoded RNAi libraries include the ability to reach poorly-transfected cell types, achieve long-term expression, and mark transduced cells. Encoded libraries also permit conditional expression when configured with regulatable promoters [43]. But because they are much easier to screen than encoded shRNAs, synthetic siRNAs have enjoyed considerable use in short-term functional profiling experiments conducted in easily transfected cells. In the first such study, Aza-Blanc et al. constructed a library of 1020 siRNAs mostly targeting human protein kinases and screened them in human HeLa cervical cancer cells for those impacting TRAIL-induced apoptosis [44]. Both pro- and antiapoptotic TRAIL-dependent regulators were identified among the set of genes not already associated with TRAIL dependence, including the novel proapoptotic kinase GSKα, whose RNAi-mediated depletion was shown to reduce caspase-8 processing. MacKeigan et al. constructed an siRNA library targeting 650 human protein kinases and 222 human protein phosphatases, profiling the roles their targets play in regulating apoptosis in HeLa cells treated with caspase inhibitors [45]. In so doing they, like Aza-Blanc et al., identified human kinases essential for cell survival, including a number of recognized essential kinases, such as the cyclin-dependent kinase CDK6 and the survival kinases SGK and ATK2. This study also suggested a prominent role for phosphatases in regulating cell survival, as depletion of over 30% of the phosphatases targeted lead to accelerated cell death. Similar studies have employed subgenomic synthetic siRNA libraries to explore cellular processes like endocytosis and insulin signaling [46,47].

Two studies demonstrated the utility of genome-scale libraries of synthetic siRNAs. Bartz et al. used a library of approximately 60,000 siRNAs targeting 20,000 human genes to seek genes that when suppressed can synergize the cytotoxic action of chemotherapeutic agents in transformed cell lines [48]. After progression through a series of filters, a list of 53 candidate enhancers of the DNA crosslinking drug cisplatin were identified and carefully validated, including many uncharacterized genes and members of two networks responsible for monitoring and correcting DNA damage, the BRCA network and the RAD network. Mukherji et al. used a library of 58,746 siRNAs targeting 24,373 unique human genes to systematically catalog key cell cycle regulators by high-content imaging [49]. Eight cellular and subcellular parameters characteristic of G1, S, or G2/M progression were independently measured and analyzed. Even employing conservative metrics, over 1000 genes were implicated in processes governing cell cycle progression. The activities of genes known to play roles in regulating specific phases of the cell cycle were reconfirmed, such as cyclin-dependent kinase CDK1 and polokinase PLK1. Some genes known to regulate progression at one point appeared as hits impacting other points, underscoring the ability of unbiased genomic functional profiling to unmask new functions for well-characterized genes. For example, 15 components of the TNF signaling cascade were shown to be required for successful passage through G2/M phase, despite their well-characterized roles in governing G1/S progression. Still other genes had either no previously reported role in regulating cell cycle progression, although they may have reported function in another cellular process, or were entirely without functional attribution, underscoring the power of functional profiling to supply unbiased information on the function of both annotated and unknown genes. Fully 27% of the 1152 genes picked up as key cell cycle regulators in the screen were devoid of any Gene Ontology functional annotation or Interpro protein domains.

4.3 GENERAL CONSIDERATIONS WHEN APPLYING FUNCTIONAL PROFILING STRATEGIES

Any comprehensive review of the use of genomic functional profiling should discuss its general methodologies. Understanding the inputs, similarities, differences, advantages, and liabilities of these methodologies will be key to exploring potential application to the study of stem cells. Considerations fall into two categories: those of concern to all high-throughput profiling approaches and those pertaining to specific advantages, disadvantages, and peculiarities of individual approaches, assays, or libraries. From the general perspective of high-throughput screening, these concerns revolve around the rates at which false positives and false negatives are encountered and the applicability of certain libraries in desired cellular contexts. Genomic functional profiling seeks to identify rate-limiting components of interesting cellular processes. From a biological standpoint, how rate-limiting must a component be before it can be detected by such methods, and how frequently do hits from the screening deck prove false and why? It is illustrative for a number of reasons to consider by way of example the metabolic enzyme phenylalanine hydroxylase (PAH), the rate-limiting enzyme in phenylalanine catabolism. Studies have demonstrated that as little as 5% PAH activity can metabolize the phenylalanine in the normal human diet [50]. Since exceptionally few RNAi constructs have been shown capable of reducing target mRNA levels in excess of 95% [39,40,51] and since PAH is present in great excess, PAH's role in phenylalanine catabolism isn't likely to be revealed by genomic functional profiling strategies employing an unchallenged cellular model of phenylalanine metabolism, regardless of whether a gain- or loss-of-function approach is employed. Many assay strategies are in fact designed to maximize sensitivity to experimental perturbation of known critical components [44,45]. Nevertheless, the principal conclusion to be drawn from this example is that one should expect to encounter cases when overexpression or RNAi-mediated depletion of real components will fail to elicit a measurable effect on the process under investigation (false negatives). The high reconfirmation rate for known components in most screens does suggest a relatively low false-negative rate, however. What is surprising is not the occurrence of false negatives, which are common in other high-throughput techniques as well, but the large numbers of true activities obtained. It is also comforting to observe that genomic functional profiling does not appear to favor genes having particular molecular functionality; gene families whose activities have been identified include, among others, kinases [15], cytokines [17], cell surface receptors [18], proteases [24], signaling modulators [12], adaptors [25], phosphatases [29], tumor suppressors [33], oncogenes [42], and many genes heretofore unclassifiable.

In addition to false negatives, genomics-profiling experiments can also report false positives. In overexpression experiments, false positives usually occur as a consequence of either phenocopying, when one protein assumes the role of another, or spatiotemporal misexpression, either of which can generate biologically irrelevant responses. These false positives are often detected and discriminated from biologically relevant hits through careful follow-up experiments. What is seldom in doubt is the sequence of the entities responsible for the observed activities, particularly in unique-clone array experiments where clone identity is often known in advance or easily derived after screening. Just the opposite is true for RNAi-based profiling experiments. Responses in RNAi-based experiments are assuredly biologically relevant because RNAi works by depleting endogenous cellular targets. False positives occur when an RNAi construct achieves this response by depleting an RNA other than the one against which it was designed, a phenomenon known as *off-target activity*. To avoid host innate immune responses to double-stranded RNA and exploit the RNAi machinery's preference for dsRNA of specific lengths, RNAi constructs must typically be between 19 and 27 ribonucleotides in length [52–54]. With a centrally located seed region 9–12 nucleotides long conferring most of their specificity [55–57], RNAi constructs are capable of acting on unintended off-targets that share homology in this region [57–59]. In a particularly thorough study demonstrating the capacity of RNAi constructs to achieve biological effects through off-target silencing, siRNAs targeting human GPCR kinase GRK4 and hexokinase HK were shown to efficiently deplete their intended mRNAs and elicit what was presumed to be an associated hypoxic phenotype. Careful follow-on experiments subsequently attributed the associated phenotype not to knockdown of GRK4 or HK, but to off-target knockdown of HIFA through conserved, seven nucleotide segments in the seed regions of the original GRK4- and HK-directed siRNAs [60]. Additional siRNAs to GRK4 and HK depleted their intended targets but not HIFA, and failed to elicit the hypoxic phenotype. As a consequence of these and similar findings, it is generally considered prudent to disregard genotype–phenotype relationships on the bases of activities of single RNAi constructs; only when multiple, independent RNAi constructs elicit a phenotype can depletion of intended targets be safely presumed to be responsible [58,61–63].

Clearly many factors must be weighed when deciding which approach, library, cell type, and assay endpoint to employ in a functional genomic profiling experiment. Some of these factors will be beyond the experimenter's control; few will have access to more than a limited set of libraries, and many will be forced to restrict experimental methodologies to the reagents at hand. In general, selection of library and approach—overexpression or RNAi-mediated depletion—are dictated by assay duration, assay endpoint, and transfectability of the chosen model cell type. Phenotypes resulting from profound, continual, long-term changes in cellular morphology or fate may require assays of considerable duration [17], whereas signaling events whose occurrence can be readily and immediately detected can employ assays as short as 24–48 h [12]. Phenotypes requiring long-term expression or suppression of target components are best approached using libraries of constructs encoded by viruses that integrate into host cellular genomes so that as cells divide, each daughter will retain the profiling construct. Short-term assays can be prosecuted transiently using transfected cDNAs, shRNAs, or siRNAs. The choice between overexpression and RNAi-mediated approaches is less straightforward. It may be helpful in general to equate overexpression with gain of function, where a phenotype in a few wells or cells occurs above a background of nonevents, and to equate RNAi-mediated depletion with loss of function, where the rare disappearance of a phenotype is sought. In reality both require the ability to detect rare events achieved through any number of integrated cellular mechanisms and components, some of which will be positive in nature and others negative. Adipogenesis of mouse 3T3-L1 cells can be promoted by overexpression of the peroxisome proliferator PPARγ and the transcription factor C/EBP but antagonized by overexpression of the proinflammatory cytokine TNFα [64–66]. In general it is useful to configure assays and readouts to seek positive rather than negative events, regardless of whether the underlying source is due to gain or loss of target function, since seeking signal loss in a background of elevated signal due to nonevents can lead to excessive false-positive rates. Rare positive events above a generally low signal background may be more difficult to achieve. Processes such as oncogenesis resulting from misexpression or dysregulation of host cellular proteins have been productively approached using both overexpression [12,19,24] and RNAi-based approaches, [33,34], although both assay types were configured so that mediators generated positive responses.

Assays in easily transfected cells requiring only transient gene overexpression or depletion can exploit naked cDNAs, shRNAs, or synthetic siRNAs. More complex profiling assays may require image-based cell-autonomous endpoints, long-term expression unsuited to transient transfection, or cell types refractory to lipid-mediated transfection, all of which will dictate the use of virally encoded cDNA or shRNA RNAi libraries. Genome-scale shRNA RNAi libraries have been constructed in oncoretroviral and HIV-based lentiviral vectors, each capable of infecting cells, integrating into the host cell genome, and effecting stable, long-term expression of virally encoded shRNA constructs [39–41,51]. Sub-genome-scale overexpression cDNA libraries have been constructed in an adenoviral vector [17]. The choice between lentiviral, retroviral, and adenoviral libraries is dictated by whether target cells are dividing or nondividing. Lenti viral vectors can infect and integrate into the genomes of nondividing cells whereas efficient retroviral transduction requires the nuclear envelope breakdown accompanying cell division and adenoviral vectors do not integrate at all, though they can frequently infect and remain epiosomal [67,68]. Viral tropism can also have a large impact in selection of a particular library [69], and this is no less likely to be true in stem cells than in non–stem cells [70]. Those in possession of viral shRNA libraries frequently test a variety of viral envelopes to empirically match optimal coat protein to experimental cell type.

If an RNAi-dependent assay is required in easily transfected cells, the use of synthetic siRNAs offers considerable advantages over plasmid-based libraries. Because they are synthesized and not cloned, siRNAs don't require cumbersome DNA preparation or viral production steps, are usually available in assay-friendly formats ready for dilution for screening, and can often be purchased at scales sufficient to permit execution of multiple screens in 96- or 384-well format. Transfection efficiencies are usually higher compared to naked, nonviral plasmids and well-to-well concentrations more even than with untitered virus. Dependence on lipid-based transfection does restrict the use of synthetic siRNAs in functional profiling experiments to efficiently transfected cell types, however, and many interesting cell types including some stem cells are not efficiently transfected, [71]. For assays whose measured endpoint is the sum of activities in all cells in a well (lysis-dependent luciferase assays represent a common example), transfection efficiencies must be very high indeed for many phenotypes to be detected; if only 50% of cells are transfected, then RNAi-dependent suppression of targets must be virtually 100% in successfully transfected cells to achieve an average knockdown of 50% across all cells in the well. The lack of published findings reporting the successful detection of phenotypes in poorly transfected cell types suggests that synthetic siRNA transfection rates in excess of 50–70% are required before depletion-dependent phenotypes can be reliably detected.

4.4 POTENTIAL APPLICATION OF GENOMIC FUNCTIONAL PROFILING TO STEM CELL BIOLOGY

To what degree has genomic functional profiling been helpful in studying stem cell biology, and how it might be exploited in the future? To date, genomic functional profiling has been applied mostly to fully differentiated cell lines except per its application to the study of osteogenesis in human mesenchymal stem cells using cDNAs in adenoviral vector [17]. Lack of progress is likely due to both the rapidly evolving nature of the stem cell field and overweening reliance on the use of reagents and cells suited to transient transfection. While the wide diversity of stem cell types renders comprehensive introduction to their genetic manipulation difficult, Table 4.1 lists several different categories of stem cells, the efficiency with which they can be transfected with cDNAs, and viral vectors used successfully to deliver overexpressed cDNAs. Table 4.2 does the same for RNAi constructs, either synthetic or encoded. Stem cells of all types are generally recalcitrant to efficient lipid-mediated transfection, with several exceptions [71]. Effective siRNA transfection protocols have been reported for human mesenchymal stem cells, achieving lipid-mediated transfection efficiencies of 50–90% [72,73], and for Schwann cell precursors of myofibroblasts and melanoctyes [74]. Nevertheless, transfection appears to be too inefficient in general to support functional profiling in 96- and 384-well formats in most stem cell types. In contrast, highly efficient infection with individual viral vectors is well established in virtually all stem cell types. Over 50% transduction efficiencies with lenti/retro viral vectors have been achieved in human ES cells where efficient transfection has yet to be reported, and viral transduction of stem cells and terminally differentiated cells remains the technical basis for most gene therapy applications, whether somatic or *in vivo* [75].

The convergence of more recent developments in genomic functional profiling technology promises to make its use in stem cells more tractable. The major development concerns the deployment of viral vector systems that will likely be able

TABLE 4.1 Common Stem Cell Types and Methods for Introducing and Overexpressing cDNAs

Stem Cell Type	Source	cDNA Lipid Transfection	Standard Viral Delivery Vectors	References
Hematopoetic	Adult	No	Adeno, retro, lenti	85–90
Stromal/ mesenchymal	Adult	Inefficient–efficient	Adeno, retro, lenti	91–95
Neural	Adult	Inefficient	Adeno, retro, lenti	96–100
Epithelial	Adult	No	Retro, lenti	101–103
Dermal	Adult	No	Retro, lenti	104–106
Embryonic	Embryo	Inefficient	AAV, adeno, retro, lenti	107–110

TABLE 4.2 Common Stem Cell Types and Methods for Introducing Synthetic or Encoded RNAi Constructs

Stem Cell Type	Source	siRNA Lipid Transfection	Standard Viral Delivery Vectors[a]	References
Hematopoetic	Adult	No	Lenti	111–114
Stromal; mesenchymal	Adult	Efficient	Adeno, lenti	72,115–118
Neural	Adult	Inefficient–efficient	Lenti	74,119
Epithelial	Adult	Unknown	Adeno	120
Dermal	Adult	Efficient	—	121
Embryonic	Embryo	Inefficient	Lenti, retro	122–124

[a]Note that shRNAs are essentially a different type of cDNA, and therefore all viral delivery methods described in Table 4.1 should function for shRNA delivery.

to infect stem cells in microwell format [17,39–41,51]. For reasons unrelated to infectivity, integrative viral vectors offer even greater advantages over transfection-based approaches for stem cell applications. Retroviral and lentiviral vectors are distinguished from adenoviral vectors by their capacity to stably integrate into the host cell genome and provide long-term expression of encoded products. Long-term expression in both dividing and nondividing cells has been reported with adenoviral vector systems (up to 15 months), but since self-renewal requires cell division, which dilutes nonintegrated genetic material over time, adeno transduction should be limited to stem cell phenotypes developing over short timeframes [76,77]. Long-term expression of at least three genes, the transcription factors Nanog and Oct4 and the peptide hormone activin A, appears to be required for human ES cell self-renewal, for example, [1,78–80], and thus genomic functional profiling studies of factors contributing to self-renewal may depend on achieving long-term expression using host-integrating lentiviral or retroviral transduction. Differentiation, which can be achieved in human ES cells through transient interruption of Nanog and Oct4 expression [1], may be amenable to study using either adenoviral or siRNA-based systems, presuming that high-throughput siRNA delivery issues can be solved.

Because current viral vectors incorporate separately expressed selection markers, transduction can be tracked in successfully transduced cells regardless of efficiency, in contrast to siRNAs, where effective use of cell-autonomous markers for transfection has yet to be reported. Under the best of circumstances it may be difficult to transduce stem cells, leading to a few successfully transduced cells in a background of uninfected cells. In addition, many phenotypes occur too quickly to employ outgrowth endpoints, necessitating the ability to rapidly distinguish transduced from uninfected cells. In both cases, automated microscopic imaging systems equipped with algorithms for automatically discriminating marked stem cells and their cellular and subcellular parameters are likely to prove increasingly useful when combined with viral transduction [81–83]. Transducing stem cells with viruses expressing fluorescent markers followed by high-content imaging would permit cell-autonomous analysis of function, eliminating the need to achieve high transduction efficiencies and enabling detection of short- and long-term phenotypes in only successfully transduced cells. Similar studies in differentiated cells have already been performed successfully [28,29]. Viral libraries incorporating puromycin or other selectable markers, such as the TRC lentiviral shRNA library, are likely to play equally important roles in the application of functional profiling to stem cells for similar reasons. For differentiation assays, where stem cell populations may require long-term culture before detectable lineage markers become apparent, selection against nontransduced cells can ensure apt comparison of genes by removing signals associated with the false-negative, nontransduced cells.

Genome-scale analysis of stem cell biology has to date been restricted to differential expression analysis. While this has been tremendously useful in gathering baseline data and establishing the cascade of effects associated with differentiation, these findings do not prove causation. Increases in expression of key genes may be the result of earlier differentiation events, for example, rather than the cause of those or subsequent events; causation can be best determined through genuine experimental manipulation. Genomic functional profiling offers the opportunity to test the contribution of individual genes to any stem cell event that can be measured in a 96- or 384-well plate, in relatively straightforward high-throughput experiments. Phosphorylation status of the nuclear corepressor N-CoR has been shown to influence neural stem cell differentiation [84]. An understanding of the role kinases play in regulating differentiation or self-renewal could be achieved by systematically depleting all protein kinases using virally encoded shRNAs, for example, and reagents and methods exist to profile contribution of each gene in the human genome in human mesenchymal stem cells to adipogenesis or chondrogenesis. By offering the ability to individually interrogate genes for their impact on key aspects of stem cell function without regard to what we already know about them, genomic functional profiling will enable researchers to rapidly expand our knowledge of programs for self-renewal and differentiation on a global level, to compare these findings to those generated for differentiated or transformed cells, and to permit greater understanding of common mechanisms guiding cellular fates.

REFERENCES

[1] Zaehres H, Lensch MW, Daheron L, Stewart SA, Itskovitz-Eldor J, Daley GQ: High-efficiency RNA interference in human embryonic stem cells, *Stem Cells* **23**:299–305 (2005).

[2] Sperger JM, Chen X, Draper JS, Antosiewicz JE, Chon CH, Jones SB, Brooks JD, Andrews PW, Brown PO, Thomson JA: Gene expression patterns in human embryonic stem cells and human pluripotent germ cell tumors, *Proc Natl Acad Sci USA* **100**:13350–13355 (2003).

[3] Park IK, He Y, Lin F, Laerum OD, Tian Q, Bumgarner R, Klug CA, Li K, Kuhr C, Doyle MJ et al: Differential gene expression profiling of adult murine hematopoietic stem cells, *Blood* **99**:488–498 (2002).

[4] Maglott D, Ostell J, Pruitt KD, Tatusova T: Entrez gene: Gene-centered information at NCBI, *Nucl Acids Res* **35**:D26–31 (2007).

[5] Hoffmann R, Valencia, A: Life cycles of successful genes, *Trends Genet* **19**:79–81 (2003).

[6] Strausberg RL, Feingold EA, Klausner RD, Collins FS: The mammalian gene collection, *Science* **286**:455–457 (1999).

[7] Su AI, Wiltshire T, Batalov S, Lapp H, Ching KA, Block D, Zhang J, Soden R, Hayakawa M, Kreiman G et al: A gene atlas of the mouse and human protein-encoding transcriptomes, *Proc Natl Acad Sci USA* **101**:6062–6067 (2004).

[8] Sonnhammer EL, Eddy SR, Durbin R: Pfam: A comprehensive database of protein domain families based on seed alignments, *Proteins* **28**:405–420 (1997).

[9] Manning G, Whyte DB, Martinez R, Hunter T, Sudarsanam S: The protein kinase complement of the human genome, *Science* **298**:1912–1934 (2002).

[10] Venter JC, Adams MD, Myers EW, Li PW, Mural RJ, Sutton GG, Smith HO, Yandell M, Evans CA, Holt RA et al: The sequence of the human genome, *Science* **291**:1304–1351 (2001).

[11] Imanishi T, Itoh T, Suzuki Y, O'Donovan C, Fukuchi S, Koyanagi KO, Barrero RA, Tamura T, Yamaguchi-Kabata Y, Tanino M et al: Integrative annotation of 21,037 human genes validated by full-length cDNA clones, *PLoS Biol* **2**:e162(2004).

[12] Chanda SK, White S, Orth AP, Reisdorph R, Miraglia L, Thomas RS, DeJesus P, Mason DE, Huang Q, Vega R et al: Genome-scale functional profiling of the mammalian AP-1 signaling pathway, *Proc Natl Acad Sci USA* **100**:12153–12158 (2003).

[13] Adams MD, Celniker SE, Holt RA, Evans CA, Gocayne JD, Amanatides PG, Scherer SE, Li PW, Hoskins RA, Galle RF et al: The genome sequence of Drosophila melanogaster, *Science* **287**:2185–2195 (2000).

[14] Lander ES, Linton LM, Birren B, Nusbaum C, Zody MC, Baldwin J, Devon K, Dewar K, Doyle M, FitzHugh W et al: Initial sequencing and analysis of the human genome, *Nature* **409**:860–921 (2001).

[15] Ziauddin J, Sabatini DM: Microarrays of cells expressing defined cDNAs, *Nature* **411**:107–110 (2001).

[16] Albayrak T, Grimm S: A high-throughput screen for single gene activities: Isolation of apoptosis inducers, *Biochem Biophys Res Commun* **304**:772–776 (2003).

[17] Michiels F, van Es H, van Rompaey L, Merchiers P, Francken B, Pittois K, van der Schueren J, Brys R, Vandersmissen J, Beirinckx F et al: Arrayed adenoviral expression libraries for functional screening, *Nat Biotechnol* **20**:1154–1157 (2002).

[18] Matsuda A, Suzuki Y, Honda G, Muramatsu S, Matsuzaki O, Nagano Y, Doi T, Shimotohno K, Harada T, Nishida E et al: Large-scale identification and characterization of human genes that activate NF-kappaB and MAPK signaling pathways, *Oncogene* **22**:3307–3318 (2003).

[19] Huang Q, Raya A, DeJesus P, Chao SH, Quon KC, Caldwell JS, Chanda SK, Izpisua-Belmonte JC, Schultz PG: Identification of p53 regulators by genome-wide functional analysis, *Proc Natl Acad Sci USA* **101**:3456–3461 (2004).

[20] Liu J, Bang AG, Kintner C, Orth AP, Chanda SK, Ding S, Schultz PG: Identification of the Wnt signaling activator leucine-rich repeat in flightless interaction protein 2 by a genome-wide functional analysis, *Proc Natl Acad Sci USA* **102**:1927–1932 (2005).

[21] Conkright MD, Canettieri G, Screaton R, Guzman E, Miraglia L, Hogenesch JB, Montminy M: TORCs: Transducers of regulated CREB activity, *Mol Cell* **12**:413–423 (2003).

[22] Bittinger MA, McWhinnie E, Meltzer J, Iourgenko V, Latario B, Liu X, Chen CH, Song C, Garza D, Labow M: Activation of cAMP response element-mediated gene expression by regulated nuclear transport of TORC proteins, *Curr Biol* **14**:2156–2161 (2004).

[23] Luesch H, Chanda SK, Raya RM, DeJesus PD, Orth AP, Walker JR, Izpisua Belmonte JC, Schultz PG: A functional genomics approach to the mode of action of apratoxin A, *Nat Chem Biol* **2**:158–167 (2006).

[24] Koenig-Hoffmann K, Bonin-Debs AL, Boche I, Gawin B, Gnirke A, Hergersberg C, Madeo F, Kazinski M, Klein M, Korherr C et al: High throughput functional genomics: Identification of novel genes with tumor suppressor phenotypes, *Int J Cancer* **113**:434–439 (2005).

[25] Fiscella M, Perry JW, Teng B, Bloom M, Zhang C, Leung K, Pukac L, Florence K, Concepcion A, Liu B et al: TIP, a T-cell factor identified using high-throughput screening increases survival in a graft-versus-host disease model, *Nat Biotechnol* **21**:302–307 (2003).

[26] Chen C, Grzegorzewski KJ, Barash S, Zhao Q, Schneider H, Wang Q, Singh M, Pukac L, Bell AC, Duan R et al: An integrated functional genomics screening program reveals a role for BMP-9 in glucose homeostasis, *Nat Biotechnol* **21**:294–301 (2003).

[27] Korherr C, Gille H, Schafer R, Koenig-Hoffmann K, Dixelius J, Egland KA, Pastan I, Brinkmann U: Identification of proangiogenic genes and pathways by high-throughput functional genomics: TBK1 and the IRF3 pathway, *Proc Natl Acad Sci USA* **103**: 4240–4245 (2006).

[28] Harada JN, Bower KE, Orth AP, Callaway S, Nelson CG, Laris C, Hogenesch JB, Vogt PK, Chanda SK: Identification of novel mammalian growth regulatory factors by genome-scale quantitative image analysis, *Genome Res* **15**:1136–1144 (2005).

[29] Cho CY, Koo SH, Wang Y, Callaway S, Hedrick S, Mak PA, Orth AP, Peters EC, Saez E, Montminy M et al: Identification of the tyrosine phosphatase PTP-MEG2 as an antagonist of hepatic insulin signaling, *Cell Metab* **3**:367–378 (2006).

[30] Brummelkamp TR, Nijman SM, Dirac AM, Bernards R: Loss of the cylindromatosis tumour suppressor inhibits apoptosis by activating NF-kappaB, *Nature* **424**:797–801 (2003).

[31] Silva JM, Mizuno H, Brady A, Lucito R, Hannon GJ: RNA interference microarrays: high-throughput loss-of-function genetics in mammalian cells, *Proc Natl Acad Sci USA* **101**:6548–6552 (2004).

[32] Berns K, Hijmans EM, Mullenders J, Brummelkamp TR, Velds A, Heimerikx M, Kerkhoven RM, Madiredjo M, Nijkamp W, Weigelt B et al: A large-scale RNAi screen in human cells identifies new components of the p53 pathway, *Nature* **428**:431–437 (2004).

[33] Kolfschoten IG, van Leeuwen B, Berns K, Mullenders J, Beijersbergen RL, Bernards R, Voorhoeve PM, Agami R: A genetic screen identifies PITX1 as a suppressor of RAS activity and tumorigenicity, *Cell* **121**:849–858 (2005).

[34] Westbrook TF, Martin ES, Schlabach MR, Leng Y, Liang AC, Feng B, Zhao JJ, Roberts TM, Mandel G, Hannon GJ et al: A genetic screen for candidate tumor suppressors identifies REST, *Cell* **121**:837–848 (2005).

[35] Brummelkamp TR, Fabius AW, Mullenders J, Madiredjo M, Velds A, Kerkhoven RM, Bernards R, Beijersbergen RL: An shRNA barcode screen provides insight into cancer cell vulnerability to MDM2 inhibitors, *Nat Chem Biol* **2**:202–206 (2006).

[36] Pollard KS, Salama SR, Lambert N, Lambot MA, Coppens S, Pedersen JS, Katzman S, King B, Onodera C, Siepel A et al: An RNA gene expressed during cortical development evolved rapidly in humans, *Nature* **443**:167–172 (2006).

[37] Willingham AT, Orth AP, Batalov S, Peters EC, Wen BG, Aza-Blanc P, Hogenesch JB, Schultz PG: A strategy for probing the function of noncoding RNAs finds a repressor of NFAT, *Science* **309**:1570–1573 (2005).

[38] Voorhoeve PM, le Sage C, Schrier M, Gillis AJ, Stoop H, Nagel R, Liu YP, van Duijse J, Drost J, Griekspoor A et al: A genetic screen implicates miRNA-372 and miRNA-373 as oncogenes in testicular germ cell tumors, *Cell* **124**:1169–1181 (2006).

[39] Silva JM, Li MZ, Chang K, Ge W, Golding MC, Rickles RJ, Siolas D, Hu G, Paddison PJ, Schlabach MR et al: Second-generation shRNA libraries covering the mouse and human genomes, *Nat Genet* **37**:1281–1288 (2005).

[40] Moffat J, Grueneberg DA, Yang X, Kim SY, Kloepfer AM, Hinkle G, Piqani B, Eisenhaure TM, Luo B, Grenier JK et al: A lentiviral RNAi library for human and mouse genes applied to an arrayed viral high-content screen, *Cell* **124**:1283–1298 (2006).

[41] Root DE, Hacohen N, Hahn WC, Lander ES, Sabatini DM: Genome-scale loss-of-function screening with a lentiviral RNAi library, *Nat Meth* **3**:715–719 (2006).

[42] Ngo VN, Davis RE, Lamy L, Yu X, Zhao H, Lenz G, Lam LT, Dave S, Yang L, Powell J et al: A loss-of-function RNA interference screen for molecular targets in cancer, *Nature* **441**:106–110 (2006).

[43] Wiznerowicz M, Szulc J, Trono D: Tuning silence: Conditional systems for RNA interference, *Nat Meth* **3**:682–688 (2006).

[44] Aza-Blanc P, Cooper CL, Wagner K, Batalov S, Deveraux QL, Cooke MP: Identification of modulators of TRAIL-induced apoptosis via RNAi-based phenotypic screening, *Mol Cell* **12**:627–637 (2003).

[45] MacKeigan JP, Murphy LO, Blenis J: Sensitized RNAi screen of human kinases and phosphatases identifies new regulators of apoptosis and chemoresistance, *Nat Cell Biol* **7**:591–600 (2005).

[46] Pelkmans L, Fava E, Grabner H, Hannus M, Habermann B, Krausz E, Zerial M: Genome-wide analysis of human kinases in clathrin- and caveolae/raft-mediated endocytosis, *Nature* **436**:78–86 (2005).

[47] Tang X, Guilherme A, Chakladar A, Powelka AM, Konda S, Virbasius JV, Nicoloro SM, Straubhaar J, Czech MP: An RNA interference-based screen identifies MAP4K4/NIK as a negative regulator of PPARgamma, adipogenesis, and insulin-responsive hexose transport, *Proc Natl Acad Sci USA* **103**:2087–2092 (2006).

[48] Bartz SR, Zhang Z, Burchard J, Imakura M, Martin M, Palmieri A, Needham R, Guo J, Gordon M, Chung N et al: Small interfering RNA screens reveal enhanced cisplatin cytotoxicity in tumor cells having both BRCA network and TP53 disruptions, *Mol Cell Biol* **26**:9377–9386 (2006).

[49] Mukherji M, Bell R, Supekova L, Wang Y, Orth AP, Batalov S, Miraglia L, Huesken D, Lange J, Martin C et al: Genome-wide functional analysis of human cell-cycle regulators, *Proc Natl Acad Sci USA* **103**:14819–14824 (2006).

[50] Treacy EP, Delente JJ, Elkas G, Carter K, Lambert M, Waters PJ, Scriver CR: Analysis of phenylalanine hydroxylase genotypes and hyperphenylalaninemia phenotypes using L-[1-13C]phenylalanine oxidation rates *in vivo*: A pilot study, *Pediatr Res* **42**:430–435 (1997).

[51] Paddison PJ, Silva JM, Conklin DS, Schlabach M, Li M, Aruleba S, Balija V, O'Shaughnessy A, Gnoj L, Scobie K et al: A resource for large-scale RNA-interference-based screens in mammals, *Nature* **428**:427–431 (2004).

[52] Elbashir SM, Lendeckel W, Tuschl T: RNA interference is mediated by 21- and 22-nucleotide RNAs, *Genes Dev* **15**:188–200 (2001).

[53] Sledz CA, Holko M, de Veer MJ, Silverman RH, Williams BR: Activation of the interferon system by short-interfering RNAs, *Nat Cell Biol* **5**:834–839 (2003).

[54] Kim DH, Behlke MA, Rose SD, Chang MS, Choi S, Rossi JJ: Synthetic dsRNA Dicer substrates enhance RNAi potency and efficacy, *Nat Biotechnol* **23**:222–226 (2005).

[55] Elbashir SM, Martinez J, Patkaniowska A, Lendeckel W, Tuschl T: Functional anatomy of siRNAs for mediating efficient RNAi in Drosophila melanogaster embryo lysate, *Embo J* **20**:6877–6888 (2001).

[56] Holen T, Amarzguioui M, Babaie E, Prydz H: Similar behaviour of single-strand and double-strand siRNAs suggests they act through a common RNAi pathway, *Nucl Acids Res* **31**:2401–2407 (2003).

[57] Du Q, Thonberg H, Wang J, Wahlestedt C, Liang Z: A systematic analysis of the silencing effects of an active siRNA at all single-nucleotide mismatched target sites, *Nucl Acids Res* **33**:1671–1677 (2005).

[58] Jackson AL, Bartz SR, Schelter J, Kobayashi SV, Burchard J, Mao M, Li B, Cavet G, Linsley PS: Expression profiling reveals off-target gene regulation by RNAi, *Nat Biotechnol* **21**:635–637 (2003).

[59] Scacheri PC, Rozenblatt-Rosen O, Caplen NJ, Wolfsberg TG, Umayam L, Lee JC, Hughes CM, Shanmugam KS, Bhattacharjee A, Meyerson M et al: Short interfering RNAs can induce unexpected and divergent changes in the levels of untargeted proteins in mammalian cells, *Proc Natl Acad Sci USA* **101**:1892–1897 (2004).

[60] Wang JL, Zheng YL, Ma R, Wang BL, Guo AG, Jiang YQ: Disulfide-stabilized single-chain antibody-targeted superantigen: Construction of a prokaryotic expression system and its functional analysis, *World J Gastroenterol* **11**:4899–4903 (2005).

[61] Whither RNAi? *Nat Cell Biol* **5**:489–490 (2003).

[62] Cullen BR: Enhancing and confirming the specificity of RNAi experiments, *Nat Meth* **3**:677–681 (2006).

[63] Echeverri CJ, Beachy PA, Baum B, Boutros M, Buchholz F, Chanda SK, Downward J, Ellenberg J, Fraser AG, Hacohen N et al: Minimizing the risk of reporting false positives in large-scale RNAi screens, *Nat Meth* **3**:777–779 (2006).

[64] Tontonoz P, Hu E, Spiegelman BM: Stimulation of adipogenesis in fibroblasts by PPAR gamma 2, a lipid-activated transcription factor, *Cell* **79**:1147–1156 (1994).

[65] Wu Z, Xie Y, Bucher NL, Farmer SR: Conditional ectopic expression of C/EBP beta in NIH-3T3 cells induces PPAR gamma and stimulates adipogenesis, *Genes Dev* **9**:2350–2363 (1995).

[66] Xing H, Northrop JP, Grove JR, Kilpatrick KE, Su JL, Ringold GM: TNF alpha-mediated inhibition and reversal of adipocyte differentiation is accompanied by suppressed

expression of PPARgamma without effects on Pref-1 expression, *Endocrinology* **138**:2776–2783 (1997).

[67] Lewis PF, Emerman M: Passage through mitosis is required for oncoretroviruses but not for the human immunodeficiency virus, *J Virol* **68**:510–516 (1994).

[68] Yamashita M, Emerman M: Retroviral infection of non-dividing cells: Old and new perspectives, *Virology* **344**:88–93 (2006).

[69] Osten P, Grinevich V, Cetin A: Viral vectors: A wide range of choices and high levels of service, *Handb Exp Pharmacol* **178**:177–202 (2007).

[70] Zhang XY, La Russa VF, Reiser J: Transduction of bone-marrow-derived mesenchymal stem cells by using lentivirus vectors pseudotyped with modified RD114 envelope glycoproteins, *J Virol* **78**:1219–1229 (2004).

[71] Ovcharenko D, Jarvis R, Hunicke-Smith S, Kelnar K, Brown D: High-throughput RNAi screening *in vitro*: From cell lines to primary cells, *Rna* **11**:985–993 (2005).

[72] Hoelters J, Ciccarella M, Drechsel M, Geissler C, Gulkan H, Bocker W, Schieker M, Jochum M, Neth P: Nonviral genetic modification mediates effective transgene expression and functional RNA interference in human mesenchymal stem cells, *J Gene Med* **7**:718–728 (2005).

[73] Xu Y, Mirmalek-Sani SH, Yang X, Zhang J, Oreffo RO: The use of small interfering RNAs to inhibit adipocyte differentiation in human preadipocytes and fetal-femur-derived mesenchymal cells, *Exp Cell Res* **312**:1856–1864 (2006).

[74] Roh J, Cho EA, Seong I, Limb JK, Lee S, Han SJ, Kim J: Down-regulation of Sox10 with specific small interfering RNA promotes transdifferentiation of Schwannoma cells into myofibroblasts, *Differentiation* **74**:542–551 (2006).

[75] O'Connor TP, Crystal RG: Genetic medicines: Treatment strategies for hereditary disorders, *Nat Rev Genet* **7**:261–276 (2006).

[76] Macq AF, Czech C, Essalmani R, Brion JP, Maron A, Mercken L, Pradier L, Octave JN: The long term adenoviral expression of the human amyloid precursor protein shows different secretase activities in rat cortical neurons and astrocytes, *J Biol Chem* **273**:28931–28936 (1998).

[77] Li JZ, Holman D, Li H, Liu AH, Beres B, Hankins GR, Helm GA: Long-term tracing of adenoviral expression in rat and rabbit using luciferase imaging, *J Gene Med* **7**:792–802 (2005).

[78] Yasuda SY, Tsuneyoshi N, Sumi T, Hasegawa K, Tada T, Nakatsuji N, Suemori H: NANOG maintains self-renewal of primate ES cells in the absence of a feeder layer, *Genes Cells* **11**:1115–1123 (2006).

[79] Gerrard L, Zhao D, Clark AJ, Cui W: Stably transfected human embryonic stem cell clones express OCT4-specific green fluorescent protein and maintain self-renewal and pluripotency, *Stem Cells* **23**:124–133 (2005).

[80] Xiao L, Yuan X, Sharkis SJ: Activin A maintains self-renewal and regulates fibroblast growth factor, Wnt, and bone morphogenic protein pathways in human embryonic stem cells, *Stem Cells* **24**:1476–1486 (2006).

[81] Ghosh RN, Lapets O, Haskins JR: Characteristics and value of directed algorithms in high content screening, *Meth Mol Biol* **356**:63–81 (2007).

[82] Carpenter AE, Jones TR, Lamprecht MR, Clarke C, Kang IH, Friman O, Guertin DA, Chang JH, Lindquist RA, Moffat J et al: CellProfiler: Image analysis software for identifying and quantifying cell phenotypes, *Genome Biol* **7**:R100(2006).

[83] Garippa RJ, Hoffman AF, Gradl G, Kirsch A: High-throughput confocal microscopy for beta-arrestin-green fluorescent protein translocation G protein-coupled receptor assays using the Evotec Opera, *Meth Enzymol* **414**:99–120 (2006).

[84] Park DM, Li J, Okamoto H, Akeju O, Kim SH, Lubensky I, Vortmeyer A, Dambrosia J, Weil RJ, Oldfield EH et al: N-CoR pathway targeting induces glioblastoma derived cancer stem cell differentiation, *Cell Cycle* **6**:(2007).

[85] Barrette S, Douglas J, Orlic D, Anderson SM, Seidel NE, Miller AD, Bodine DM: Superior transduction of mouse hematopoietic stem cells with 10A1 and VSV-G pseudotyped retrovirus vectors, *Mol Ther* **1**:330–338 (2000).

[86] Austin TW, Salimi S, Veres G, Morel F, Ilves H, Scollay R, Plavec I: Long-term multilineage expression in peripheral blood from a Moloney murine leukemia virus vector after serial transplantation of transduced bone marrow cells, *Blood* **95**:829–836 (2000).

[87] Barrette S, Douglas JL, Seidel NE, Bodine DM: Lentivirus-based vectors transduce mouse hematopoietic stem cells with similar efficiency to moloney murine leukemia virus-based vectors, *Blood* **96**:3385–3391 (2000).

[88] Worsham DN, Schuesler T, von Kalle C, Pan D: *In vivo* gene transfer into adult stem cells in unconditioned mice by in situ delivery of a lentiviral vector, *Mol Ther* **14**:514–524 (2006).

[89] Takahashi T, Yamada K, Tanaka T, Kumano K, Kurokawa M, Hirano N, Honda H, Chiba S, Tsuji K, Yazaki Y et al: A potential molecular approach to *ex vivo* hematopoietic expansion with recombinant epidermal growth factor receptor-expressing adenovirus vector, *Blood* **91**:4509–4515 (1998).

[90] Kahl CA, Pollok K, Haneline LS, Cornetta K: Lentiviral vectors pseudotyped with glycoproteins from Ross River and vesicular stomatitis viruses: Variable transduction related to cell type and culture conditions, *Mol Ther* **11**:470–482 (2005).

[91] Yoshida T, Kawai-Kowase K, Owens GK: Forced expression of myocardin is not sufficient for induction of smooth muscle differentiation in multipotential embryonic cells, *Arterioscler Thromb Vasc Biol* **24**:1596–1601 (2004).

[92] Brunelli S, Tagliafico E, De Angelis FG, Tonlorenzi R, Baesso S, Ferrari S, Niinobe M, Yoshikawa K, Schwartz RJ, Bozzoni I et al: Msx2 and necdin combined activities are required for smooth muscle differentiation in mesoangioblast stem cells, *Circ Res* **94**:1571–1578 (2004).

[93] McMahon JM, Conroy S, Lyons M, Greiser U, O'Shea C, Strappe P, Howard L, Murphy M, Barry F, O'Brien T: Gene transfer into rat mesenchymal stem cells: A comparative study of viral and nonviral vectors, *Stem Cells Dev* **15**:87–96 (2006).

[94] Anjos-Afonso F, Siapati EK, Bonnet D: *In vivo* contribution of murine mesenchymal stem cells into multiple cell-types under minimal damage conditions, *J Cell Sci* **117**:5655–5664 (2004).

[95] Gordon EM, Skotzko M, Kundu RK, Han B, Andrades J, Nimni M, Anderson WF, Hall FL: Capture and expansion of bone marrow-derived mesenchymal progenitor cells with a transforming growth factor-beta1-von Willebrand's factor fusion protein for retrovirus-mediated delivery of coagulation factor IX, *Hum Gene Ther* **8**:1385–1394 (1997).

[96] Geraerts M, Eggermont K, Hernandez-Acosta P, Garcia-Verdugo JM, Baekelandt V, Debyser Z: Lentiviral vectors mediate efficient and stable gene transfer in adult neural stem cells *in vivo*, *Hum Gene Ther* **17**:635–650 (2006).

[97] Derrington EA, Lopez-Lastra M, Darlix JL: Dicistronic MLV-retroviral vectors transduce neural precursors *in vivo* and co-express two genes in their differentiated neuronal progeny, *Retrovirology* **2**:60(2005).

[98] Consiglio A, Gritti A, Dolcetta D, Follenzi A, Bordignon C, Gage FH, Vescovi AL, Naldini L: Robust *in vivo* gene transfer into adult mammalian neural stem cells by lentiviral vectors.*Proc Natl Acad Sci USA* **101**:14835–14840 (2004).

[99] Falk A, Holmstrom N, Carlen M, Cassidy R, Lundberg C, Frisen J: Gene delivery to adult neural stem cells, *Exp Cell Res* **279**:34–39 (2002).

[100] Cao F, Hata R, Zhu P, Ma YJ, Tanaka J, Hanakawa Y, Hashimoto K, Niinobe M, Yoshikawa K, Sakanaka M: Overexpression of SOCS3 inhibits astrogliogenesis and promotes maintenance of neural stem cells, *J Neurochem* **98**:459–470 (2006).

[101] Beausejour CM, Krtolica A, Galimi F, Narita M, Lowe SW, Yaswen P, Campisi J: Reversal of human cellular senescence: roles of the p53 and p16 pathways, *Embo J* **22**:4212–4222 (2003).

[102] Sugiyama-Nakagiri Y, Akiyama M, Shimizu H: Hair follicle stem cell-targeted gene transfer and reconstitution system, *Gene Ther* **13**:732–737 (2006).

[103] Milosevic A, Goldman JE: Potential of progenitors from postnatal cerebellar neuroepithelium and white matter: lineage specified vs. multipotent fate, *Mol Cell Neurosci* **26**:342–353 (2004).

[104] Ghazizadeh S, Harrington R, Taichman L: *In vivo* transduction of mouse epidermis with recombinant retroviral vectors: implications for cutaneous gene therapy, *Gene Ther* **6**:1267–1275 (1999).

[105] Tunici P, Bulte JW, Bruzzone MG, Poliani PL, Cajola L, Grisoli M, Douglas T, Finocchiaro G: Brain engraftment and therapeutic potential of stem/progenitor cells derived from mouse skin, *J Gene Med* **8**:506–513 (2006).

[106] Gache Y, Baldeschi C, Del Rio M, Gagnoux-Palacios L, Larcher F, Lacour JP, Meneguzzi G: Construction of skin equivalents for gene therapy of recessive dystrophic epidermolysis bullosa, *Hum Gene Ther* **15**:921–933 (2004).

[107] Smith-Arica JR, Thomson AJ, Ansell R, Chiorini J, Davidson B, McWhir J: Infection efficiency of human and mouse embryonic stem cells using adenoviral and adeno-associated viral vectors, *Clon Stem Cells* **5**:51–62 (2003).

[108] Oka M, Chang LJ, Costantini F, Terada N: Lentiviral vector-mediated gene transfer in embryonic stem cells, *Meth Mol Biol* **329**:273–281 (2006).

[109] Hamaguchi I, Woods NB, Panagopoulos I, Andersson E, Mikkola H, Fahlman C, Zufferey R, Carlsson L, Trono D, Karlsson S: Lentivirus vector gene expression during ES cell-derived hematopoietic development in vitro, *J Virol* **74**:10778–10784 (2000).

[110] Psarras S, Karagianni N, Kellendonk C, Tronche F, Cosset FL, Stocking C, Schirrmacher V, Boehmer Hv H, Khazaie K: Gene transfer and genetic modification of embryonic stem cells by Cre- and Cre-PR-expressing MESV-based retroviral vectors, *J Gene Med* **6**: 32–42 (2004).

[111] Banerjea A, Li MJ, Bauer G, Remling L, Lee NS, Rossi J, Akkina R: Inhibition of HIV-1 by lentiviral vector-transduced siRNAs in T lymphocytes differentiated in SCID-hu mice and CD34+ progenitor cell-derived macrophages, *Mol Ther* **8**:62–71 (2003).

[112] Scherr M, Battmer K, Dallmann I, Ganser A, Eder M: Inhibition of GM-CSF receptor function by stable RNA interference in a NOD/SCID mouse hematopoietic stem cell transplantation model, *Oligonucleotides* **13**:353–363 (2003).

[113] Schomber T, Kalberer CP, Wodnar-Filipowicz A, Skoda RC: Gene silencing by lentivirus-mediated delivery of siRNA in human CD34+ cells, *Blood* **103**:4511–4513 (2004).

[114] Samakoglu S, Lisowski L, Budak-Alpdogan T, Usachenko Y, Acuto S, Di Marzo R, Maggio A, Zhu P, Tisdale JF, Riviere I et al: A genetic strategy to treat sickle cell anemia by coregulating globin transgene expression and RNA interference, *Nat Biotechnol* **24**:89–94 (2006).

[115] Salasznyk RM, Klees RF, Boskey A, Plopper GE: Activation of FAK is necessary for the osteogenic differentiation of human mesenchymal stem cells on laminin-5, *J Cell Biochem* **100**:499–514 (2007).

[116] Inoue I, Ikeda R, Tsukahara S: Current topics in pharmacological research on bone metabolism: Promyelotic leukemia zinc finger (PLZF) and tumor necrosis factor-alpha-stimulated gene 6 (TSG-6) identified by gene expression analysis play roles in the pathogenesis of ossification of the posterior longitudinal ligament, *J Pharmacol Sci* **100**:205–210 (2006).

[117] Piersanti S, Sacchetti B, Funari A, Di Cesare S, Bonci D, Cherubini G, Peschle C, Riminucci M, Bianco P, Saggio I: Lentiviral transduction of human postnatal skeletal (stromal, mesenchymal) stem cells: *in vivo* transplantation and gene silencing, *Calcif Tissue Int* **78**:372–384 (2006).

[118] Kafienah W, Mistry S, Williams C, Hollander AP: Nucleostemin is a marker of proliferating stromal stem cells in adult human bone marrow, *Stem Cells* **24**:1113–1120 (2006).

[119] Kimura A, Ohmori T, Ohkawa R, Madoiwa S, Mimuro J, Murakami T, Kobayashi E, Hoshino Y, Yatomi Y, Sakata Y: Essential roles of sphingosine 1-phosphate/S1P1 receptor axis in the migration of neural stem cells toward a site of spinal cord injury, *Stem Cells* **25**:115–124 (2007).

[120] Jiang M, Wang B, Wang C, He B, Fan H, Guo TB, Shao Q, Gao L, Liu Y: Inhibition of hypoxia-inducible factor-1alpha and endothelial progenitor cell differentiation by adenoviral transfer of small interfering RNA *in vitro*, *J Vasc Res* **43**:511–521 (2006).

[121] Mildner M, Ballaun C, Stichenwirth M, Bauer R, Gmeiner R, Buchberger M, Mlitz V, Tschachler E: Gene silencing in a human organotypic skin model, *Biochem Biophys Res Commun* **348**:76–82 (2006).

[122] Zou GM, Wu W, Chen J, Rowley JD: Duplexes of 21-nucleotide RNAs mediate RNA interference in differentiated mouse ES cells, *Biol Cell* **95**:365–371 (2003).

[123] Rubinson DA, Dillon CP, Kwiatkowski AV, Sievers C, Yang L, Kopinja J, Rooney DL, Ihrig MM, McManus MT, Gertler FB et al: A lentivirus-based system to functionally silence genes in primary mammalian cells, stem cells and transgenic mice by RNA interference, *Nat Genet* **33**:401–406 (2003).

[124] Conti MA, Even-Ram S, Liu C, Yamada KM, Adelstein RS: Defects in cell adhesion and the visceral endoderm following ablation of nonmuscle myosin heavy chain II-A in mice, *J Biol Chem* **279**:41263–41266 (2004).

5

CHEMICAL TECHNOLOGIES: PROBING BIOLOGY WITH SMALL MOLECULES

NICOLAS WINSSINGER, ZBIGNIEW PIANOWSKI, AND SOFIA BARLUENGA

Organic and Bioorganic Chemistry Laboratory, Institut de Science et Ingénierie Supramoléculaires, Université Louis Pasteur, Strasbourg, France

The use of small molecules to probe biological events has a long history, starting with bioactive natural products that are being progressively replaced with synthetic compounds. Compared to traditional genetic approaches where the function of a protein is analyzed by knocking out its corresponding gene or knocking down its translation by RNA-interference-based methods, small molecules allow for a rapid and often reversible modulation of biological function with temporal control. More importantly, small molecules have the potential to discriminate between different posttranslational forms or different conformations of a protein and as such can be used to dissect the various functions of multifunctional proteins. This pharmacological or chemical genetics approach is thus complementary to a genetic approach and can be particularly valuable to dissect dynamic biological processes. At the core of this approach is the capacity to make small molecule libraries, screen these libraries, and identify the target of the small molecules. The main features of this process are reviewed herein, starting with a section on enabling technologies for high-throughput synthesis, library synthesis [split-and-mix (split–pool) synthesis and dynamic combinatorial synthesis], a section on chemical diversity, with a discussion highlighting several reactions that can be used to create molecular diversity, and case studies of several libraries presented as a vignette of the state of the art in library synthesis. Screening technologies will not be discussed as

Chemical and Functional Genomic Approaches to Stem Cell Biology and Regenerative Medicine
Edited by Sheng Ding

high-throughput screening is reviewed in Chapter 4. However, the use of small-molecule microarrays as a new screening platform is briefly discussed. Finally, a section on target identification covers available techniques to assign the cellular target of a small molecule responsible for a phenotype.

5.1 COMBINATORIAL SYNTHESIS

Several technologies have been developed to facilitate the isolation of a reaction product from reagents enabling automation of the chemistry and the production of chemical libraries. A pivotal advance was the pioneering work of Merrifield's solid-phase synthesis of peptides [1]. By attaching the first amino acid residue to an insoluble matrix such as polystyrene, a large excess of reagents can be used to promote amino acid couplings and deprotections; these excesses are removed by a simple filtration. This technique lends itself to automation and has been expanded to oligonucleotides, carbohydrates, and the synthesis of drug-like nonoligomeric small molecules (Figure 5.1a). Conversely, reagents can be immobilized to an insoluble matrix or scavenged by a reactive functional group on solid phase such that an excess of reagents can be used to achieve the desired transformations [2,3]. Alternatively, the first intermediate in a synthesis can be tethered to a tag with unique physical properties such that it can be selectively isolated following a reaction. A successful implementation of this strategy relies on fluorocarbon tags, which have a high affinity for fluorinated phases (either fluorinated solvent or fluorinated silica) and can be isolated in an automated fashion [4,5]. These synthetic techniques can be easily parallelized to produce hundreds of diverse compounds in a matter of days.

These methods for parallel synthesis have significantly increased the speed at which libraries can be prepared. However, libraries in the thousands of compounds are still laborious or require expensive robotics. Another important development was a technique called "split and mix" or "split–pool" synthesis [6]. In this case, a batch of resin is divided into several pools, and the first synthetic intermediate or element of diversity is introduced (Figure 5.1b). The pools are then recombined, mixed, and redistributed into several new pools. If a large excess of solid support beads is used per pool, each new pool will now contain a statistical distribution of all the first synthetic intermediates. Reaction with a different element of diversity (reagent or synthetic intermediate) in every pool will yield all possible combinatorial permutations of products. An important feature of this method, compared to parallel synthesis, is that the number of products generated increases exponentially with every step. For example, a library of $20 \times 20 \times 20$ (all permutations of a tripeptide) would yield 8000 compounds but require only 60 individual reactions (omitting protecting group manipulations)! While this scheme is extremely powerful to synthesize libraries, the structure of the product coming from an individual bead is unknown unless an encoding method is used to track the path of this bead through the splits and the mixes. Elegant tracking or encoding methods [7] have been developed on the basis of chemical tags [8] with optimal analytical profile or physical tags appended to a container for the resin (optical encoding with 2D barcodes [9] or radiofrequency encoding with Rf tags [10])

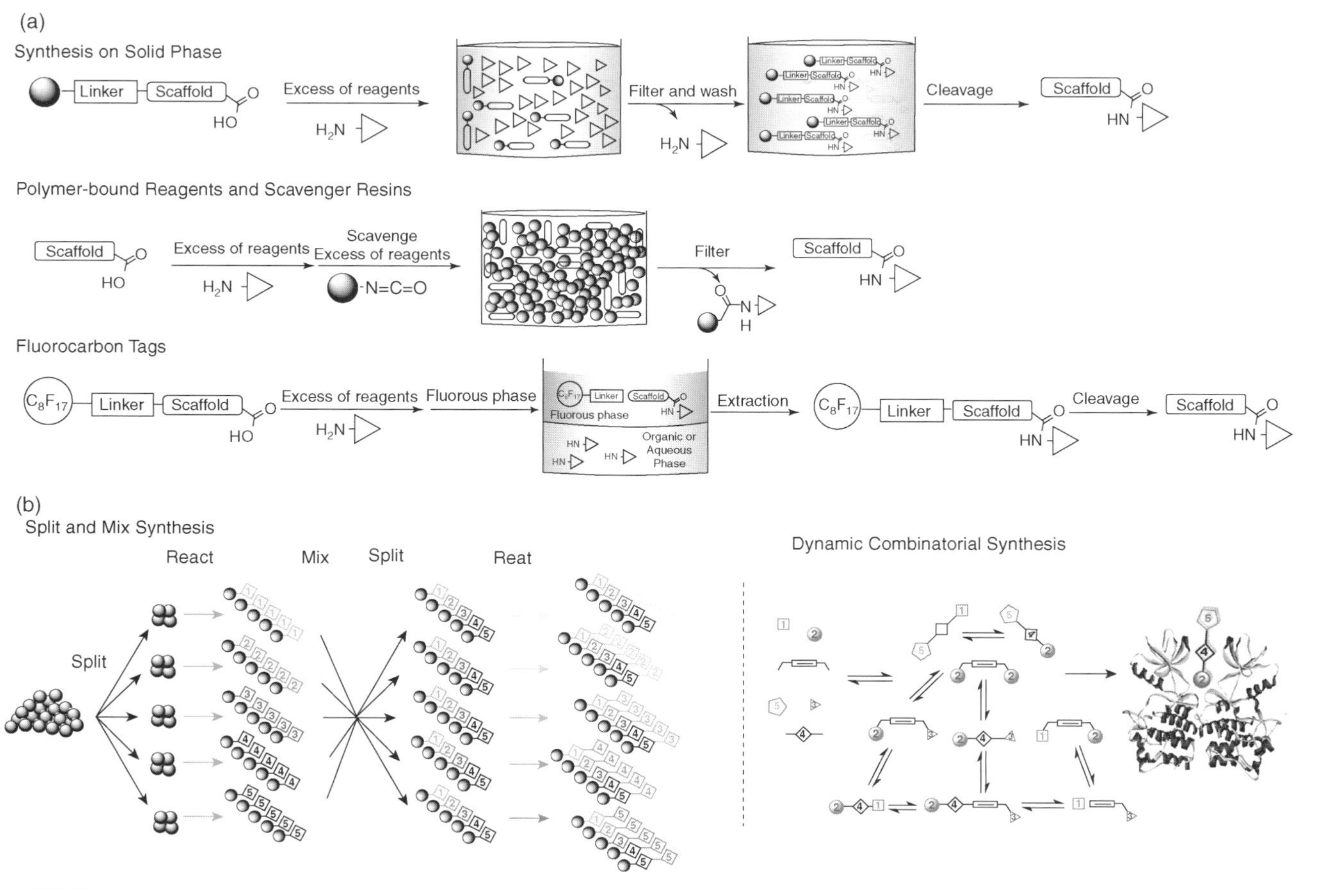

FIGURE 5.1 Enabling technologies for automated synthesis (a) and library synthesis (b). (See insert for color representation.)

that can be read in an automated fashion. A significant advantage of the latter is the fact that the final library can be arrayed in microtiter plates and that the structures of all compounds are known according to their location. The container can carry 5–100 mg of resin providing sufficient compound for thousands of assays.

In a conceptually different approach, it has been shown that the fittest ligand for a target can be selected from a virtual dynamic library prepared from smaller sets of synthetic intermediates containing reactive moieties (Figure 5.1b). Reactions between the synthetic intermediates produce more complex molecules in virtually every possible permutation, thereby affording a library for *in situ* screening [11–13]. If the reactions leading to the more complex molecules are reversible, the presence of the target in the reaction mixture will enrich the fittest compound relative to other possible library members. Conversely, if the reaction between the synthetic intermediates is slow, the target may accelerate the formation of the fittest ligand by bringing the reactive species in close proximity [14–16]. While this approach is attractive, it is, of course, limited to the use of reactions that are compatible with the target and its aqueous environment.

5.2 MEASURING AND DESIGNING DIVERSITY

With the advent of high-throughput screening in the 1980s, it appeared that the number of hits in a screen was directly proportional to the number of compounds being screened. Consequently, the first combinatorial libraries in the 1990s were designed to produce the largest number of compounds possible. It quickly became clear that simply maximizing the number of compounds without consideration of the diversity space—a space created by mapping the principal components of a molecule's spatial and physicochemical properties such as molecular weight, charges, number of hydrogen bond donors and acceptors, surface area, and log P [17]—would in fact result in libraries containing many similar compounds that are clustered in this diversity space. However, historical libraries of large pharmaceutical companies were composed of natural products from diverse sources and small libraries based on the diverse medicinal chemistry efforts that took place over the years resulting in a diversity that could not be matched by a single scaffold or single combinatorial library. It has been estimated that the number of possible small organic molecules exceeds 10^{60}, but regardless of the exact number, it is quite clear that we will be able to explore only small segments of this chemical space at any time [18]. Despite its vastness, there are many indications that biologically active compounds cluster into small pockets of diversity space; the challenge is thus to identify those regions rather than individual scattered compounds (Figure 5.2) [19,20]. Some scaffolds such as the benzodiazepines have been identified as privileged structures, based on an intrinsic ability to bind structurally or functionally unrelated classes of proteins [21]. As bioavailability of small molecules is a significant hurdle in drug discovery, researchers at Pfizer formulated a rule to predict bioavailability based on an analysis of all approved drugs listed in the World Drug Index [22]. This empirical rule, known as "Lipinski's rule-of-five," limits dramatically the subset of potential drug candidate. However, it is important to keep in mind that it is based on a dataset of limited diversity. Many

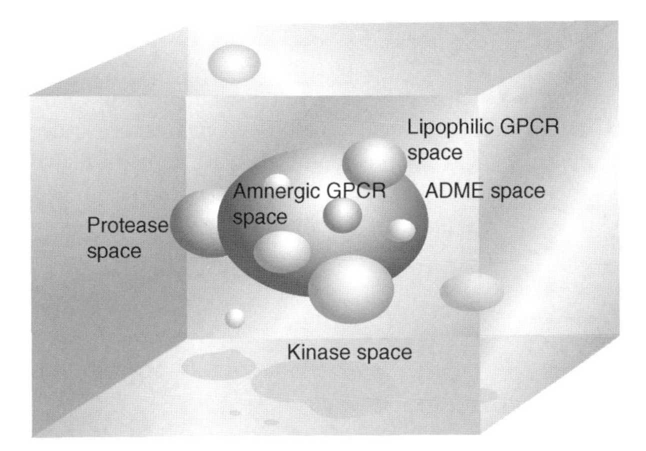

FIGURE 5.2 Cartoon representation of the chemical space. Representation obtained by plotting the principal component of physicochemical properties on *x,y,z* axis. Biologically active compounds tend to cluster into discrete areas (represented by the colored spheres). (Reprinted with permission from Ref [20]). (See insert for color representation.)

biologically active natural products, for example, do not obey this rule. Natural products, which have evolved for a specific function—typically to bind with high affinity to a target and perturb a biological process in competing organisms—may be considered as validated starting points for the design of focused libraries. As a number of protein domains are redundantly present in functionally distinct protein families, natural-product-based libraries should provide protein ligands even against new targets with enhanced quality and probability compared to random libraries [23,24]. The fact that libraries of biologically active molecules should give a higher hit rate in a screen against a new target than random libraries has been demonstrated. A library of 2036 biologically active molecules with a well-characterized mechanism of action was screened against an engineered human tumor cell line produced by genetically introducing four oncogenic proteins (two simian viral oncogenes, telomerase and Ras). The result was compared to the same screen with a commercially available diverse library (Comgenex). While neither libraries had any known ligand for the oncogenic proteins, the library of biologically active compounds had a 10-fold higher hit rate (1% vs 0.1%) than did the diverse library in a screen for compounds that were selectively cytotoxic for the transformed cell line [25]. On the other hand, a number of libraries based on elegant and efficient chemistries affording seemingly complex and diverse products failed to give significant hits.

5.3 DIVERSITY-BUILDING REACTIONS

A major focus of organic synthesis in the twentieth century has been the total synthesis of natural products and other designed molecules where a highly convergent route to a target was deemed most attractive. The synthesis of libraries consisting of diverse molecules obviously has different requirements and limitations. While there are many techniques to remove excess reagents used in the reactions, it is extremely cumbersome to purify each product in a final library, and it is important that the chemistry that is used has nearly quantitative efficiency for every transformation. In reaction-driven approaches to generate molecular diversity, multicomponent and cascade reactions have proved particularly suitable [26]. While a comprehensive review is far beyond the scope of this chapter, selected examples are presented to illustrate some efficient reactions that have been used to synthesize libraries (Figure 5.3). The four-component Ugi reaction

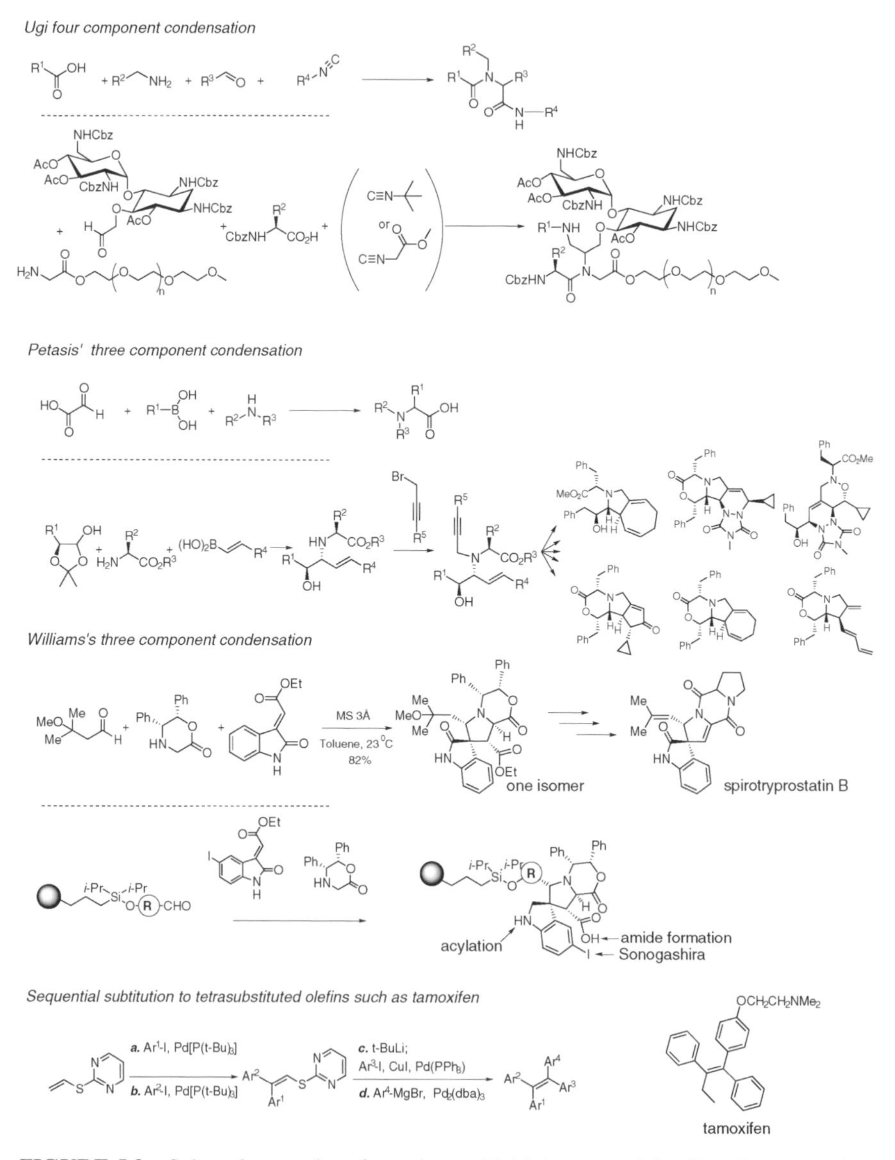

FIGURE 5.3 Selected examples of reactions with high potential for diversity generation.

[27,28] is the condensation of an aldehyde, an amine, a carboxylic acid, and an isocyanide to yield an α-aminoacyl amide derivative. The reaction is very general, and hundreds of each component are commercially available, making a library based on this chemistry within easy reach. To improve the purity of the final product, one of the components can be immobilized on a solid phase such that the other three are used in excess [29]. Release of the compound from the resin affords the final product in excellent yields. Furthermore, the intermediates can be further elaborated to obtain diverse scaffolds [30]. For example, the Ugi reaction has been used to generate a library based on neomycin B, a natural product currently used as an antibiotic, leading to the discovery of inhibitors with improved activity [31]. The library was synthesized using a neamine-derived aldehyde, *tert*-butyl isocyanide or isocyanoacetic acid methyl ester, a glycine-conjugated polyethylene glycol (PEG) methyl ester, and several Cbz-*N*-protected amino acids. Another example of a useful multicomponent reaction yielding highly substituted amines and α-amino acids is the Petasis reaction [32,33], where an imine formed *in situ* is engaged in a reaction with a boronic acid acting as the nucleophile. This reaction has been used to prepare a library of 44 products using propargylation of the amine resulting from the Petasis reaction to generate a library of β-amino alcohols strategically functionalized for further diversifications in divergent pathways [34]. The Petasis reaction is stereoselective, thus the diversity generated with this strategy is not only a skeletal or scaffold diversity but also a stereochemical diversity. (Four possible stereoisomers are accessible in their pure form by using the *R* or *S* lactols and the *R* or *S* amino acids). The different series of skeletal diversification reactions included a cycloisomerization catalyzed by Pd, a [5 + 2] reaction catalyzed by Ru, a Pauson–Khand multicomponent reaction, and an enyne metathesis followed by a Diels–Alder cycloaddition. Another example of an elegant reaction to assemble complex molecular scaffolds in a single operation is the Williams three-component reaction [35,36] where an aldehyde and 5,6-diphenylmorpholino-2-one are reacted *in situ* to generate a dipolarophile that can be engaged in a [3 + 2] cycloaddition. This reaction was exploited to generate a library of 3000 spirooxindoles [37]. The development of new enabling synthetic protocols can also extend the types of molecules accessible in a combinatorial fashion. For example, a library of tamoxifen analogs (one of the most important drug in breast cancer prevention) was prepared using four successive Pd-catalyzed cross-couplings starting from 2-pyrimidyl sulfide [38]. These examples illustrate the development of reaction sequences that are particularly efficient to synthesize libraries in a so-called diversity-oriented synthesis [39]. Of course, while the chemistry may efficiently generate large libraries, the pharmacophore may not be validated.

5.4 SELECTED EXAMPLES OF LIBRARIES THAT HAVE YIELDED INHIBITORS USEFUL IN EXPLORING BIOLOGY

5.4.1 Structure-Based Libraries Targeting Kinases and Other Purine-Dependent Enzymes

The purine scaffold is a key structural element of the substrates and ligands in many biosynthetic, regulatory, and signal transduction proteins, including G-coupled proteins, kinases, motor proteins, and polymerases. It follows that libraries based on

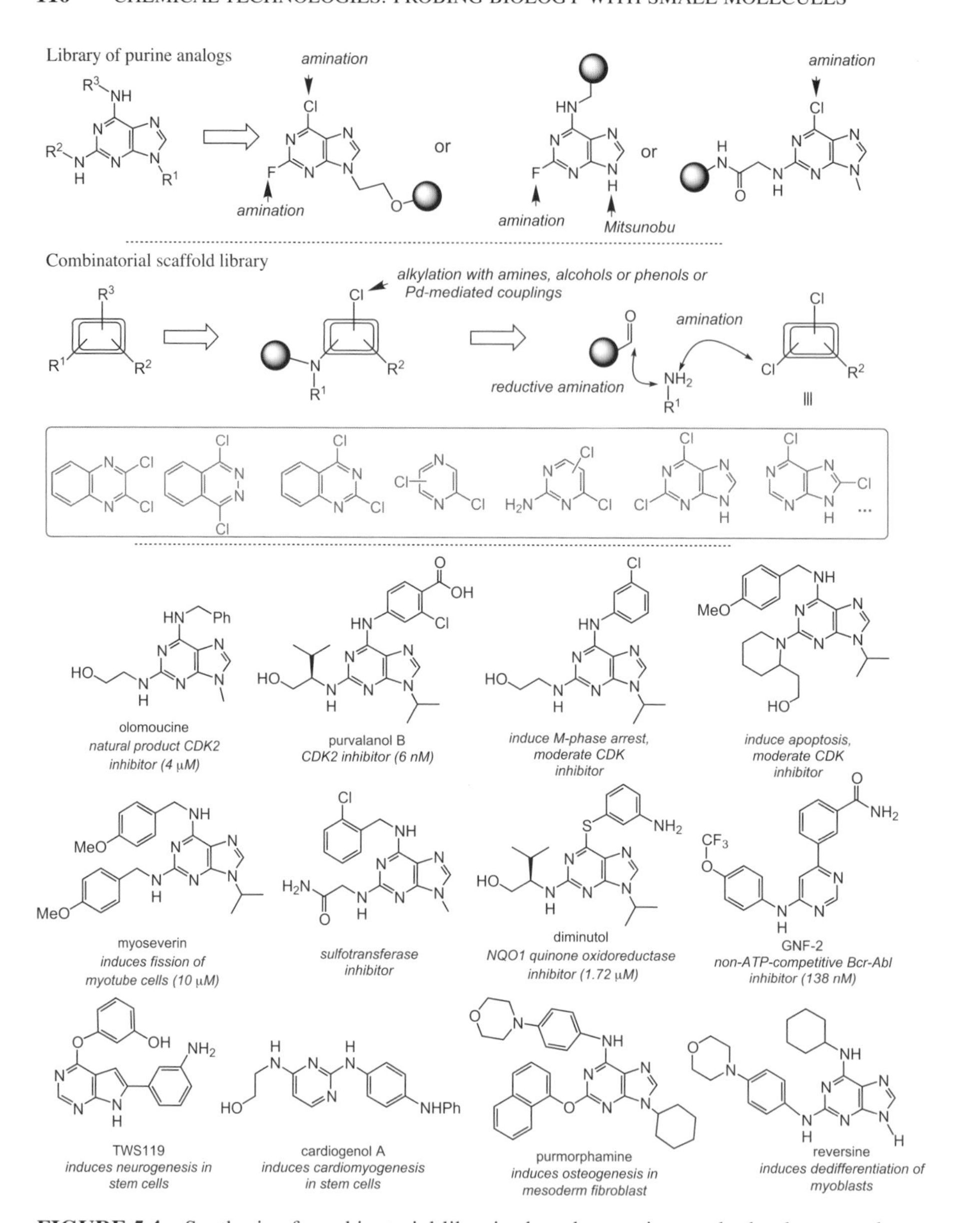

FIGURE 5.4 Synthesis of combinatorial libraries based on purines and other heterocycles and discovery of kinase inhibitors, sulfotransferase inhibitors, a tubulin binder, and cell fate modulators.

this scaffold may provide versatile leads to probe signaling and metabolic pathways. Olomoucine (Figure 5.4), a natural derivative of purine that selectively inhibits CDK2, indicated that despite the ubiquitous role of purines, selective inhibition was possible. Schultz and coworkers reported an initial library of 348 purine derivatives in 1996 that led to the discovery of an improved and selective CDK2 inhibitor named purvalanol B (6 nM, 1000-fold improvement over olomoucine) [40]. Interestingly, a crystal structure revealed that this compound had a different binding mode than the adenosine of ATP. Larger libraries extending the scope of the chemistry yielded inhibitors targeting other cell cycle regulators. Notably, a compound that induced apoptosis irrespective of cell cycle progression and a second compound that induce cell cycle arrest at the *M* phase were identified, suggesting that closely related analogs target different cell cycle regulators [41]. Beyond kinases, this library also afforded inhibitors of carbohydrate sulfotransferases, which is not surprising since such enzymes derive the sulfate group from 3′-phosphoadenosine 5′-phosphosulfate (PAPS) [42], and estrogen sulfotransferase inhibitors [43]. A screen against the malaria parasite *Plasmodium falciparum* identified two subfamilies of purines with good activities (nanomolar) against the parasite but modest CDK activities raising the possibility to achieve a good therapeutic window [44]. Purine libraries were also screened in a phenotypic assay for myotube fission—myotubes are multinucleated cells that eventually develop into mature muscle fibers and have the capacity to regenerate wounded muscles—leading to the discovery of a new purine analog named *myoseverin* (Figure 5.4) [45]. Further experiments using affinity chromatography revealed that myoseverin interacts with microtubulin. Interestingly, gene expression profiling of myoseverin-treated myotubes revealed that myoseverin induced a profile similar to that of tissue injury, suggesting the possibility of using such molecules to prime the natural regenerative capacity of muscles. Later, it was observed that while myoseverin's inhibition of microtubule polymerization induced the reversion of terminally differentiated myotubes to mononucleated cells, it lacked the toxicity of other tubulin modulators such as taxol, vinblastine, nocodazole, and colchicine. Myoseverin reverted terminal muscle-differentiated cells to a state that was responsive to environmental cues [46]. In an extension of this study, a library of purines was screened in a phenotypic assay for perturbation of mitotic spindle assembly in *Xenopus* egg extracts. Out of a collection of 1561 compounds, 15 compounds destabilized microtubules without targeting tubulin directly. Affinity chromatography with one compound, diminutol, suggested NQO1, an NADP-dependent oxidoreductase, as the target that was confirmed through immunodepletion studies [47]. The chemistry developed to access libraries of purines was extended to other heterocycles, including pyrimidines, quinazolines, pyrazines, phtalazines, pyridazines, and quinoxalines. A library of over 45,000 compounds was prepared using a combinatorial scaffold approach. For a library of this size, a split–pool synthesis was necessary; however, the authors used the optical encoding system [9] which directs the path of small reactors harboring 5–10 mg of resin through the split and pool cycles and allows a reformatting of the final library into a spatially addressable format for the cleavage of the library compounds [48]. This larger library and focused libraries derived from library hits have yielded an impressive collection of biological probes and potential therapeutics, including a number of compounds that control stem cell fate [49–52] or even reverted differentiated cells [53]. The importance of these discoveries for stem cell biology is discussed in more detail in Chapter 4. More recently, several non-ATP-competitive inhibitors of Bcr-Abl-kinase were identified and shown to bind to an allosteric site distant from the ATP-binding site.

These compounds are synergic with imatinib, the Bcr-abl inhibitor currently used for the treatment of chronic myelogenous leukaemia [54].

5.4.2 Libraries Based on a Privileged Scaffold—Discovery of Fexaramine

The benzopyran motif is found in more than 4000 natural products and designed analogs including many biologically active compounds. This observation prompted Nicolaou and coworkers to develop a highly efficient and divergent synthesis based on the benzopyran core motif which was formed by cycloaddition reaction upon loading on a selenium-based resin (Figure 5.5). This chemistry was used to prepare over 10,000 compounds, using split–pool synthesis and optical encoding with directed sorting [9,55]. The automated sorting was essential in this library synthesis as not all intermediates followed the same synthetic paths. The final library was further diversified by reaction on the pyran moiety [56]. The library was first screened for antibiotic activity leading to the identification of several benzopyrans with activity against methycillin-resistant bacterias at a comparable level to vancomycin [57]. In collaboration with Evans and coworkers, the library was screened for FXR agonists [58], a nuclear receptor involved in the regulation of cholesterol biosynthesis. The efficient regulation of cholesterol metabolism is essential for mammals as its misregulation leads to arteriosclerosis and heart diseases. It is controlled through a complex feedback loop consisting of LXR and FXR nuclear receptors. The LXR (liver X receptors) are activated by oxysterols (early cholesterol metabolites) leading to upregulation of CYP7A1, the enzyme catalyzing the rate-limiting step in the conversion of cholesterol to bile acids. The latter ones (e.g., chenodeoxycholic acid CDCA) are, however, ligands for FXR (farnesoid X receptors) whose activation downregulates the CYP7A1, and closes the feedback circuit. Additionally, both LXR and FXR are involved in the regulation of several gene products responsible for cholesterol absorption, metabolism, and transport. Selective small-molecule FXR agonists would be powerful tools to dissect FXR's physiological function and may provide lead structure for potential therapeutics. The benzopyran library was screened using a cell-based reporter assay in which a FXR-responsive promoter was linked to a luciferase reporter. The initial hits were further optimized in terms of potency and pharmacological properties affording a FXR ligand, fexaramine, with 25 nM affinity. Expression profiling experiments were then carried out to evaluate the effects of fexaramine-induced FXR activation on a genomic scale [59]. Notably, fexaramine did not agonize other nuclear hormone receptors, showing high specificity in contrast to bile acid. Fexaramine is thus a useful tool to dissect the FXR genetic network from the bile acid network. A cocrystal structure of FXR–fexaramine complex was obtained providing insightful structural information and suggesting a mechanism for the initial steps of bile acid signaling.

5.4.3 Unbiased Library—Discovery of Uretupamines

With the goal of developing unbiased libraries reminiscent of natural products, Schreiber and coworkers reported the diversity-oriented synthesis of a 1,3-dioxane library (Figure 5.6) [60]. Using polystyrene macrobeads (500-μm rather than the

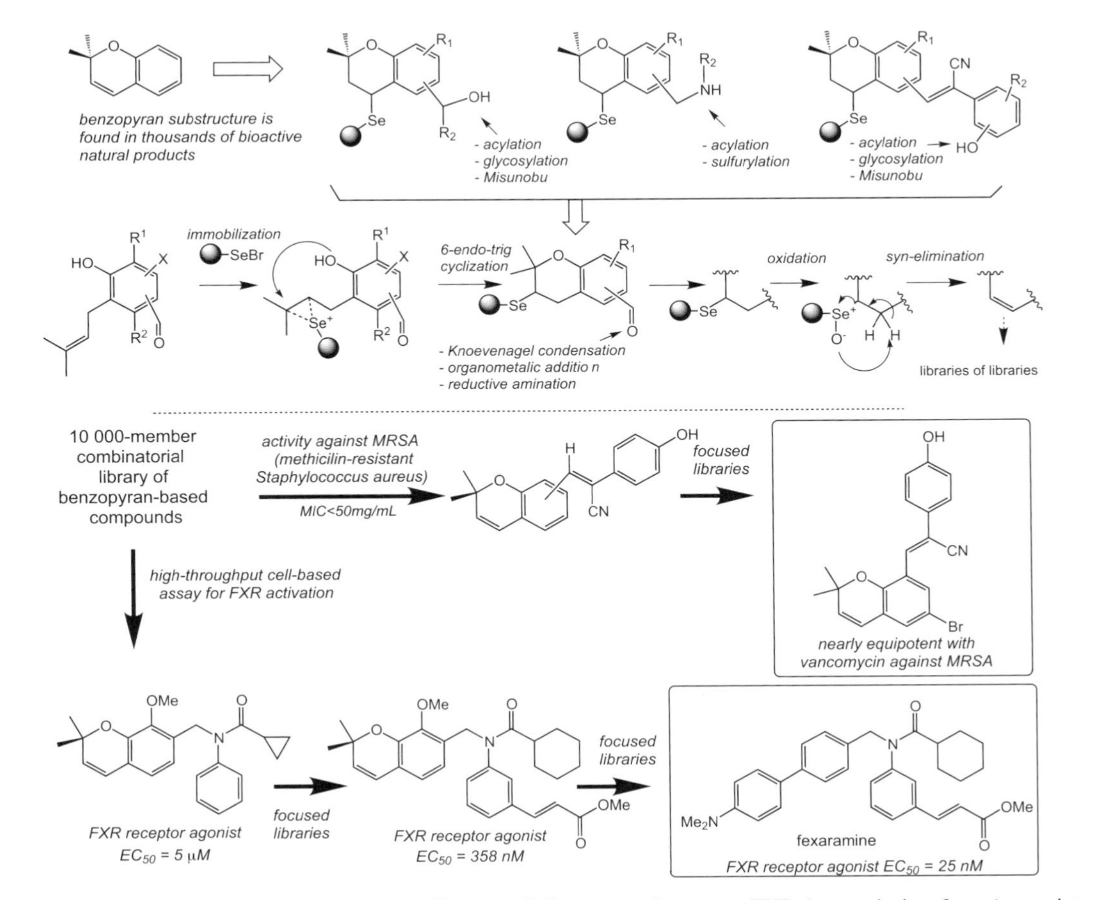

FIGURE 5.5 Synthesis of a benzopyran library and discovery of a potent FXR (transcription factor) agonist.

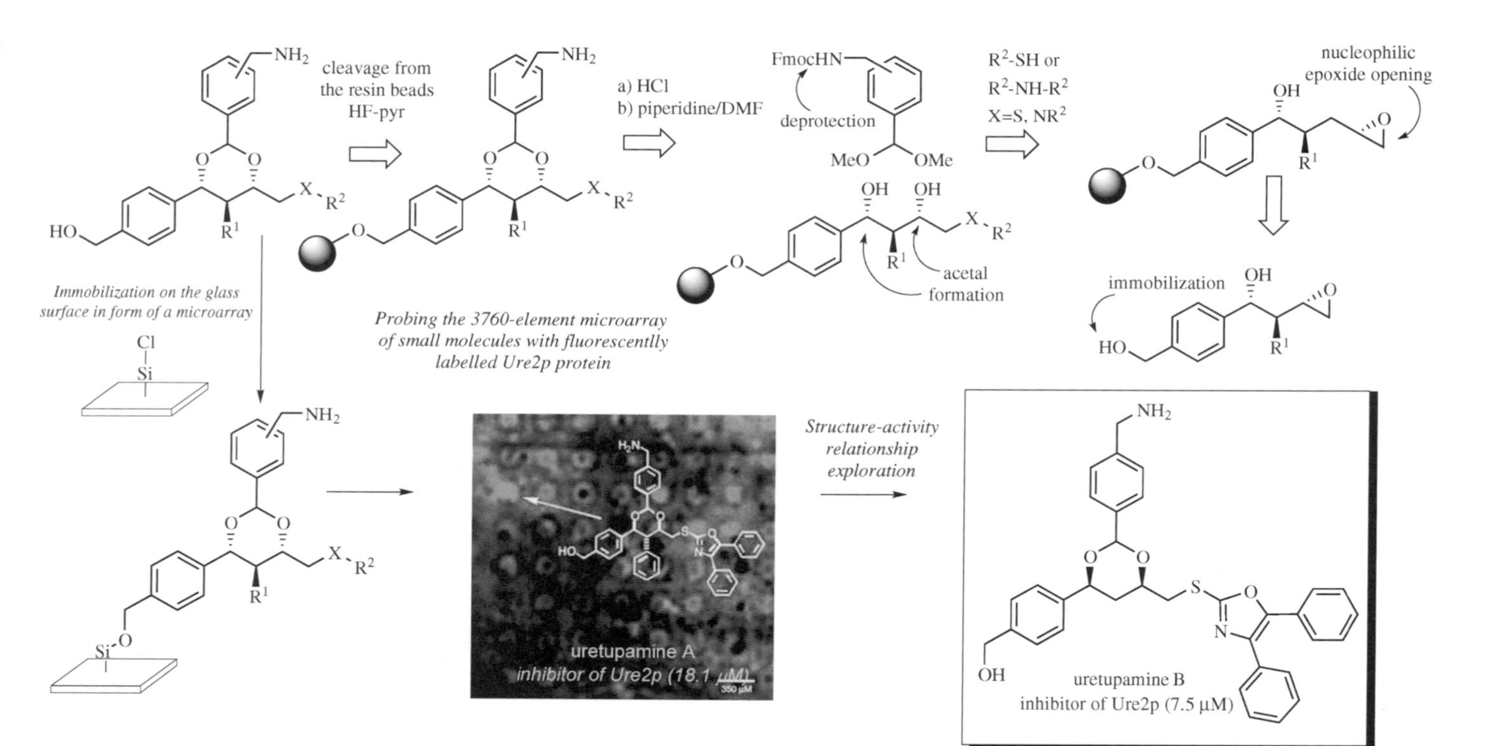

FIGURE 5.6 Synthesis and microarraying of a 1,3-dioxane library. Discovery of a selective Ure2p (transcription factor) modulator. (See insert for color representation.)

regular 100-μm beads), a sufficient amount of each compound was obtained from individual beads for multiple biological assays. A first library of 1800 compounds was prepared by split–pool synthesis and distributed in a microtiter plate format for cleavage. For the purpose of smaller libraries where every compound within a synthetic pool has a unique mass, the identity of an active compound can be obtained by LC-MS analysis of the stock solution. A library based on the 1,3-dioxanes was microarrayed (~800 spots/cm^2) on slilyl chloride slides [61] which reacted covalently with the primary alcohol present in the whole library [62]. The microarrayed compounds were screened as potential ligands of the yeast transcription factor Ure2p using a fluorescently labeled Ure2p leading to the identification of eight compounds that showed reproducible binding [63]. Although this protein has been widely studied as a central repressor of genes involved in nitrogen metabolism [64] and as part of the signaling cascade downstream of the Tor proteins [65], there was no reported ligand that could modulate its function. To determine the cellular activity of identified ligands, the molecules were resynthesized and tested for modulation of endogenous Ure2p function using a reporter assay. One of the eight compounds—namely uretupamine A—gave a concentration-dependent dose response approaching the level of a positive control using rapamycin. A focused library was then prepared to optimize the activity, yielding a more potent analog, uretupamine B (K_d 7.5 μM vs. 18.1 μM). Neither compound had significant effects on a strain lacking the gene encoding Ure2p, suggesting that despite their moderate affinity, both compounds were selective. Analysis of whole-genome transcription profiles revealed that both compounds upregulate only a subset of genes known to be under the control of Ure2p (like PUT1, PUT2, PRB1, NIL1, and UGA1). Ure2p-controlled genes were thought to be responsive to nitrogen quality; however, uretupamine induced the expression of only glucose-sensitive genes in a NIL1-dependent manner and not in GLN3. Interestingly, when quality of energy source decreased (from glucose to acetate or glycerol), Ure2p was dephosphorylated, which was not the case with decreasing quality of nitrogen source (from ammonium sulfate to proline) nor in case of other stresses known to upregulate Ure2p-dependent genes, like heatshock, pH 9.5, 1 M NaCl, or 1 M sorbitol. Together, these results suggested that Ure2p is part of a signaling pathway specific to glucose and that uretupamines modulate the glucose-sensitive subset of genes downstream of Ure2p. In a larger context, this example illustrates the utility of small molecules in dissecting multifunctional proteins.

5.4.4 Inhibitors of Histone Deacetylases (HDAC)

Histone deacetylases (HDACs) are zinc hydrolases that remove the lysine's ε-*N*-acetyl group from proteins, thereby modulating their function. In the case of histones, this deacetylation restores their cationic character and increases their affinity for DNA, thus downregulating DNA transcription. Nine human HDACs have been identified thus far, and while several inhibitors are known, none are selective for individual HDAC and thus have limited utility to deconvolute the function of individual HDAC. Interestingly, it has been found that inhibition of

HDACs results in hyperacetylation of α-tubulin, suggesting that HDACs modulate the function of proteins beyond the histones. Inspired by the structure of a natural product, trichostatin A, containing a hydroxamic acid known to chelate zinc. Schreiber and coworkers designed a library based on the previously made 1,3-dioxanes targeting the HDAC by incorporating a hydroxamic acid or a 2-amino-carboxyamide functionality (Figure 5.7). A library of 7200 compounds was prepared on the aforementioned polystyrene macrobeads using split–pool synthesis and chemical encoding [60]. After the synthesis, the macrobeads were separated into individual wells of a microtiter plate, and the compounds were cleaved from the resin to obtain a stock solution that was used in biological assays (one bead–one stock solution) [66,67]. On identification of an active compound, the chemical tag remaining on the bead was oxidatively cleaved and analyzed with gas chromatography to determine the structure of the compound. A cell-based cytoblot assay to measure histone and tubulin acetylation levels in the presence of inhibitors followed by secondary fluorescence assays led to the identification of tubacin—a compound that selectively induces α-tubulin hyperacetylation ($EC_{50} = 2.9\,\mu M$) and histacin—a compound that selectively induces histone hyperacetylation ($EC_{50} = 34\,\mu M$) [68]. Tubacin, in contrast to trichostatin A, had no effect on gene expression and did not affect cell cycle progression. Interestingly, it was shown that tubacin inhibited only one of the HDAC6's two deacteylase domains. While α-tubulin hyperacetylation had no impact on microtubule dynamics, it was shown to reduce cellular motility, consistent with the fact that overexpression of HDAC6 had previously been shown to increase cell motility [69]. Further experiments highlighted the role of α-tubulin acetylation in mediating the localization of microtubule-associated proteins such as p58, a protein that mediates binding of microtubules to the Golgis [70].

5.4.5 Secramine

Galanthamine is a natural alkaloid and a potent acetylcholinesterase inhibitor. Recognizing the potential to access this rigid polycyclic core through a biomimetic oxidative coupling/Michael addition, Shair and coworkers developed a highly efficient diversity-oriented synthesis based on this scaffold (Figure 5.8) [71]. A library of 2527 compounds was prepared on the aforementioned macrobeads by split–pool synthesis and distributed in microtiter plates for cleavage. The structure of individual compounds was obtained by mass spectroscopic analysis of the library. The library was screened in a phenotypic assay for protein trafficking—translocation of proteins from the endoplasmatic reticulum via the Golgi apparatus to the plasma membrane—using a fluorescent fusion protein between VSVG, which is targeted toward the cell membrane, and GFP, leading to the identification of an inhibitor, namely secramine. It blocked protein trafficking out of the Golgi at $2\,\mu M$ concentration, whereas galanthamine had no effect up to $100\,\mu M$. Further experiments showed that secramine inhibited the activation of Rho GTPase Cdc42 and thus inhibited Cdc42-dependent functions. As this inhibition is independent of the prenylation state of Cdc42, secramine is fast-acting compared to prenylation inhibitors [72,73].

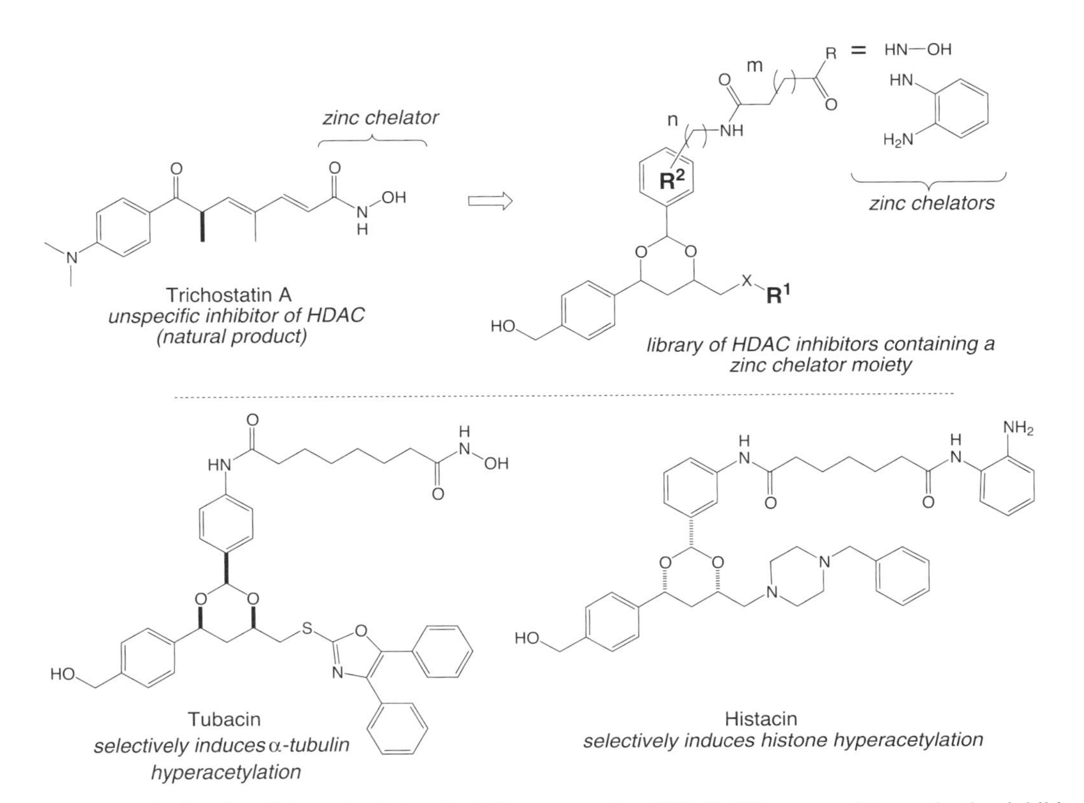

FIGURE 5.7 Synthesis of a trichostatin A–inspired library targeting HDAC. Discovery of two selective inhibitors modulating different HDAC functions.

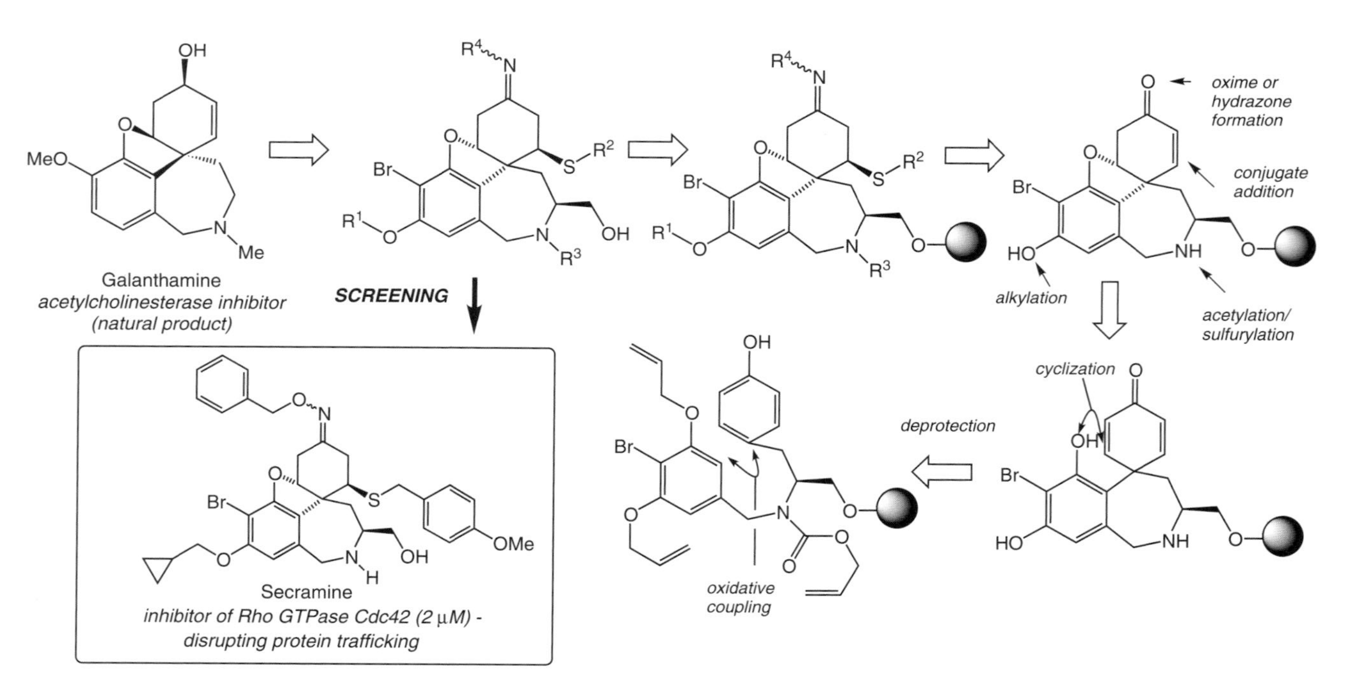

FIGURE 5.8 Synthesis of a galanthamine-inspired library and discovery of secramine, an inhibitor of protein trafficking.

5.4.6 Small-Molecule Antagonists of Protein–Protein Interactions

Protein–protein interactions regulate numerous biological pathways, and their dysfunctions have accordingly been implicated in an equally large number of pathologies. Needless to say, there is tremendous interest in small molecules that could agonize or antagonize such interactions not only as potential therapeutics but also as tools to gain insight into the function of specific interactions. However, in contrast to an enzyme active site, which is typically a small cavity bearing lipophilic groups well suited to bind small molecules with high affinity, protein–protein interactions are typically the product of interactions of larger, discontinuous open surfaces that lend themselves poorly to competitive binding by small molecules [74]. Boger and coworkers have designed a series of libraries prepared using solution-phase chemistry aimed at disrupting protein–protein interactions and protein–DNA interactions (Figure 5.9), including EPO/EPOR and Myc-Max [75]. The EPO glycoprotein is the principal factor regulating red blood cell fate through homodimerization of EPO receptor (EPOR)—a class I cytokine receptor. Inspired by the discovery of a peptidic agonist of EPO from phage libraries, investigators prepared solution-phase libraries derived from *N*-Boc-iminodiacetic acid anhydride containing roughly 100,000 compounds and screened them for inhibition of ^{125}I-EPO binding to EPOR. Several hits (1–80 μM) were discovered and covalently linked to afford C_2-symmetric ligands (Figure 5.9), which were found to be EPO agonists at 5–10 μM. A second successful application of the solution-phase libraries has been the discovery of a small-molecule antagonist of Myc-Max dimerization. Myc is a transcriptional regulator of the basic helix–loop–helix leucine zipper (bHLHLZ) protein family overexpressed in many human cancers. Its oncogenic properties come from the formation of a heterodimer with ubiquitous Max—another member of the bHLHLZ protein family leading to the transcription of Myc target genes. To screen for compounds that could disrupt this interaction and thus prevent transcription, an assay based on fluorescence–resonance energy transfer between a Myc-GFP fusion protein and Max-YFP fusion protein was used [76]. A isoindoline library containing 240 compounds (Figure 5.9) yielded several candidate inhibitors that were subsequently validated by ELISA and electrophoretic mobility-shift assay. Two of these compounds were also shown to be effective *in vivo* and interfered with Myc-induced oncogenesis of chicken embryo fibroblast.

5.4.7 Phosphatase Inhibitors

Protein phosphatases specifically hydrolyze phosphate groups from proteins and act as key regulators in biological pathways, often in equilibrium with kinase phosphorylation. They represent a growing class of therapeutic targets such as for insulin signaling (PTP1B), tuberculosis (MptpA and MptpB) and numerous other infections (Shp2), cell adhesion and angiogenesis (VE-PTP), the cell cycle (Cdc25A), and signaling by kinase dephosphorylation (VHR) [77]. Recognizing the necessity for selective phosphatase inhibitors, Waldmann and coworkers have developed a series of libraries based on natural products targeting phosphatases [78–80]. Searching for new scaffolds to be used for the synthesis of focused libraries, a small library of natural products was

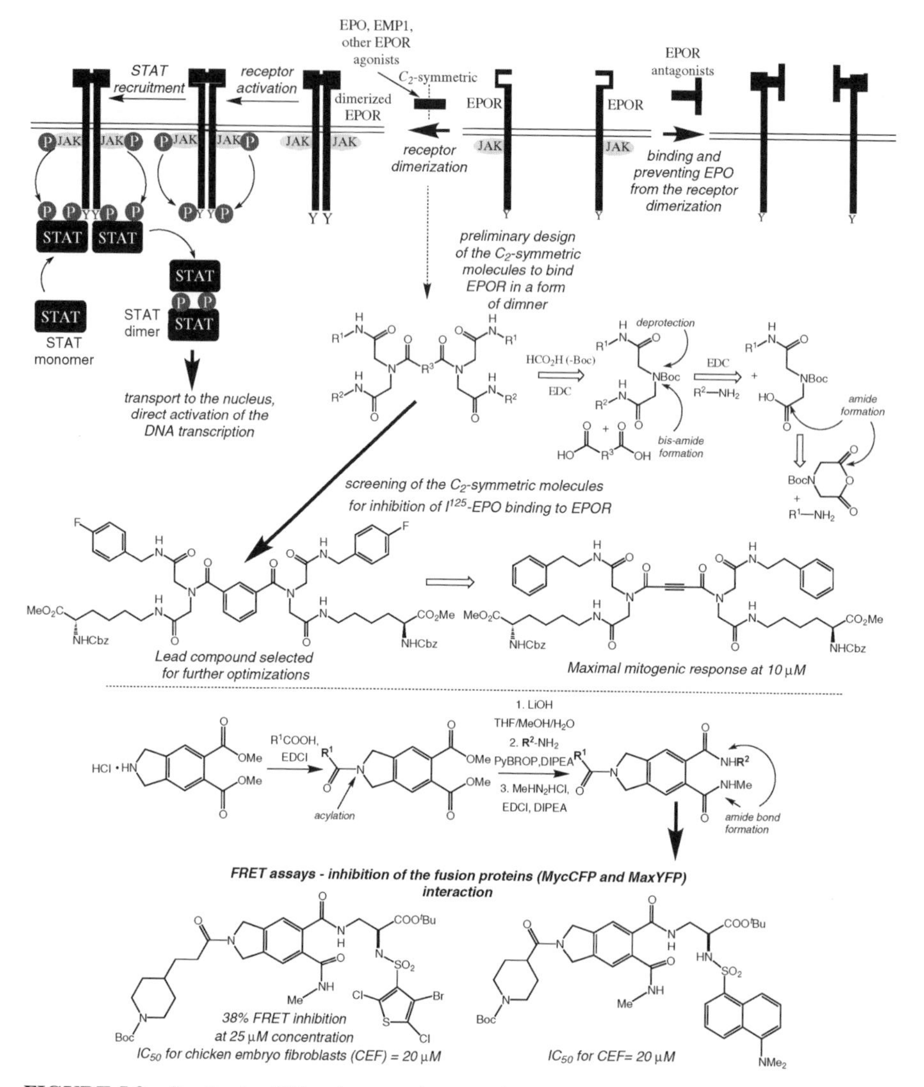

FIGURE 5.9 Synthesis of libraries targeting protein–protein interactions. Libraries based on C_2-symmetric ligands led to the discovery of an EPO agonist while library of isoindolinones afforded a Myc-Max protein–protein interaction inhibitor. (See insert for color representation.)

screened against a panel of seven phosphatases (PTP1B, Shp2, VE-PTP, MptpA, MptpB, and the dual-specificity phosphatases Cdc25A and VHR), leading to the identification of three yohimbine alkaloids (yohimbine, ajmalicine, and reserpine) with weak inhibition of Cdc25A (Figure 5.10) [81]. After detailed structural analysis, it was speculated that simplified tetracyclic and tricyclic analogues should retain most of the biological activity. A collection of 450 diastereomerically pure indoloquinolizidines was prepared by a divergent solid-phase synthesis and screened against seven

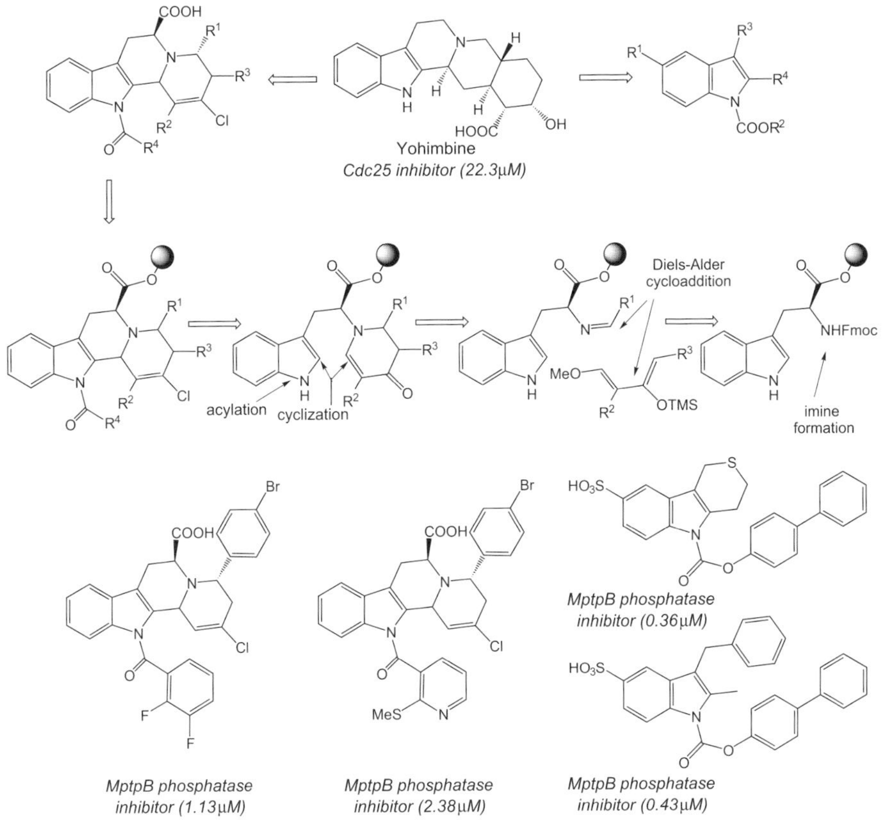

FIGURE 5.10 Phosphatase inhibitors derived from natural-product like libraries based on simplified yohimbine analogs. Discovery of selective sub-μM phosphatase inhibitors.

phosphatases revealing two weak Cdc25A inhibitors (comparable to the yohimbines despite their simplified structure), but also 11 compounds with $IC_{50} < 10\,\mu M$ against the MptpB (nine of them did not inhibit any other phosphatase at the $100\,\mu M$ concentration), thus reporting the first inhibitors for this phosphatase.

5.5 MICROARRAY TECHNOLOGIES

The success of DNA microarrays to measure the expression level of thousands of genes simultaneously has inspired researchers in bioorganic chemistry to explore this format [82]. From an analytical perspective, the ability of screening thousands of analytes in a few microliters is attractive and results in a significant miniaturization as compared to HTS. Small molecules from libraries can be immobilized in a microarray format by contact printing using the same robotic equipment as for DNA microarray preparation by exploiting a chemoselective reaction such as thiols reacting with a maleimide surface [83] (Figure 5.11), a Staudinger ligation [84], or an oxime reacting with an aldehyde surface [85], affording the immobilized molecule with controlled orientation. Of course, this methodology requires that the whole library bear the appropriate functionality for immobilization. Alternatively, two methods have been reported to immobilize small molecules in an indiscriminate fashion using either photocrosslinking [86] (where the microarray surface is derivatized with a photocrosslinker) or an isocyanate [87] that reacts promiscuously with heteroatoms. Two applications of small-molecule microarrays have already yielded important results beyond the proof of concept: ligand discovery and enzyme profiling. To probe ligand–protein interaction, a small-molecule microarray was used to discover a new ligand for a transcription factor (Ure2p, see Section 5.4.3) using only 4 μg of the protein [63]. Several reports have also highlighted the utility of carbohydrate microarrays notably to probe the recognition motif of therapeutically important antibodies such as 2G12, a human antibody that neutralizes a broad range of HIV-1 [88] or to study the importance of sulfation pattern on carbohydrates [89,90]. A second important application of small-molecule microarrays has been to profile the substrate specificity of given enzymes or measure the activity of enzymes from crude cell lysates [91]. This has been demonstrated for kinases and proteases (Figure 5.12). For kinase profiling, microarrays containing 700–1300 kinase substrates (13-mer peptides) identified bioinformatically from sequence analysis of the human genome was used to define the preferred substrate of several kinases. Using [γ^{33}P]-ATP and autoradiography or phospho-imaging detection, the authors showed that these arrays could be used to identify the preferred substrate of a given kinase as exemplified by profiling two kinases: protein kinase A (PKA) and 3-phosphoinositide-dependent protein kinase (PDK1) [92]. Rather than using specific peptide sequences of predicted phosphorylation sites from database analysis, it was also shown that the preferred substrate of Abl could be inferred from the phosphorylation of a random array of 1433 peptides using the weight matrix–nearest-neighbour algorithm [93]. The same strategy was used to identify the preferred substrates of CK2, a serine/threonine kinase [94]. For proteases, it has also been shown that enzymatic activity could be detected specifically using immobilized fluorogenic substrates [95]. Using an alternative approach where the substrates were linked to an oligonucleotide tag (PNA) such that a library in solution could be arrayed by hybridization to a DNA microarray, the preferred substrate of cysteine and serine protease could be defined [96]. This method was shown to be sufficiently sensitive to measure the difference in proteolytic activity (caspase 3) between lysates of apoptotic cells versus healthy cells.

5.6 TARGET IDENTIFICATION

There is unfortunately not a general streamlined approach to target identification, and it can be a laborious process that often remains a major bottleneck. This bottleneck has been accentuated by the development of phenotypic screens or pathway-specific high-throughput screens of small-molecule libraries for the discovery of small molecule modulators of biological processes. Traditionally, the target of a small molecule was

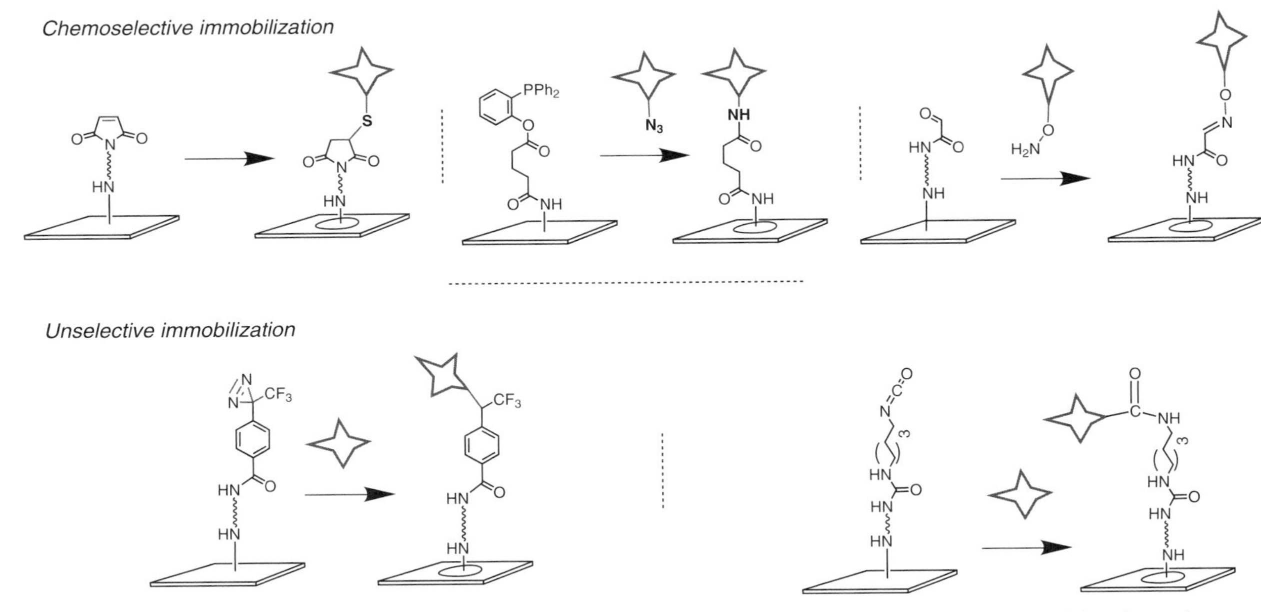

FIGURE 5.11 Immobilization of small molecules in a microarray format using chemoselective or unbiased reactions.

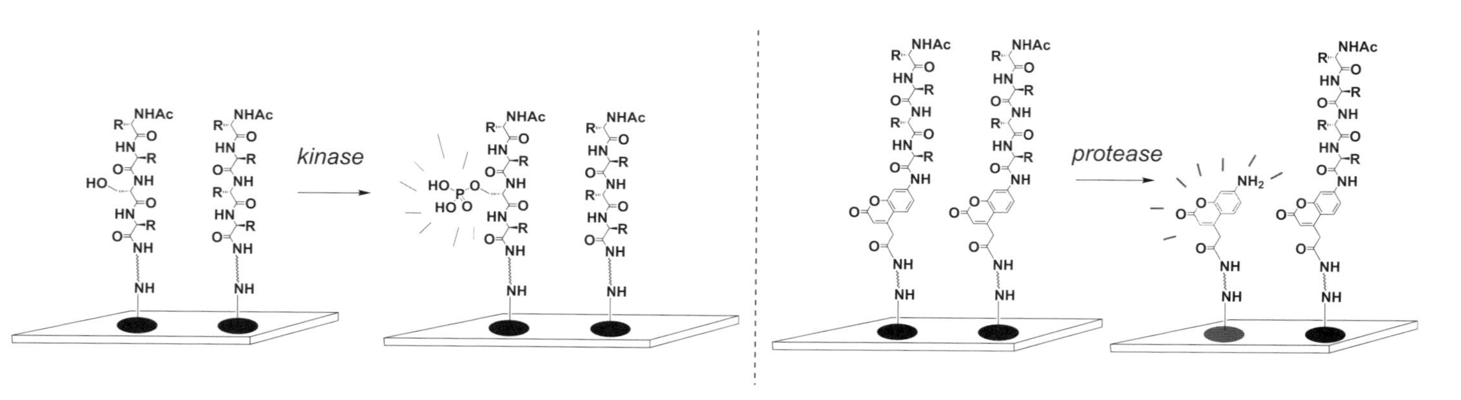

FIGURE 5.12 Profiling substrate specificity of kinases and proteases.

identified using *in vitro* biochemical approaches ranging from radiolabeling of the active molecule to affinity-based methods with or without photocrosslinking. Following the sequencing of the human and other genomes, a number of systematic approaches capitalizing on genomic technologies have been developed. Several examples of successful target identification are presented to illustrate the repertoire of available techniques.

5.6.1 Affinity Labeling

Colchicine had been known to inhibit mitosis and spindle organization since the beginning of the twentieth century. However, the precise target of colchicine was identified only in the 1960s using column fractionation by measuring the specific colchicine binding activity of each fraction with [^{3}H]-labeled colchicine (Figure 5.13) [97]. This example remains important and relevant as hydrogen–tritium substitution is a very sensitive method to label a small molecule without affecting its biological activity. However, it should be noted that tubulin is one of the most abundant proteins and that colchicine has a picomolar affinity for its target. It is unlikely that this approach would have been successful for lower-affinity molecules targeting low-abundance proteins. If enough structure–activity information is available about an active molecule to know where such molecule can be substituted with an affinity label, affinity purification can be highly informative. For example, the antiangiogenic activity [98] discovered for fumagillin in 1990 generated tremendous interest in the mode of action of this natural product, which was first isolated in the 1950s. From available structure–activity data, it was known that the aliphatic tail (Figure 5.13) was not essential for activity and could be substituted for an affinity tag. A biotinylated analog of fumagillin was incubated with epithelial cells and following lysis, the crude proteomic mixture was separated by gel and visualized with avidin–horseradish peroxidase, leading to the identification of a metalloprotease, MetAP2, as the unique target of fumagillin [99]. The purification of the fumagillin's target was possible by virtue of the reactive epoxide, which forms an irreversible adduct with the protease. Alternatively, a nondenaturing gel to separate crude proteomic mixtures can be blotted with the labeled small molecule of interest as was successfully done to identify radicicol's target. Radicicol was also reported in the 1950s but did not originally attract attention until it was reported to inhibit Src and other oncogenic kinases in cellular assays [100]. However, this inhibitory activity could not be replicated with purified proteins. Because the ketone could be converted to an oxime without ablating its biological activity, a biotinylated version of radicicol was prepared and used in a Western blot format with crude proteomic mixture from lysates (despite the epoxide and the conjugate olefin, radicicol is not an irreversible inhibitor) [101]. This led to the identification of HSP90, a molecular chaperone necessary for the folding and maturation of a number of oncogenic kinases, as the target of radicicol. In the absence of HSP90's function, the misfolded kinases are unfunctional and targeted to the proteasome, which explained the observed antikinase activities in cellular assays but not *in vitro*. Conversely, the small molecule inhibitors can be immobilized to a matrix for affinity purification. This strategy has been successfully used in the context of the

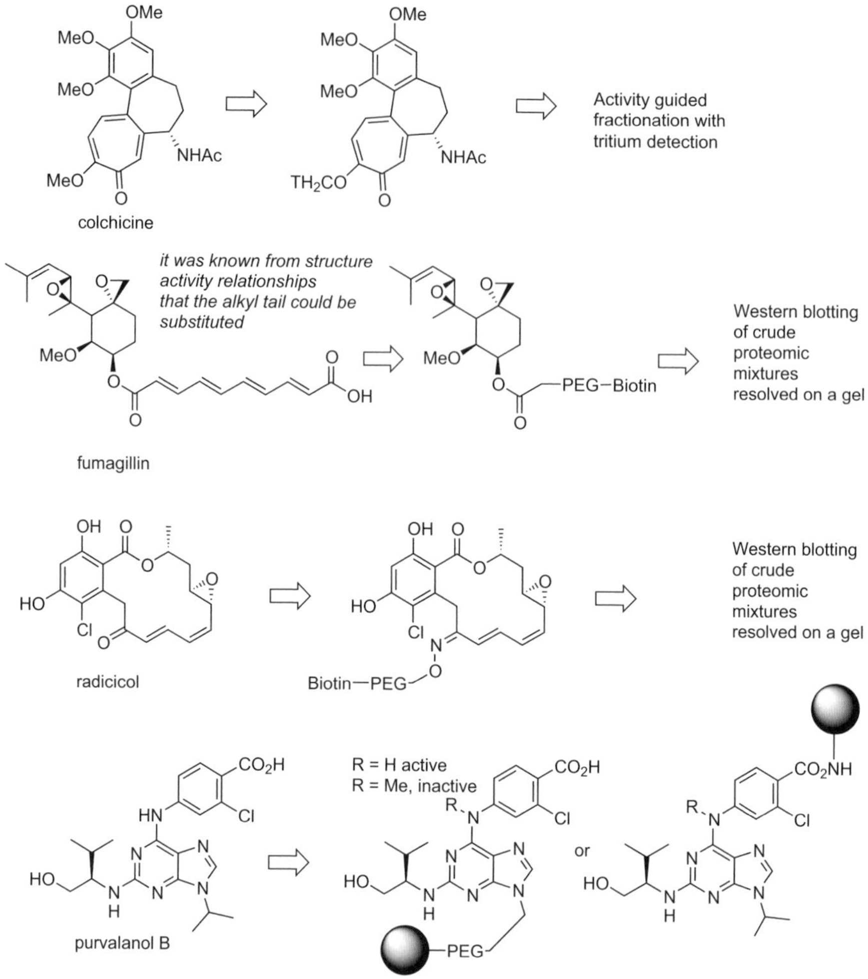

colchicine

Activity guided fractionation with tritium detection

it was known from structure activity relationships that the alkyl tail could be substituted

fumagillin

Western blotting of crude proteomic mixtures resolved on a gel

radicicol

Western blotting of crude proteomic mixtures resolved on a gel

purvalanol B

R = H active
R = Me, inactive

FIGURE 5.13 Chemical structure of selected molecules used for target identification by fractionation and affinity purification.

purvalanols which were discovered from combinatorial libraries based on the purine scaffold (see discussion in Section 5.4.1). Purvalanol B in particular was found to be a potent inhibitor of CDK1 and selective with respect to 25 tested kinases [41]. While a number of factors were consistent with the fact that purvalanol B is a CDK1 inhibitor [40], its actual intracellular target remained unverified. To this end, an affinity matrix with purvalanol B and a second matrix with an inactive analog as a control were prepared to identify all intracellular proteins having an affinity for purvalanol. Interestingly, seven proteins were selectively retained on the matrix containing purvalanol B including CDK1 and several other kinases. However, it was shown that purvalanol had at least 1000 times lower IC_{50} for CDK1 than for the other isolated kinases, suggesting that their isolation may be a product of relative abundance. When the same experiment was carried out with a series of human oncogenic cell lines, p42/p44 MAP kinase was isolated in addition to CDK1 [102]. It was then shown that the *in vivo* inhibition of both CDK1 and p42/p44 MAP kinase were likely to be relevant to purvalanol B's antimitotic activity and its potential as a chemotherapeutic compound. In a more general context this study demonstrated the ability of using the affinity matrix to identify unexpected targets [103]. While the initial characterization of proteins isolated from the affinity matrix was achieved using microsequencing, technological advances in protein separation and characterization will undoubtedly accelerate the throughput of this approach.

More recently, the potential of yeast three-hybrid (Y3H) was explored to identify small-molecule targets particularly in the context of kinase inhibition. Y3H is an extension of Fields and Song's two-hybrid method (Y2H) [104], where the association between two proteins is linked to the transcription of a specific reporter such as His-3, which is necessary for growth on a histidine-deficient medium. The potential of Y3H was first demonstrated in a proof-of-concept experiment with a well-characterized small molecule (FK506) known to induce dimerization between distinct proteins [105]. To demonstrate the applicability of this approach to target identification, six known inhibitors including purvalanol B and other kinase inhibitors conjugated to methotrexate (MTX) were used (Figure 5.14) [106]. A yeast strain with dihydrofolate (DHFR) conjugated to a DNA-binding domain (LexA-DBD) was transformed with a

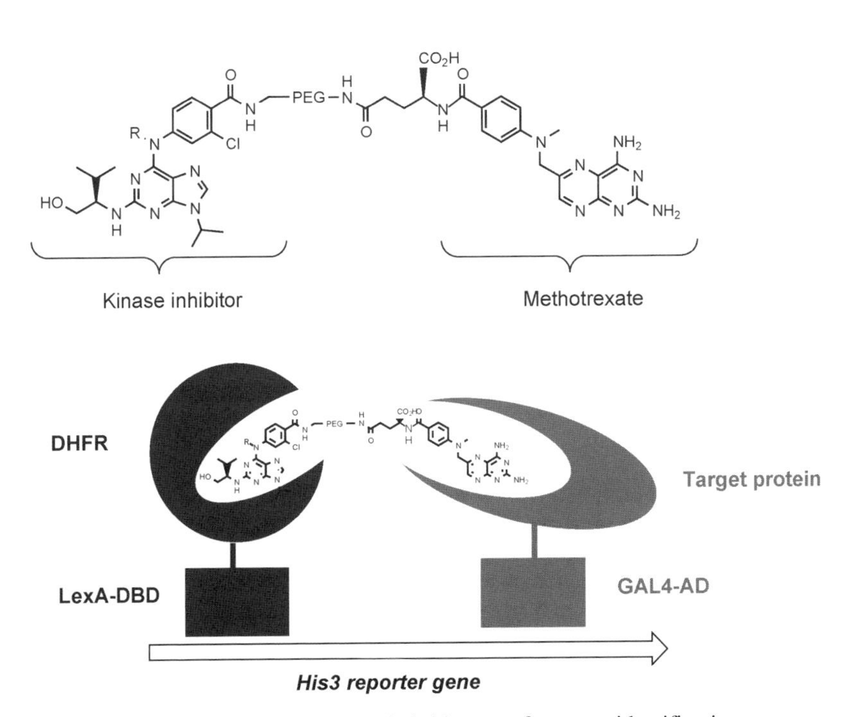

FIGURE 5.14 Yeast–three-hybrid screen for target identification.

library of cDNA fused to a transcriptional activation domain (Gal-4AD) for His3 expression. Incubation of the yeast with the inhibitors-MTX adducts on hisitdine-deficient media should result in the survival of only the strain with the target of the small protein fused to the transcriptional activation domain. The known targets of purvalanol (CDK1, CDK2, and casein) were indeed successfully identified by this technique, suggesting that it should be applicable to a broader spectrum of small molecules. More recently, the same group used this technique to identify the targets of a new pyrazolopirimidinone that exhibits potent cyctotoxicity on noncycling tumor cells in addition to be antimitotic [107]. This analysis led to the identification of CRKs (CDK-related kinases) as putative targets. A potential advantage of Y3H systems over the previously described affinity methods is that the interaction of the small molecule with its target takes place in intact cells rather than *in vitro* following a lysis. Limitations are the use of yeast versus mammalian cells and the use of engineered fusion proteins at concentrations that may not reflect the relative abundance of proteins in the mammalian cells. The traditional affinity purification and Y3H are thus complementary but are both limited by the necessity to modify the small molecule of interest.

5.6.2 Unbiased Genomic Approaches with Unlabeled Molecules

Synthetic lethality screens, where single viable mutants are paired to identify combinations that are no longer viable, have been used in genetics to identify genes that are linked in biochemical pathways [108]. By the same principle, lowering the dosage of a single gene from two copies to one copy in diploid wild-type yeast results in a heterozygote that is sensitized to any drug that acts on the product of this gene. This haploinsufficient phenotype thereby identifies the gene product as the likely drug target. During the construction of the yeast deletion allele, unique oligonucleotide sequences flanked by common primers were introduced as genetic barcodes of the deleted gene such that PCR amplification of the barcodes could be directly quantified by microarray hybridization to measure the concentration of each mutant simultaneously (Figure 5.15a). The proof of concept behind haploinsufficiency was demonstrated with the identification of the correct target for tunicamycin, a known natural product that inhibits glycosylation [109]. More recently, this approach has been used to profile the target(s) of 78 therapeutically relevant small molecules with a pool of 3503 haploinsufficient yeast strains (roughly half of the yeast genome) [110]. Impressively, targets were easily identifiable for 58 of the compounds, and out of the 20 compounds with known targets, the correct target was identified for 11 of them. Interestingly, several well-characterized small molecules were found to have additional targets. Hypersensitivity for 5-fluorouracil (5FU), a drug frequently prescribed for the treatment of solid tumor known to inhibit thymidylate synthase, was found for strains encoding genes for ribosomal RNA processing. This observation corroborates growing arguments that the therapeutic effect of 5FU also comes from its inhibition of ribosomal assembly. Likewise, molsidomine, a potent vasodilator currently in use, induced hypersensitivity for haploids missing a gene-encoding lanosterol synthase, thereby providing a rationale for the observed cholesterol reduction in patients taking the drug. In an concurrent study, a set of 5916 heterozygous strains were tested against

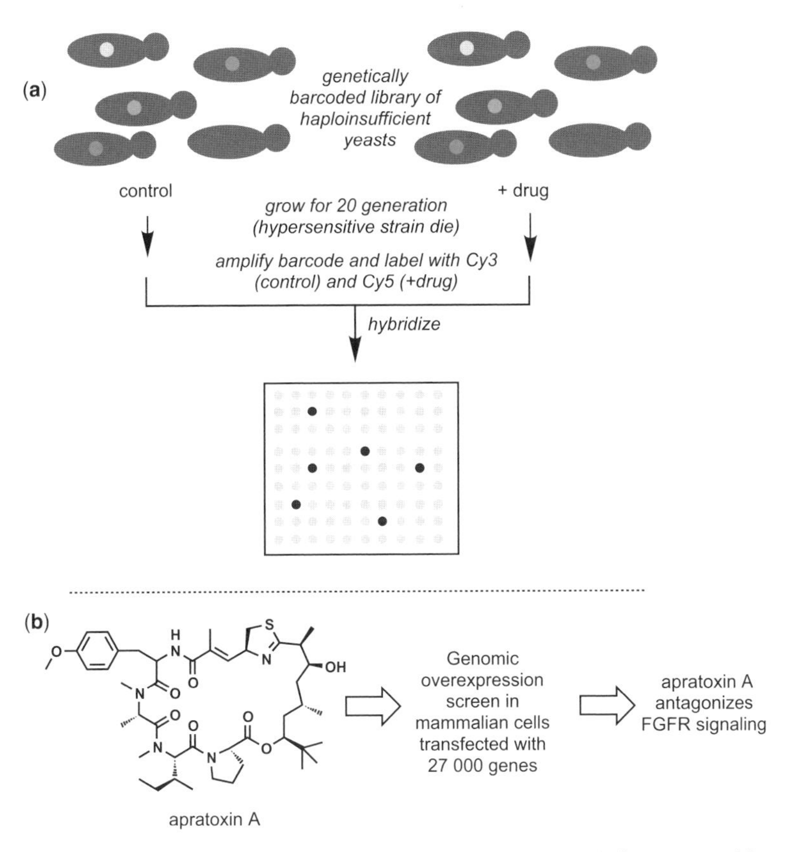

FIGURE 5.15 Genomic approaches to target identification: (a) lethality screen with genetically tagged haploinsufficient yeasts; (b) discovery of the mode of action of apratoxin A. (See insert for color representation.)

10 diverse compounds [111]. This study also found evidence suggesting that 5FU targets ribosomal assembly and processing. Of course, such a lethality screen will also give positive results for genes involved in the bioavailability and metabolism of the drug, and only 30% of the genes implicated in human diseases have yeast orthologs. Nevertheless, this approach can be carried out on a genomic scale and does not require any form of labeling. For cytotoxic compounds, the reverse strategy can be taken, specifically, screening for resistance of yeast strains overexpressing a target [112]. This was successfully achieved using a library of 1536 transformed yeasts arranged in a microarray format harboring a multicopy of a gene. This technique was used to identify the target of a novel cytotoxic compound derived from combinatorial libraries targeting kinases. This enabled the identification of Pkc1 as the target, which was confirmed by affinity purification and biochemical assays. The array format used in the latter study can dramatically increase the throughput as the readout does not require sequencing and allows easy subtraction of nonspecific genes related to drug metabolism and bioavailability.

Most recently, a genomic overexpression screen has been adapted to mammalian cells to investigate the mode of action of apratoxin A (Figure 5.15b), a novel natural product with a potent cytotoxicity toward tumor cell lines [113]. A screen of apratoxin A in the NIC-60 cancer cell line revealed a unique pattern of cytoxicity which could not be correlated to previously profiled antineoplastic compounds. The use of mammalian cell lines in the resistance screen was indispensable in this case as the compound was inactive in yeast. Osteosarcoma cells were transiently transfected with a cDNA collection (27,000 genes) along with a reporter plasmid (luciferase) in a 384-well plate format. The cells were then screened for resistance to apratoxin A, which led to the identification of 46 genes attenuating cytotoxicity by two- to threefold. Retesting identified 24 of these genes as encoding proteins that either inhibited apoptosis or caused various cell cycle arrests and thus were independent of apratoxin A's target. Out of the 22 remaining potential targets, five genes encoded FGFRs. Interestingly, cell lines that were found to be least sensitive among the NIC-60 cell line panel showed a high expression level of FGFR when profiled, corroborating the results of the genomewide overexpression screen. Further biochemical assays established that the phosphorylation level of Stat3, a transcription factor regulated by the FGFR pathway, was reduced in a dose-dependent manner by apratoxin A treatment. It has also been shown that a genetic suppressor screen can be used to identify the target of a small molecule identified from phenotypic assays in *Caenorhabditis elegans* [114]. In this study, 180,000 randomly mutated wild-type genomes were screened for suppression of the small-molecule-induced phenotype (induction of population growth defect), leading to the identification of *elg*-19, a gene that encodes L-type calcium channel a1 subunit.

A different approach has been developed by researchers at NCI to rapidly assign the mode of action of cytotoxic compounds from analysis of a compound's activity against a panel of cancer cell lines (Figure 5.16). To this end, an extensive database of activity against 60 different cancer cell lines (NCI-60) with compounds having a known mode of action has led to the identification of specific patterns correlating to a particular mode of action. A number of novel cytotoxics have been assigned a putative mode of action simply by matching the cytotoxicity pattern in the NIC-60 cell line against that of known inhibitors [115]. In a conceptually related approach extending beyond cytotoxic compounds, a database of the gene expression profiles corresponding to 300 diverse mutations and chemical treatments in yeast was constructed. It was demonstrated that the cellular pathway affected by an uncharacterized perturbation (pharmacological or genetic) could be determined by pattern matching, even among very subtle profiles. The utility of this approach was confirmed with the discovery of a novel target (Erg2p involved in ergosterol biosynthesis) for the commonly used drug dyclonine, a topical anesthetic [116].

5.7 CONCLUSION

Since the mid 1990s, a number of technological developments ranging from linkers for solid–phase synthesis to new synthetic methodologies for diversity-oriented synthesis and encoding technologies for split–pool synthesis have enabled the preparation of

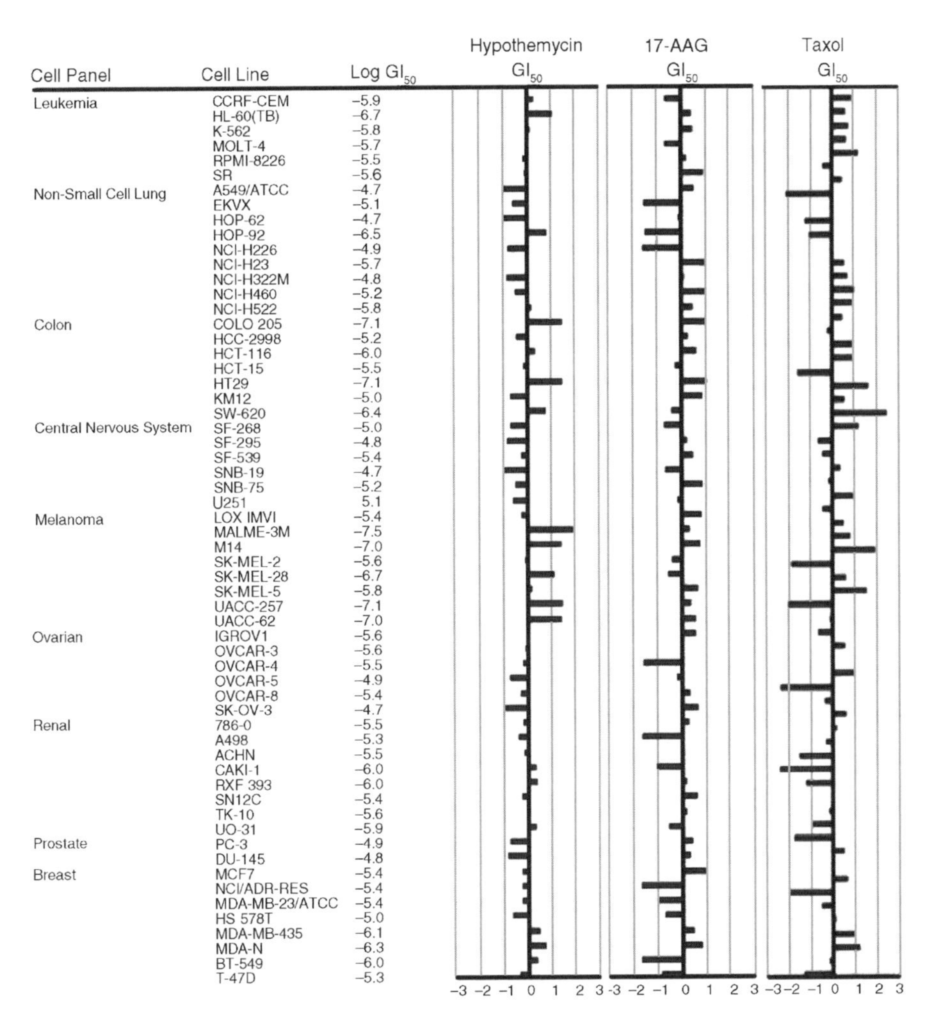

FIGURE 5.16 Activity of cytotoxic compounds (hypothemycin—MAP kinase inhibitor; 17AAG—HSP90 inhibitors; taxol—tubulin depolymerization inhibitor) against the NCI-60 cancer cell line panel. (Reproduced from www.nci.nih.gov.)

small-molecule libraries for high-throughput screening (HTS). These developments have made high-quality libraries of small molecules within reach to academic researchers. Several landmark publications have already highlighted the potential to uncover important biological mechanisms using small-molecule inhibitors discovered from combinatorial libraries. These first successes have inspired the creation of several national programs to assemble diverse libraries and establish screening centers as exemplified by the NIH molecular library initiative. High-content phenotypic assays have the greatest potential to result in the discovery of new biological mechanisms; however, identifying the target of a small molecule inducing an interesting phenotype remains a bottleneck. The advent of mass spectrometric characterization of proteins will certainly accelerate the throughput of protein identification by affinity purification. The availability of genomic information has also enabled new genomewide approaches to uncover the mode of action of small molecules that do not require any manipulation or modification of the inhibitor.

ACKNOWLEDGMENTS

We thank our collaborators who have contributed to our work in the area of chemical biology and the funding agencies (Human Frontier Science Program, Agence National de Recherche and Marie Curie and the CNRS). This review was not intended to be comprehensive but rather to highlight some recent work, and we apologize for unenviable arbitrary omissions.

REFERENCES

[1] Merrifield RB: Solid phase peptide synthesis. I. The synthesis of a tetrapeptide, *J Am Chem Soc* **85**:2149–2154 (1963).

[2] Ley SV, Baxendale IR: New tools and concepts for modern organic synthesis, *Nat Rev Drug Discov* **1**:573–586 (2002).

[3] Ley SV et al: Multi-step organic synthesis using solid-supported reagents and scavengers: A new paradigm in chemical library generation, *J Chem Soc-Perkin Trans 1* **23**:3815–4195 (2000).

[4] Studer A et al: Fluorous synthesis: A fluorous-phase strategy for improving separation efficiency in organic synthesis, *Science* **275**:823–826 (1997).

[5] Luo Z, Zhang Q, Oderaotoshi Y, Curran DP: Fluorous mixture synthesis: A fluorous-tagging strategy for the synthesis and separation of mixtures of organic compounds, *Science* **291**:1766–1769 (2001).

[6] Furka A, Sebestyen F, Asgedom M, Dibo G: General method for rapid synthesis of multicomponent peptide mixtures, *Int J Peptide Protein Res* **37**:487–493 (1991).

[7] Affleck RL: Solutions for library encoding to create collections of discrete compounds, *Curr Opin Chem Biol* **5**:257–263 (2001).

[8] Nestler HP, Bartlett PA, Still WC: A general method for molecular tagging of encoded combinatorial chemistry libraries, *J Org Chem* **59**:4723–4724 (1994).

[9] Nicolaou KC et al: Natural product-like combinatorial libraries based on privileged structures. 2. Construction of a 10 000-membered benzopyran library by directed split-and-pool chemistry using NanoKans and optical encoding, *J Am Chem Soc* **122**:9954–9967 (2000).

[10] Czarnik AW: Encoding methods for combinatorial chemistry, *Curr Opin Chem Biol* **1**: 60–66 (1997).

[11] Huc I, Lehn JM: Virtual combinatorial libraries: dynamic generation of molecular and supramolecular diversity by self-assembly, *Proc Natl Acad Sci USA* **94**:2106–2110 (1997).

[12] Lehn JM, Eliseev AV: Dynamic combinatorial chemistry, *Science* **291**:2331–2332 (2001).

[13] Ramstrom O, Lehn JM: Drug discovery by dynamic combinatorial libraries, *Nat Rev Drug Discov* **1**:26–36 (2002).

[14] Whiting M et al: Inhibitors of HIV-1 protease by using in situ click chemistry, *Angew Chem Int Ed Engl* **45**:1435–1439 (2006).

[15] Manetsch R et al: In situ click chemistry: Enzyme inhibitors made to their own specifications, *J Am Chem Soc* **126**:12809–12818 (2004).

[16] Lewis WG et al: Click chemistry in situ: Acetylcholinesterase as a reaction vessel for the selective assembly of a femtomolar inhibitor from an array of building blocks, *Angew Chem Int Ed Engl* **41**:1053–1057 (2002).

[17] Oprea TI, Gottfries J: Chemography: The art of navigating in chemical space, *J Comb Chem* **3**:157–166 (2001).

[18] Dobson CM: Chemical space and biology, *Nature* **432**:824–828 (2004).

[19] Paolini GV, Shapland RH, van Hoorn WP, Mason JS, Hopkins AL: Global mapping of pharmacological space, *Nat Biotechnol* **24**:805–815 (2006).

[20] Lipinski C, Hopkins A: Navigating chemical space for biology and medicine, *Nature* **432**:855–861 (2004).

[21] DeSimone RW, Currie KS, Mitchell SA, Darrow JW, Pippin DA: Privileged structures: Applications in drug discovery, *Comb Chem High Throughput Screen* **7**:473–493 (2004).

[22] Lipinski CA, Lombardo F, Dominy BW, Feeney PJ: Experimental and computational approaches to estimate solubility and permeability in drug discovery and development settings, *Adv Drug Deliv Rev* **23**:3–25 (1997).

[23] Breinbauer R, Vetter IR, Waldmann H: From protein domains to drug candidates—natural products as guiding principles in the design and synthesis of compound libraries, *Angew Chem Int Ed* **41**:2879–2890 (2002).

[24] Balamurugan R, Dekker FJ, Waldmann H: Design of compound libraries based on natural product scaffolds and protein structure similarity clustering (PSSC), *Mol BioSyst* **1**:36–45 (2005).

[25] Root DE, Flaherty SP, Kelley BP, Stockwell BR: Biological mechanism profiling using an annotated compound library, *Chem Biol* **10**:881–892 (2003).

[26] Armstrong RW, Combs AP, Tempest PA, Brown SD, Keating TA: Multiple-component condensation strategies for combinatorial library synthesis, *Acc Chem Res* **29**:123–131 1996).

[27] Ugi I ed: *Isonitrile Chemistry*, Academic Press (1971).

[28] Tempest PA: Recent advances in heterocycle generation using the efficient Ugi multiple-component condensation reaction, *Cur Opin Drug Discov Dev* **8**:776–788 (2005).

[29] Stocker AM, Keating TA, Tempest PA, Armstrong RW: Use of a convertible isocyanide for generation of Ugi reaction derivatives on solid support: Synthesis of a-acylamino esters and pyrroles, *Tetrahedron Lett* **37**:1149–1152 (1996).

[30] Keating TA, Armstrong RW: Postcondensation modifications of ugi four-component condensation products: 1-Isocyanocyclohexene as a convertible isocyanide. Mechanism of conversion, synthesis of diverse structures, and demonstration of resin capture, *J Am Chem Soc* **118**:2574–2583 (1996).

[31] Park WKC, Auer M, Jaksche H, Wong C-H: Rapid combinatorial synthesis of amino glycoside antibiotic mimetics: Use of a polyethylene glycol-linked amine and a neamine-derived aldehyde in multiple component condensation as a strategy for the discovery of new inhibitors of the HIV RNA Rev responsive element, *J Am Chem Soc* **118**:10150–10155 (1996).

[32] Petasis NA, Zavialov IA: A new and practical synthesis of a-amino acids from alkenyl boronic acids, *J Am Chem Soc* **119**:445–446 (1997).

[33] Prakash GK, Mandal M, Schweizer S, Petasis NA, Olah GA: A facile stereocontrolled synthesis of anti-alpha-(trifluoromethyl)-beta-amino alcohols, *Org Lett* **2**:3173–3176 (2000).

[34] Kumagai N, Muncipinto G, Schreiber SL: Short synthesis of skeletally and stereochemically diverse small molecules by coupling petasis condensation reactions to cyclization reactions, *Angew Chem Int Ed Engl* **45**:3635–3638 (2006).

[35] Williams RM, Zhai W, Aldous DJ, Aldous SC: Asymmetric [1,3]-dipolar cycloaddition reactions: Synthesis of highly substituted proline derivatives, *J Org Chem* **57**:6527–6532 (1992).

[36] Sebahar PR, Williams RM: The asymmetric total synthesis of (+)- and (−)-spirotryprostatin B, *J Am Chem Soc* **122**:5666–5667 (2000).

[37] Lo MM, Neumann CS, Nagayama S, Perlstein EO, Schreiber SL: A library of spirooxindoles based on a stereoselective three-component coupling reaction, *J Am Chem Soc* **126**:16077–16086 (2004).

[38] Itami K, Yoshida J: Platform synthesis: A useful strategy for rapid and systematic generation of molecular diversity, *Chemistry* **12**:3966–3974 (2006).

[39] Burke MD, Schreiber SL: A planning strategy for diversity-oriented synthesis, *Angew Chem Int Ed Engl* **43**:46–58 (2004).

[40] Gray NS et al: Exploiting chemical libraries, structure, and genomics in the search for kinase inhibitors, *Science* **281**:533–538 (1998).

[41] Chang YT et al: Synthesis and application of functionally diverse 2,6, 9-trisubstituted purine libraries as CDK inhibitors, *Chem Biol* **6**:361–375 (1999).

[42] Armstrong JI et al: Discovery of carbohydrate sulfotransferase inhibitors from a kinase-directed library, *Angew Chem Int Ed Engl* **39**:1303–1306 (2000).

[43] Verdugo DE et al: Discovery of estrogen sulfotransferase inhibitors from a purine library screen, *J Med Chem* **44**:2683–2686 (2001).

[44] Harmse L et al: Structure-activity relationships and inhibitory effects of various purine derivatives on the in vitro growth of Plasmodium falciparum, *Biochem Pharmacol* **62**:341–348 (2001).

[45] Rosania GR et al: Myoseverin, a microtubule-binding molecule with novel cellular effects, *Nat Biotechnol* **18**:304–308 (2000).

[46] Perez OD, Chang YT, Rosania G, Sutherlin D, Schultz PG: Inhibition and reversal of myogenic differentiation by purine-based microtubule assembly inhibitors, *Chem Biol* **9**:475–483 (2002).

[47] Wignall SM et al: Identification of a novel protein regulating microtubule stability through a chemical approach, *Chem Biol* **11**:135–146 (2004).

[48] Ding S, Gray NS, Wu X, Ding Q, Schultz PG: A combinatorial scaffold approach toward kinase-directed heterocycle libraries, *J Am Chem Soc* **124**:1594–1596 (2002).

[49] Wu X, Ding S, Ding Q, Gray NS, Schultz PG: A small molecule with osteogenesis-inducing activity in multipotent mesenchymal progenitor cells, *J Am Chem Soc* **124**:14520–14521 (2002).

[50] Ding S et al: Synthetic small molecules that control stem cell fate, *Proc Natl Acad Sci USA* **100**:7632–7637 (2003).

[51] Wu X, Ding S, Ding Q, Gray NS, Schultz PG: Small molecules that induce cardiomyogenesis in embryonic stem cells, *J Am Chem Soc* **126**:1590–1591 (2004).

[52] Warashina M et al: A synthetic small molecule that induces neuronal differentiation of adult hippocampal neural progenitor cells, *Angew Chem Int Ed Engl* **45**:591–593 (2006).

[53] Chen S, Zhang Q, Wu X, Schultz PG, Ding S: Dedifferentiation of lineage-committed cells by a small molecule, *J Am Chem Soc* **126**:410–411 (2004).

[54] Adrian FJ et al: Allosteric inhibitors of Bcr-abl-dependent cell proliferation, *Nat Chem Biol* **2**:95–102 (2006).

[55] Nicolaou K et al: Natural product-like combinatorial libraries based on privileged structures, 1. General principles and solid-phase synthesis of benzopyrans, *J Am Chem Soc* **122**:9939–9953 (2000).

[56] Nicolaou K et al: Natural product-like combinatorial libraries based on privileged structures. 3. The "libraries from libraries" principle for diversity enhancement of benzopyran libraries, *J Am Chem Soc* **122**:9968–9976 (2000).

[57] Nicolaou KC, Roecker AJ, Barluenga S, Pfefferkorn JA, Cao GQ: Discovery of novel antibacterial agents active against methicillin-resistant Staphylococcus aureus from combinatorial benzopyran libraries, *Chembiochem* **2**:460–465 (2001).

[58] Nicolaou KC et al: Discovery and optimization of non-steroidal FXR agonists from natural product-like libraries, *Org Biomol Chem* **1**:908–920 (2003).

[59] Downes M et al: A chemical, genetic, and structural analysis of the nuclear bile acid receptor FXR, *Mol Cell* **11**:1079–1092 (2003).

[60] Sternson SM, Wong JC, Grozinger CM, Schreiber SL: Synthesis of 7200 small molecules based on a substructural analysis of the histone deacetylase inhibitors trichostatin and trapoxin, *Org Lett* **3**:4239–4242 (2001).

[61] Hergenrother PJ, Depew KM, Schreiber SL: Small-molecule microarrays: Covalent attachment and screening of alcohol-containing small molecules on glass slides, *J Am Chem Soc* **122**:7849–7850 (2000).

[62] Sternson SM, Louca JB, Wong JC, Schreiber SL: Split-pool synthesis of 1,3-dioxanes leading to arrayed stock solutions of single compounds sufficient for multiple phenotypic and protein-binding assays, *J Am Chem Soc* **123**:1740–1747 (2001).

[63] Kuruvilla FG, Shamji AF, Sternson SM, Hergenrother PJ, Schreiber SL: Dissecting glucose signalling with diversity-oriented synthesis and small-molecule microarrays, *Nature* **416**:653–657 (2002).

[64] Coschigano PW, Magasanik B: The URE2 gene product of Saccharomyces cerevisiae plays an important role in the cellular response to the nitrogen source and has homology to glutathione s-transferases. Molecular and cellular biology, *Mol. Cell Biol.* **11**: 822–832 (1991).

[65] Hardwick JS, Kuruvilla FG, Tong JK, Shamji AF, Schreiber SL: Rapamycin-modulated transcription defines the subset of nutrient-sensitive signaling pathways directly controlled by the Tor proteins, *Proc Natl Acad Sci USA* **96**:14866–14870 (1999).

[66] Blackwell HE et al: A one-bead, one-stock solution approach to chemical genetics: Part 1, *Chem Biol* **8**:1167–1182 (2001).

[67] Clemons PA et al: A one-bead, one-stock solution approach to chemical genetics: Part 2, *Chem Biol* **8**:1183–1195 (2001).

[68] Haggarty SJ, Koeller KM, Wong JC, Butcher RA, Schreiber SL: Multidimensional chemical genetic analysis of diversity-oriented synthesis-derived deacetylase inhibitors using cell-based assays, *Chem Biol* **10**:383–396 (2003).

[69] Hubbert C et al: HDAC6 is a microtubule-associated deacetylase.*Nature* **417**:455–458 (2002).

[70] Haggarty SJ, Koeller KM, Wong JC, Grozinger CM, Schreiber SL: Domain-selective small-molecule inhibitor of histone deacetylase 6 (HDAC6)-mediated tubulin deacetylation, *Proc Natl Acad Sci USA* **100**:4389–4394 (2003).

[71] Pelish HE, Westwood NJ, Feng Y, Kirchhausen T, Shair MD: Use of biomimetic diversity-oriented synthesis to discover galanthamine-like molecules with biological properties beyond those of the natural product, *J Am Chem Soc* **123**:6740–6741 (2001).

[72] Pelish HE et al: Secramine inhibits Cdc42-dependent functions in cells and Cdc42 activation in vitro, *Nat Chem Bio* **2**:39–46 (2006).

[73] Pelish HE et al: The Cdc42 inhibitor secramine B prevents cAMP-induced K+ conductance in intestinal epithelial cells, *Biochem Pharmacol* **71**:1720–1726 (2006).

[74] Arkin MR, Wells JA: Small-molecule inhibitors of protein-protein interactions: Progressing towards the dream, *Nat Revi Drug Discov* **3**:301–317 (2004).

[75] Boger DL, Desharnais J, Capps K: Solution-phase combinatorial libraries: Modulating cellular signaling by targeting protein-protein or protein-DNA interactions.*Angew Chem Int Ed* **42**:4138–4176 (2003).

[76] Berg T et al: Small-molecule antagonists of Myc/max dimerization inhibit Myc-induced transformation of chicken embryo fibroblasts, *Proc Natl Acad Sci USA* **99**:3830–3835 (2002).

[77] Bialy L, Waldmann H: Inhibitors of protein tyrosine phosphatases: Next-generation drugs?*Angew Chem Int Ed Engl* **44**:3814–3839 (2005).

[78] Weide T, Arve L, Prinz H, Waldmann H, Kessler H: 3-Substituted indolizine-1-carbonitrile derivatives as phosphatase inhibitors, *Bioorg Med Chem Lett* **16**:59–63 (2006).

[79] Bialy L, Waldmann H: Total synthesis and biological evaluation of the protein phosphatase 2A inhibitor cytostatin and analogues, *Chemistry* **10**:2759–2780 (2004).

[80] Brohm D et al: Solid-phase synthesis of dysidiolide-derived protein phosphatase inhibitors, *J Am Chem Soc* **124**:13171–13178 (2002).

[81] Noeren-Mueller A et al: Discovery of protein phosphatase inhibitor classes by biology-oriented synthesis, *Proc Natl Acad Sci USA* **103**:10606–10611 (2006).

[82] Uttamchandani M, Walsh DP, Yao SQ, Chang YT: Small molecule microarrays: Recent advances and applications, *Curr Opin Chem Biol* **9**:4–13 (2005).

[83] MacBeath G, Koehler AN, Schreiber SL: Printing small molecules as microarrays and detecting protein-ligand interactions en masse, *J Am Chem Soc* **121**:7967–7968 (1999).

[84] Kohn M et al: Staudinger ligation: A new immobilization strategy for the preparation of small-molecule arrays, *Angew Chem Int Ed Engl* **42**:5830–5834 (2003).

[85] Falsey JR, Renil M, Park S, Li S, Lam KS: Peptide and small molecule microarray for high throughput cell adhesion and functional assays, *Bioconj Chem* **12**:346–353 (2001).

[86] Kanoh N et al: Immobilization of natural products on glass slides by using a photoaffinity reaction and the detection of protein-small-molecule interactions, *Angew Chem Int Ed Engl* **42**:5584–5587 (2003).

[87] Bradner JE et al: A robust small-molecule microarray platform for screening cell lysates, *Chem Biol* **13**:493–504 (2006).

[88] Calarese DA et al: Dissection of the carbohydrate specificity of the broadly neutralizing anti-HIV-1 antibody 2G12, *Proc Natl Acad Sci USA* **102**:13372–13377 (2005).

[89] de Paz JL, Noti C, Seeberger PH: Microarrays of synthetic heparin oligosaccharides, *J Am Chem Soc* **128**:2766–2767 (2006).

[90] Tully SE, Rawat M, Hsieh-Wilson LC: Discovery of a TNF-alpha antagonist using chondroitin sulfate microarrays, *J Am Chem Soc* **128**:7740–7741 (2006).

[91] Winssinger N, Harris JL: Microarray-based functional protein profiling using peptide nucleic acid-encoded libraries, *Exp Rev Proteom* **2**:937–947 (2005).

[92] Schutkowski M et al: Automated synthesis: High-content peptide microarrays for deciphering kinase specificity and biology, *Ange Chem Int Ed* **43**:2671–2674 (2004).

[93] Rychlewski L, Kschischo M, Dong L, Schutkowski M, Reimer U: Target specificity analysis of the Abl kinase using peptide microarray data, *J Mol Biol* **336**:307–311 (2004).

[94] Panse S et al: Profiling of generic anti-phosphopeptide antibodies and kinases with peptide microarrays using radioactive and fluorescence-based assays, *Mol Divers* **8**:291–299 (2004).

[95] Salisbury CM, Maly DJ, Ellman JA: Peptide microarrays for the determination of protease substrate specificity, *J Am Chem Soc* **124**:14868–14870 (2002).

[96] Winssinger N et al: PNA-encoded protease substrate microarrays, *Chem Biol* **11**:1351–1360 (2004).

[97] Weisenberg RC, Borisy GG, Taylor EW: The colchicine-binding protein of mammalian brain and its relation to microtubules, *Biochemistry* **7**:4466–4479 (1968).

[98] Ingber D et al: Synthetic analogues of fumagillin that inhibit angiogenesis and suppress tumour growth, *Nature* **348**:555–557 (1990).

[99] Sin N et al: The anti-angiogenic agent fumagillin covalently binds and inhibits the methionine aminopeptidase, MetAP-2, *Proc Natl Acad Sci USA* **94**:6099–6103 (1997).

[100] Kwon HJ, Yoshida M, Fukui Y, Horinouchi S, Beppu T: Potent and specific inhibition of p60v-src protein kinase both *in vivo* and *in vitro* by radicicol, *Cancer Res* **52**:6926–6930 (1992).

[101] Sharma SV, Agatsuma T, Nakano H: Targeting of the protein chaperone, HSP90, by the transformation suppressing agent, radicicol, *Oncogene* **16**:2639–2645 (1998).

[102] Knockaert M et al: p42/p44 MAPKs are intracellular targets of the CDK inhibitor purvalanol, *Oncogene* **21**:6413–6424 (2002).

[103] Knockaert M, Meijer L: Identifying *in vivo* targets of cyclin-dependent kinase inhibitors by affinity chromatography, *Biochem Pharmacol* **64**:819–825 (2002).

[104] Fields S, Song O: A novel genetic system to detect protein-protein interactions, *Nature* **340**:245–246 (1989).

[105] Licitra EJ, Liu JO: A three-hybrid system for detecting small ligand-protein receptor interactions, *Proc Natl Acad Sci USA* **93**:12817–12821 (1996).

[106] Caligiuri M et al: A proteome-wide CDK/CRK-specific kinase inhibitor promotes tumor cell death in the absence of cell cycle progression, *Chem Biol* **12**:1103–1115 (2005).

[107] Becker F et al: A three-hybrid approach to scanning the proteome for targets of small molecule kinase inhibitors, *Chem Biol* **11**:211–223 (2004).

[108] Winzeler EA et al: Functional characterization of the S. cerevisiae genome by gene deletion and parallel analysis, *Science* **285**:901–906 (1999).

[109] Giaever G et al: Genomic profiling of drug sensitivities via induced haploinsufficiency, *Nat Genet* **21**:278–283 (1999).

[110] Lum PY et al: Discovering modes of action for therapeutic compounds using a genome-wide screen of yeast heterozygotes, *Cell* **116**:121–137 (2004).

[111] Giaever G et al: Chemogenomic profiling: identifying the functional interactions of small molecules in yeast, *Proc Natl Acad Sci USA* **101**:793–798 (2004).

[112] Luesch H et al: A genome-wide overexpression screen in yeast for small-molecule target identification, *Chem Biol* **12**:55–63 (2005).

[113] Luesch H et al: A functional genomics approach to the mode of action of apratoxin A, *Nat Chem Biol* **2**:158–167 (2006).

[114] Kwok TC et al: A small-molecule screen in C. elegans yields a new calcium channel antagonist, *Nature* **441**:91–95 (2006).

[115] Paull KD, Lin CM, Malspeis L, Hamel E: Identification of novel antimitotic agents acting at the tubulin level by computer-assisted evaluation of differential cytotoxicity data, *Cancer Res* **52**:3892–3900 (1992).

[116] Hughes TR et al: Functional discovery via a compendium of expression profiles, *Cell* **102**:109–126 (2000).

6

PROTEIN CHARACTERIZATION BY BIOLOGICAL MASS SPECTROMETRY

VENKATESHWAR A. REDDY AND ERIC C. PETERS
Genomics Institute of the Novartis Research Foundation, San Diego, California

Numerous sophisticated approaches have been developed to study the structure and function of genes, including the mining of whole-organism genome assemblies using sophisticated gene prediction algorithms and homology models [1], global transcriptional profiling [2], and forward genetic studies [3]. However, these techniques are ultimately limited by the fact that they only assess intermediates on the way to the protein products of genes that ultimately regulate biological activity [4]. Processes such as RNA splicing, proteolytic activation, and hundreds of possible posttranslational modifications (PTMs) can result in the production of numerous proteins of unique structure and function from a limited number of genes. Additionally, biological activity often results from the assembly of numerous proteins into an active complex, the nature and composition of which can be explored only at the protein level. Therefore, proteomic studies should be able to answer many questions about cellular processes and diseases that can't be answered by genomic methods alone [5].

Since the introduction and development of new biomolecule-compatible ionization techniques such as matrix-assisted laser desorption/ionization (MALDI) [6] and electrospray ionization (ESI) [7] in the early 1990s, mass spectrometry (MS) has rapidly become one of the most important tools for the characterization of individual proteins. Additionally, constant improvements in instrumentation performance as well as data acquisition and processing strategies continue to rapidly expand the breadth of this technique for the study of numerous aspects of protein function within

Chemical and Functional Genomic Approaches to Stem Cell Biology and Regenerative Medicine
Edited by Sheng Ding

biological systems. Since a comprehensive review of this field would be prohibitively long and somewhat outdated by the time of its publication, the purpose of this chapter is to provide a general overview of the current practice and applications of modern protein mass spectrometry in order to demonstrate the potential that this technique offers for the study of stem cell biology. The various references cited are clearly not meant to be exclusive, but rather provide good starting points for obtaining additional details regarding the numerous topics and techniques mentioned throughout this review.

6.1 PROTEIN IDENTIFICATION USING MASS SPECTROMETRY

In its simplest implementation, proteomics studies require the ability to rapidly identify which proteins are present in a given sample. Historically, however, this requirement represented a significant limitation. For example, biochemists had long been able to assess changes in the expression patterns of thousands of proteins using two-dimensional gel electrophoresis (2DGE). However, the identification of those species that changed under a given set of conditions was a laborious task, requiring subjecting various proteolytic digests to high-performance liquid chromatography (HPLC) or gel electrophoresis, followed by *N*-terminal (Edman) sequencing and/or amino acid analysis of the separated peptides. This bottleneck of protein identification in 2DGE studies was effectively eliminated by the introduction of peptide mass mapping [8], which combined the emerging technique of biological mass spectrometry with the availability of protein databases of ever-increasing quality based on genomic sequencing studies.

Peptide mass mapping involves protein enzymatic digestion, mass spectrometry, and computer-facilitated data analysis to effect protein identification. In the case of 2DGE-based studies, protein spots of interest are excised and subjected to in-gel digestion using an enzyme such as trypsin [9]. The masses of the resulting peptides are experimentally measured and compared with theoretical "*in silico*" digests of all the proteins contained in a sequence database. The database proteins are then statistically evaluated and ranked according to how closely their theoretical digests match the entire set of experimental data. Clearly, the success of this or any other comparative database-searching technique requires the existence of the correct protein sequence within the database searched. However, numerous databases of continuously improving quality are now available as a result of the genomic sequencing of numerous organisms. In practice, matching five to eight different tryptic peptide masses measured using currently available instrumentation is usually sufficient to unambiguously identify a human protein with an average molecular weight of 50 kDa.

Although peptide mass mapping greatly increased the number of proteins that could readily be identified in 2DGE-based experiments, it inadvertently served to also expose some of the limitations of 2DGE [10]. Additionally, the very nature of peptide mass mapping limits its use to samples containing ideally only a single species. Given the complexity and extreme range of protein expression levels inherent in living organisms, the requirement to purify each protein to near homogeneity before its digestion and successful identification by peptide mass mapping becomes highly restrictive. By contrast, tandem mass spectrometry enables sequence information to be determined for individual peptides, regardless of the presence of other species, and thus enables the efficient analysis of protein mixtures [11].

Tandem mass spectrometry experiments yield peptide sequence information by isolating individual peptide ions within the mass spectrometer itself, physically fragmenting them using any of a number of different methodologies [12], and measuring the masses of the resulting fragment ions. Manual interpretation of tandem mass spectra can often be quite difficult because of the number of different fragmentations that can occur, not all of which yield structurally useful information. However, in analogy to peptide mass mapping experiments, the experimentally obtained fragmentation patterns can be compared to *in silico*–generated tandem mass spectra of the proteolytic peptides expected to arise from each protein sequence contained in the database being searched. Statistical evaluation of the results and scoring algorithms using search engines such as Sequest™ and MASCOT™ facilitate the identification of the best possible match.

Numerous types of MS analyzers can be employed to perform tandem MS experiments [13], but ion trap mass spectrometers remain extensively utilized because of their rapid scanning capabilities and robustness. One of the most popular means of performing such experiments on complex peptide mixtures involves the direct coupling of reverse-phase HPLC to an ion trap mass spectrometer through an ESI interface. However, other separation techniques such as capillary electrophoresis can also be employed [14]. The peptide separation serves to decrease the ion signal suppression that occurs when highly complex mixtures are directly analyzed by fractionating the peptide mixture before MS analysis. As shown in Figure 6.1, these "shotgun" proteomics studies typically employ a data-dependent analysis scheme. Specifically, the ion trap first performs a MS measurement of all the ion signals eluting from the separation column at a given time. Then, the ion trap performs three to five MS/MS experiments on individual signals detected during the initial MS scan, with the basis for selection usually being a simple criterion such as ion signal intensity A new MS scan is then performed, and additional signals are selected for tandem MS, and this cycle repeats itself throughout the course of the chromatographic run. Using this shotgun methodology, hundreds of proteins can readily be identified during the course of a typical reverse-phase chromatographic run. These types of experiments in which numerous proteins are first simultaneously digested to produce an even greater number of peptides before MS analysis are often referred to as bottom–up proteomics.

Although effective, these data-dependent schemes are relatively inefficient, in that only a small fraction of the tandem mass spectra measured actually yield useful protein identifications. In light of this and other limitations of shotgun-based analysis schemes, MALDI-LC/MS/MS analysis platforms are also being investigated [15,16]. In such systems, deposition of the effluent of the separation column(s) directly onto MALDI target plates effectively decouples the separation step from the mass spectrometer. This enables more decision-driven, targeted analyses of samples, due to the removal of time restrictions imposed on mass spectrometers by online chromatographic

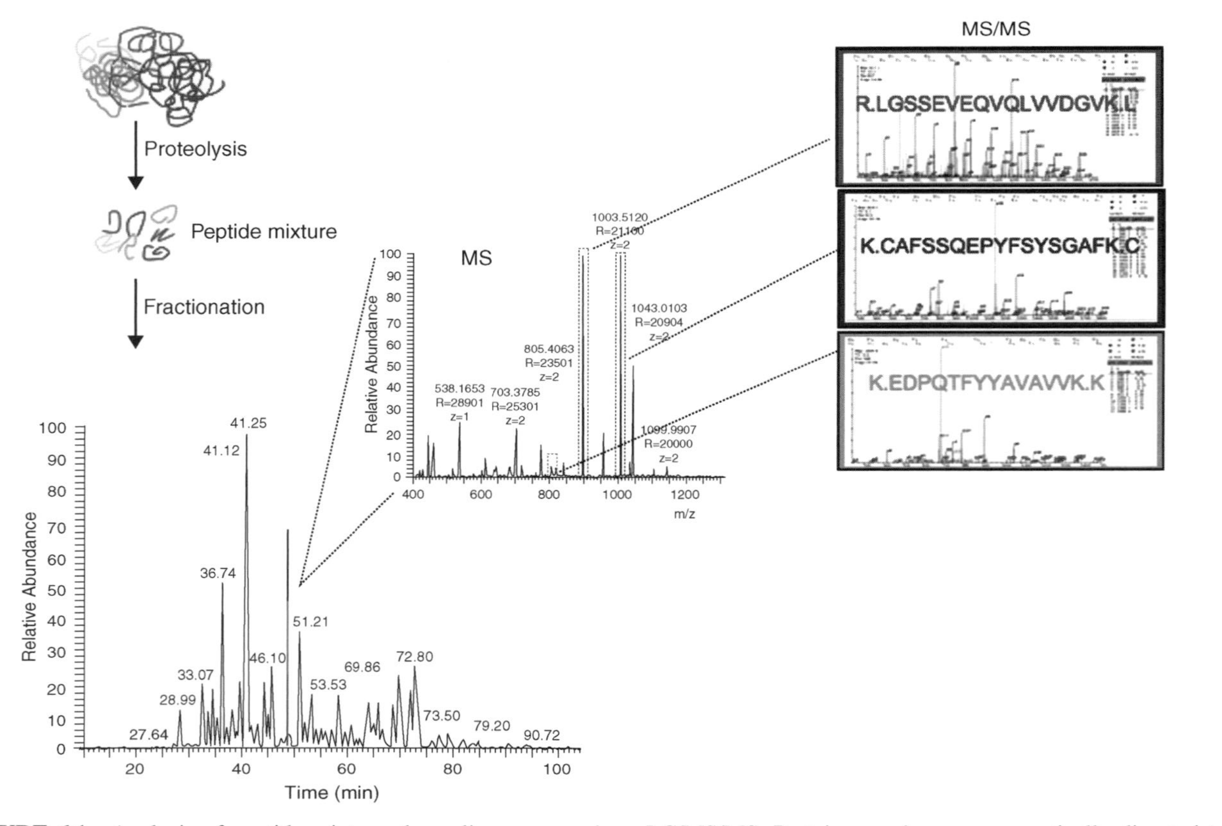

FIGURE 6.1 Analysis of peptide mixtures by online reverse-phase LC/MS/MS. Protein samples are enzymatically digested to produce a more complex peptide mixture. This mixture is subjected to reverse-phase nanoHPLC, and the peptides eluting at a given time are detected in a MS scan. Tandem MS spectra are then obtained for several of the signals detected during the MS scan. The mass and associated fragmentation pattern of each signal are subsequently searched against a sequence database in order to identify the peptide. (See insert for color representation.)

separations. Despite this and other potential benefits such as the possibility of performing parallel separations, the effective implementation of MALDI-based analysis systems requires the investigation and optimization of numerous technical issues. Thus, it remains to be seen whether such analysis platforms will become widely utilized.

Given the criterion by which ion signals are typically selected for tandem MS in data-dependent analysis schemes, it is not surprising that the protein identifications obtained are usually biased toward more highly abundant species. This limitation is of particular concern given the broad range of protein expression levels inherent in living organisms as well as the fact that most interesting classes of regulatory proteins are often expressed at low copy numbers per cell. In addition to the use of highly focused sample isolation techniques, other strategies have also been employed to increase the effective dynamic range of such analyses. Perhaps the simplest method involves additional fractionation of the initial mixture, thus further "spreading out" the sample before MS analysis. Although numerous multidimensional separation strategies have been described [17,18], the majority of such experiments reported to date employ "MuDPIT" (multidimensional protein identification technology), which combines strong cation exchange and reverse-phase chromatography of peptide mixtures [19,20]. Using this technique, thousands of unique proteins have been identified from whole-cell lysates in a single 2D LC-MS/MS experiment. Improvements in instrumentation have further increased the level of information that can be obtained from such experiments. For example, more recently introduced linear (two-dimensional) ion traps possess a significantly improved total ion trapping capacity compared to traditional three-dimensional traps, increasing their sensitivity by a factor of 10 [21]. Additionally, these new machines have vastly reduced scan times, enabling the acquisition of 5 times the number of tandem MS spectra per unit time. It should be noted that these large-scale profiling experiments require extensive computational resources to efficiently process the huge amount of data collected.

In evaluating any MS-based study that describes the detection of numerous proteins in a given sample, it is absolutely critical to understand the criteria employed for making an assignment. As typically employed in such experiments, the term *protein identification* does not imply that the protein is completely characterized in terms of its entire sequence or all of its PTMs. Rather, this term means that the search program employed matched one or more acquired tandem mass spectra to the expected spectra of one or more amino acid sequences unique to a given protein as translated from its encoding gene. In many studies reporting the presence of thousands of proteins in a given sample, the majority of proteins are identified on the basis of only one or two peptide hits. Although it is technically possible to identify the presence of a gene product from the presence of a single unique peptide, the limitations of the searching algorithms employed as well as the numerous sequences present in current databases provide ample opportunities for false identifications. The reporting of numerous incorrect identifications plagued early large-scale proteomics studies, and necessitated the introduction of strict criteria for reporting the results of MS-based protein identifications [22]. The quality of protein identifications from complex mixtures have also been greatly improved by the introduction of new hybrid MS instrumentation such as the LTQ FTICR [23] and the LTQ Orbitrap [24], which enable highly accurate mass measurements (± 5 ppm) to be routinely performed on the chromatographic timescale.

An important caveat of shotgun-based protein identification studies is that they rarely can distinguish between the numerous variations of a gene product that might exist at a given time. For example, the activities of numerous proteins, and in particular enzymes, are highly regulated through a series of protein truncations and modifications. Thus, the identification of the presence of a particular gene product need not correlate with the existence of its expected biological activity. This issue can be addressed by ABPP (activity-based protein profiling) techniques that enable the identification (and quantitation) of biologically active members of a given class of proteins [25]. ABPP reagents typically consist of a reactive group that covalently labels the active site of a specific class of enzymes, a noncovalent affinity group that helps specifically position the reactive moiety near the protein's active site, and a reporter molecule (fluorescent probe or affinity tag) that aids in the later isolation and identification of the labeled species. To date, ABPP has been applied to numerous classes of proteins, including cysteine proteases [26], tyrosine phosphatases [27], metalloproteases [28], and protein kinases [29].

Nevertheless, there are still numerous applications that focus primarily on the identification of which proteins are present in a given sample. For example, researchers are attempting to definitively catalog the protein components of various organelles [30]. Additionally, protein MS is widely used to identify proteins that interact either with other biomacromolecules in order to map functional networks [31,32] or with specific small molecules for various drug discovery applications [33,34]. Certainly, these studies will also be important for cataloging proteins that are uniquely present in embryonic stem cells during their self-renewal. However, after the identification of the protein constituents of a given system, the next step usually involves understanding how these systems change over time.

6.2 PROTEIN QUANTITATION USING MASS SPECTROMETRY

Unlike the relatively fixed nature of the genome, the proteome exists in a constant state of flux, varying over time, tissue type, and in response to external conditions. As such, understanding the overall biological significance of a particular protein requires the ability to assess changes that occur over time with respect to both the protein itself and other species in its immediate environment. Traditionally, 2DGE has been used to assess large-scale changes in protein expression levels between different samples (i.e., healthy vs. diseased samples). These experiments rely on the fact that the various chemicals used to visualize separated protein bands produce responses roughly proportional to the total level of the moiety with which they interact [35]. However, with the emergence of multidimensional LC/MS/MS methodologies, MS is increasingly used to effect simultaneous protein identification and quantification.

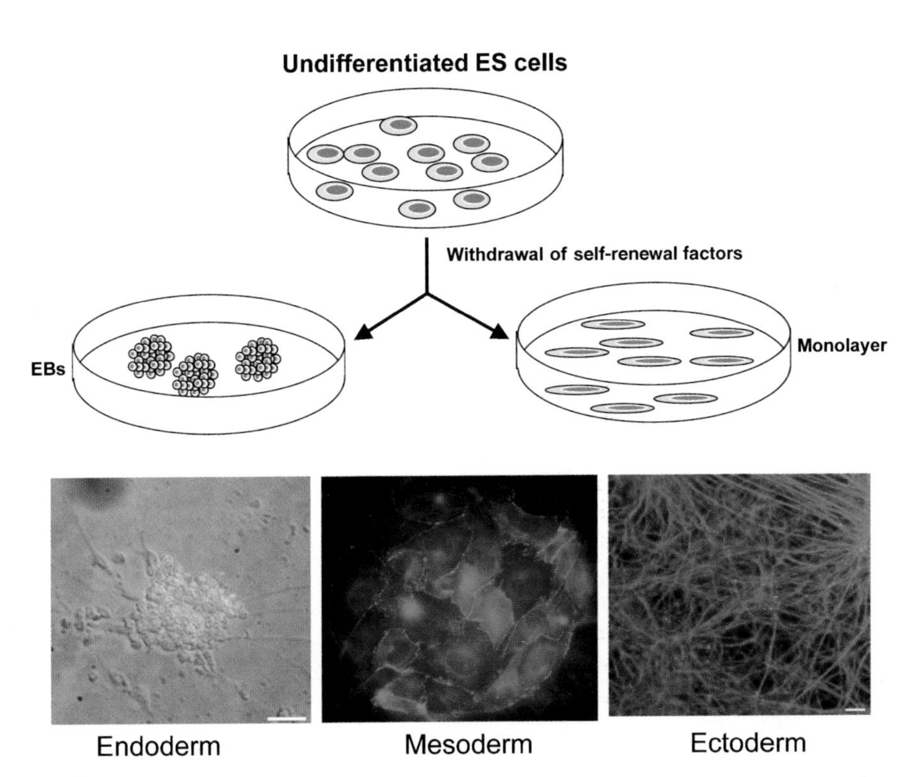

FIGURE 1.3 Differentiation potential of ES cells. ES cells can be induced to differentiate into all somatic cell lineages via the formation of three-dimensional EBs or monolayer culture.

(**a**) Tissue-specific Adult Stem Cell-Multipotent

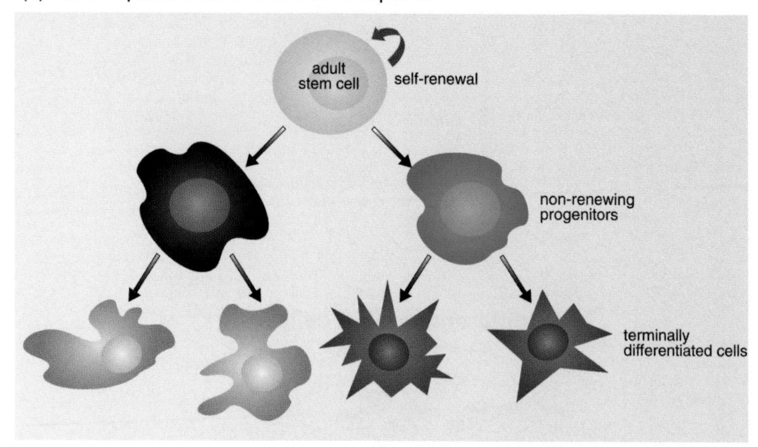

(**b**) Embryonic Stem Cell-Pluripotent

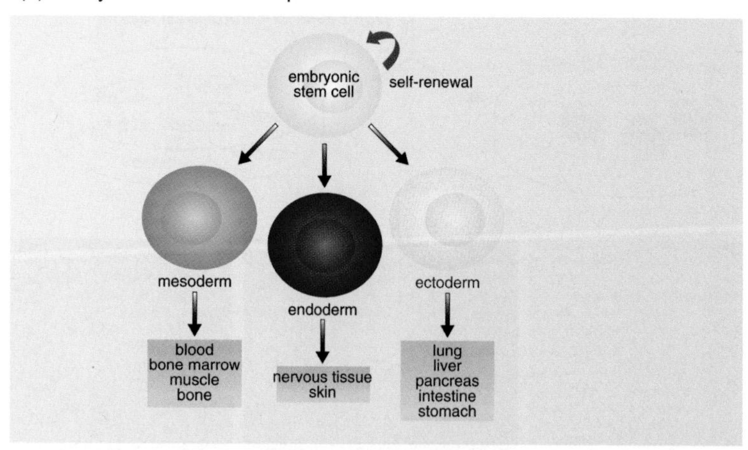

FIGURE 2.1 Adult stem cell lineages. (*See text for full caption.*)

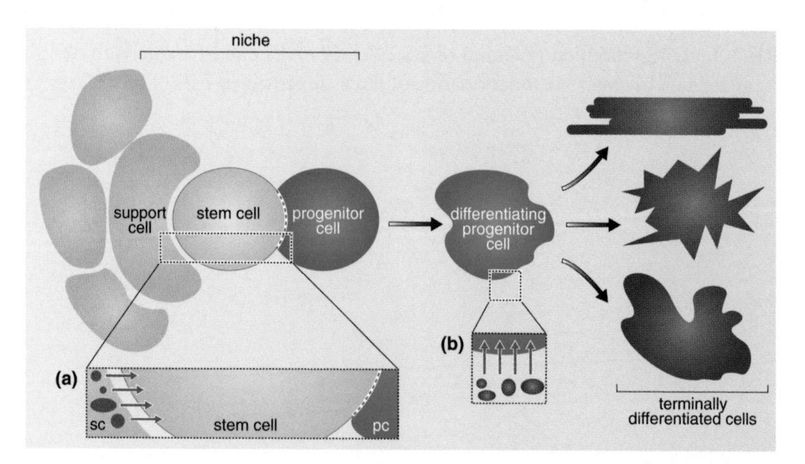

FIGURE 2.2 Adult stem cell niche. Adult stem cells are maintained in a specialized environment known as the *niche*. (*See text for full caption.*)

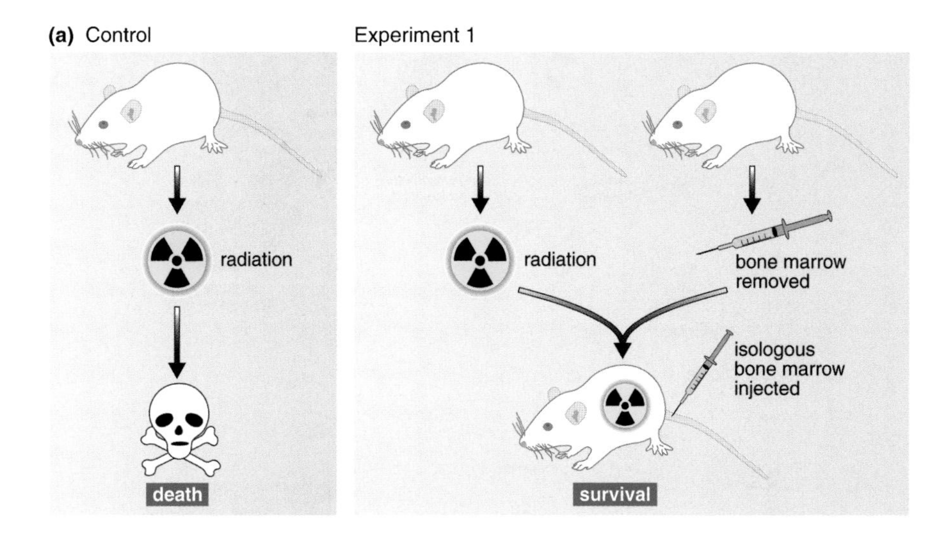

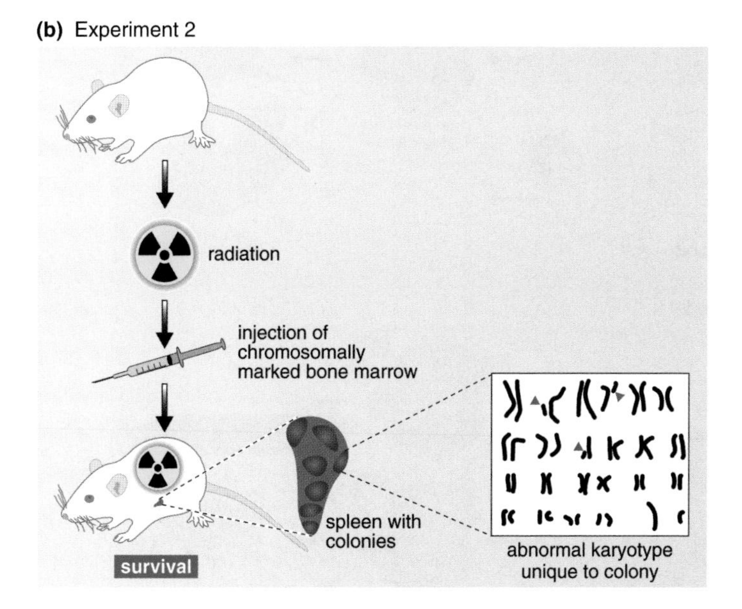

FIGURE 2.3 Discovery of the hematopoietic stem cell. (*See text for full caption.*)

(**a**) Normal Adult Stem Cell

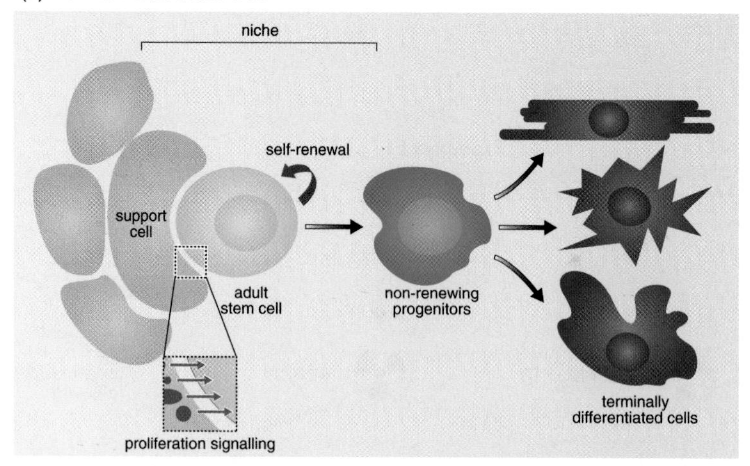

(**b**) Cancer Model I

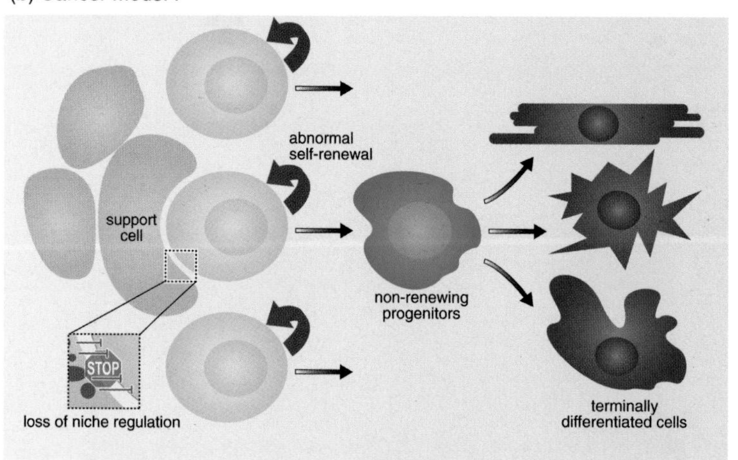

(**c**) Cancer Model II

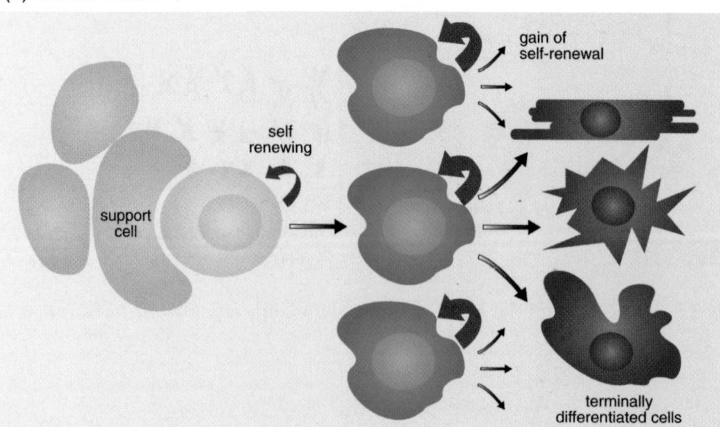

FIGURE 2.4 Cancer stem cell. Two major theories of the cancer stem cell arise from the common theme of aberrant self-renewal. (*See text for full caption.*)

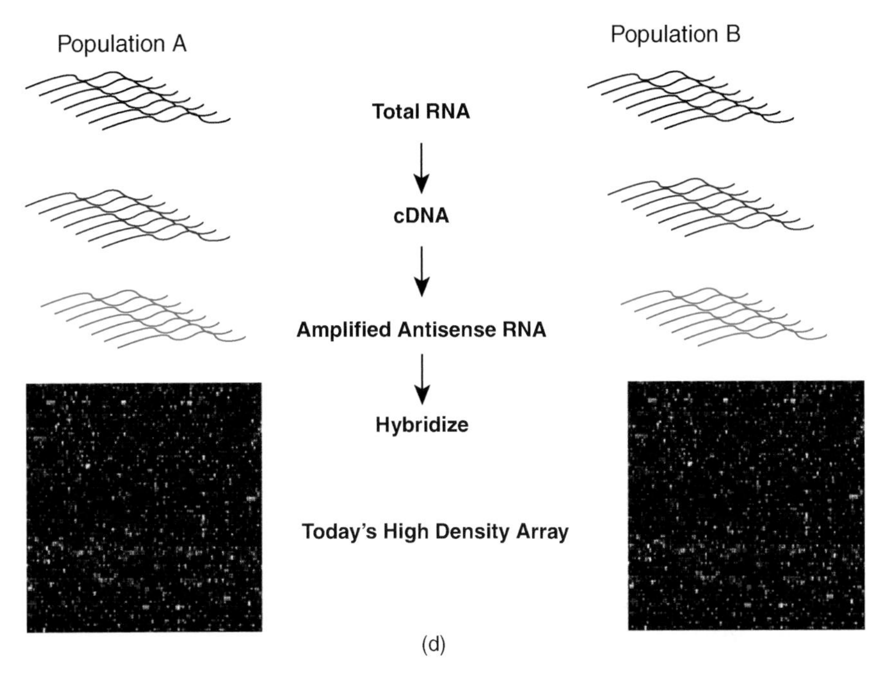

FIGURE 3.1 Differential gene expression technology progression. (*See text for full caption.*)

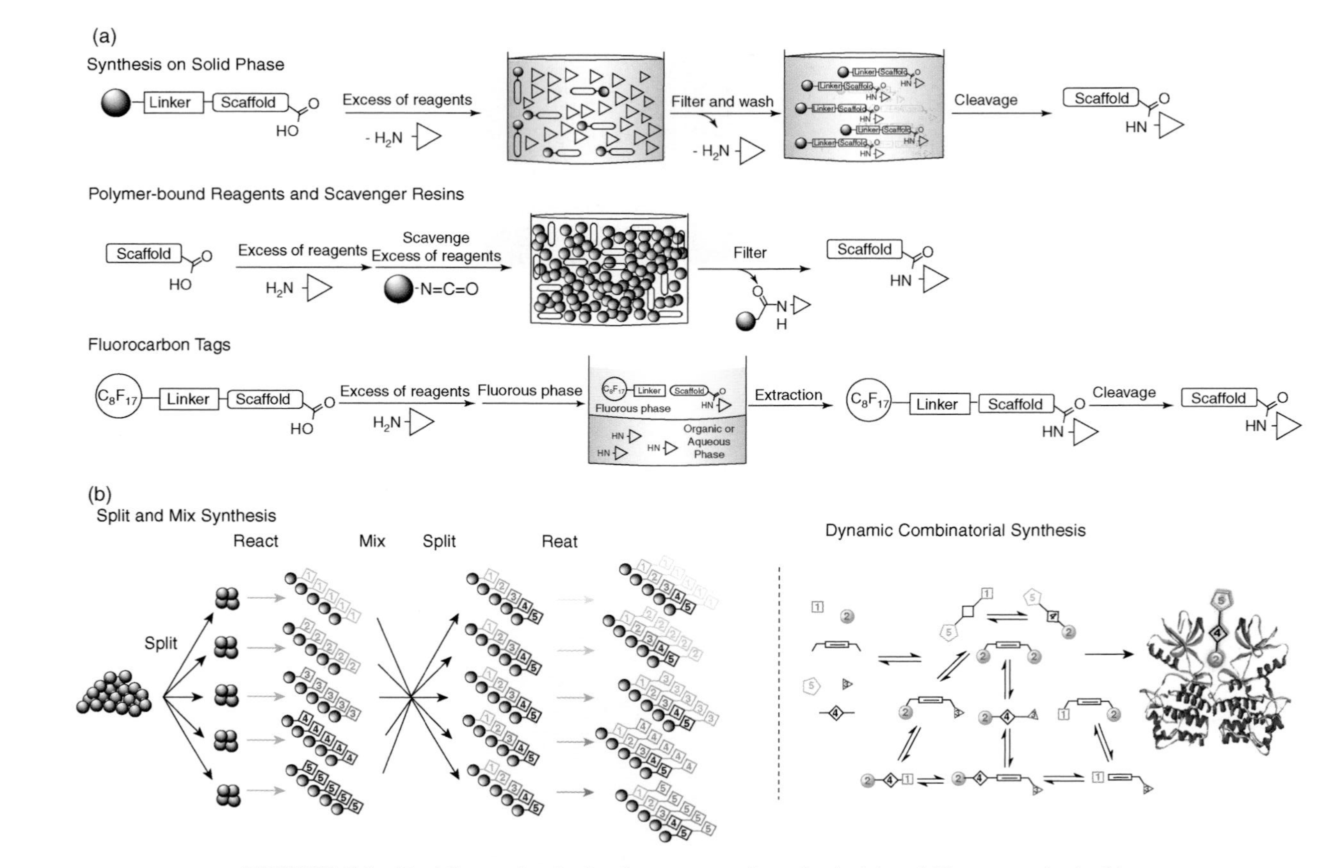

FIGURE 5.1 Enabling technologies for automated synthesis (a) and library synthesis (b).

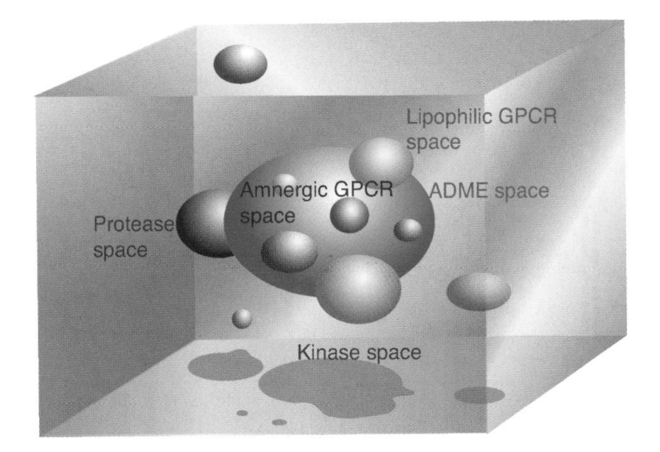

FIGURE 5.2 Cartoon representation of the chemical space. Representation obtained by plotting the principal component of physicochemical properties on x,y,z axis. Biologically active compounds tend to cluster into discrete areas (represented by the colored spheres). (Reprinted with permission from Ref [20]).

cleavage from the resin beads HF-pyr
a) HCL
b) piperidine/DMF
FmocHN
deprotection
MeO OMe
acetal formation
R^2-SH or
R^2-NH-R^2
X=S, NR^2
nucleophilic epoxide opening
immobilization
immobilization on the glass surface in form of a microarray
Probing the 3760-element microarray of small molecules with fluroescentlly labelled Ure2p protein
Structure-activity relationship exploration
uretupamine A
inhibitor of Ure2p (18.1 μM)
uretupamine B
inhibitor of Ure2p (7.5 μM)

FIGURE 5.6 Synthesis and microarraying of a 1,3-dioxane library. Discovery of a selective Ure2p (transcription factor) modulator.

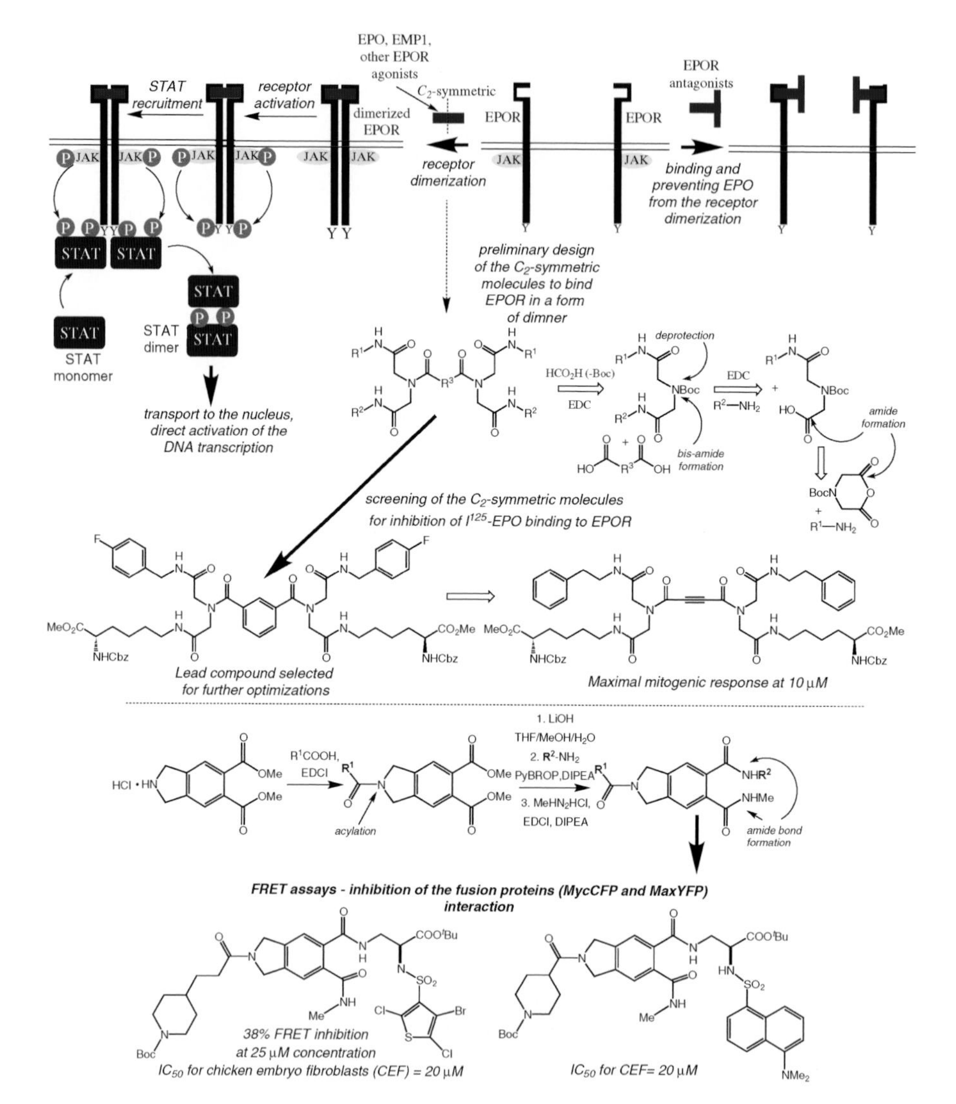

FIGURE 5.9 Synthesis of libraries targeting protein–protein interactions. Libraries based on C_2-symmetric ligands led to the discovery of an EPO agonist while library of isoindolinones afforded a Myc-Max protein–protein interaction inhibitor.

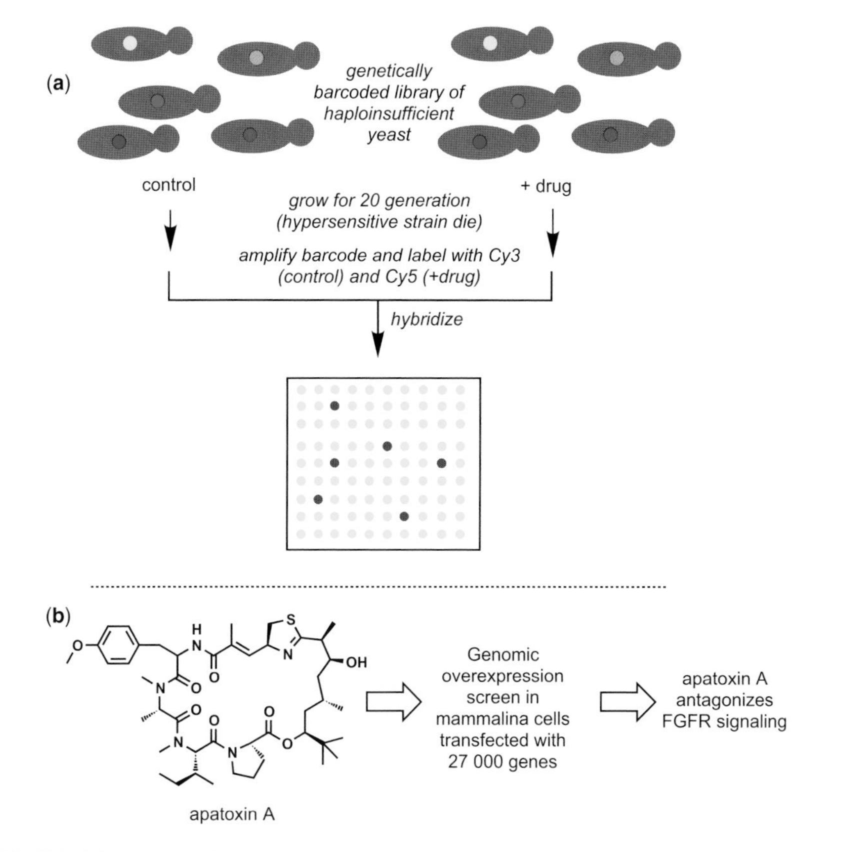

FIGURE 5.15 Genomic approaches to target identification: (a) lethality screen with genetically tagged haploinsufficient yeasts; (b) discovery of the mode of action of apatoxin A.

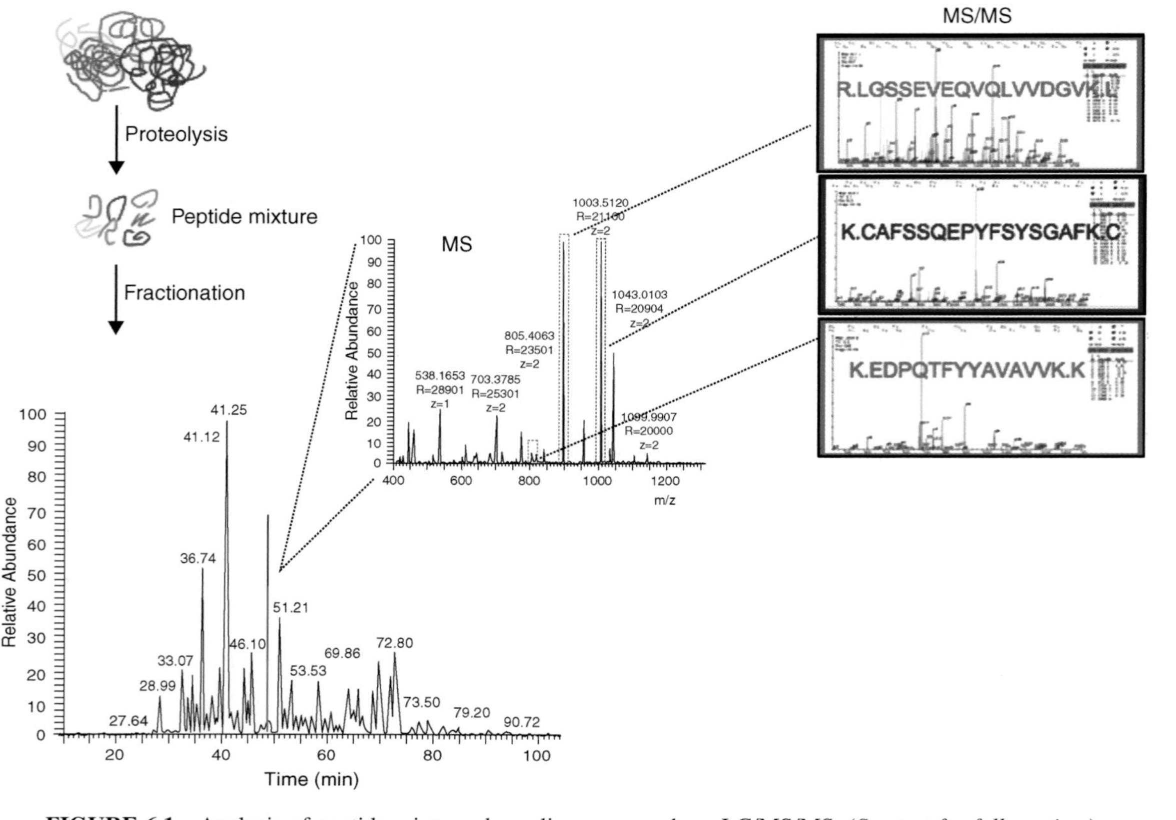

FIGURE 6.1 Analysis of peptide mixtures by online reverse-phase LC/MS/MS. (*See text for full caption.*)

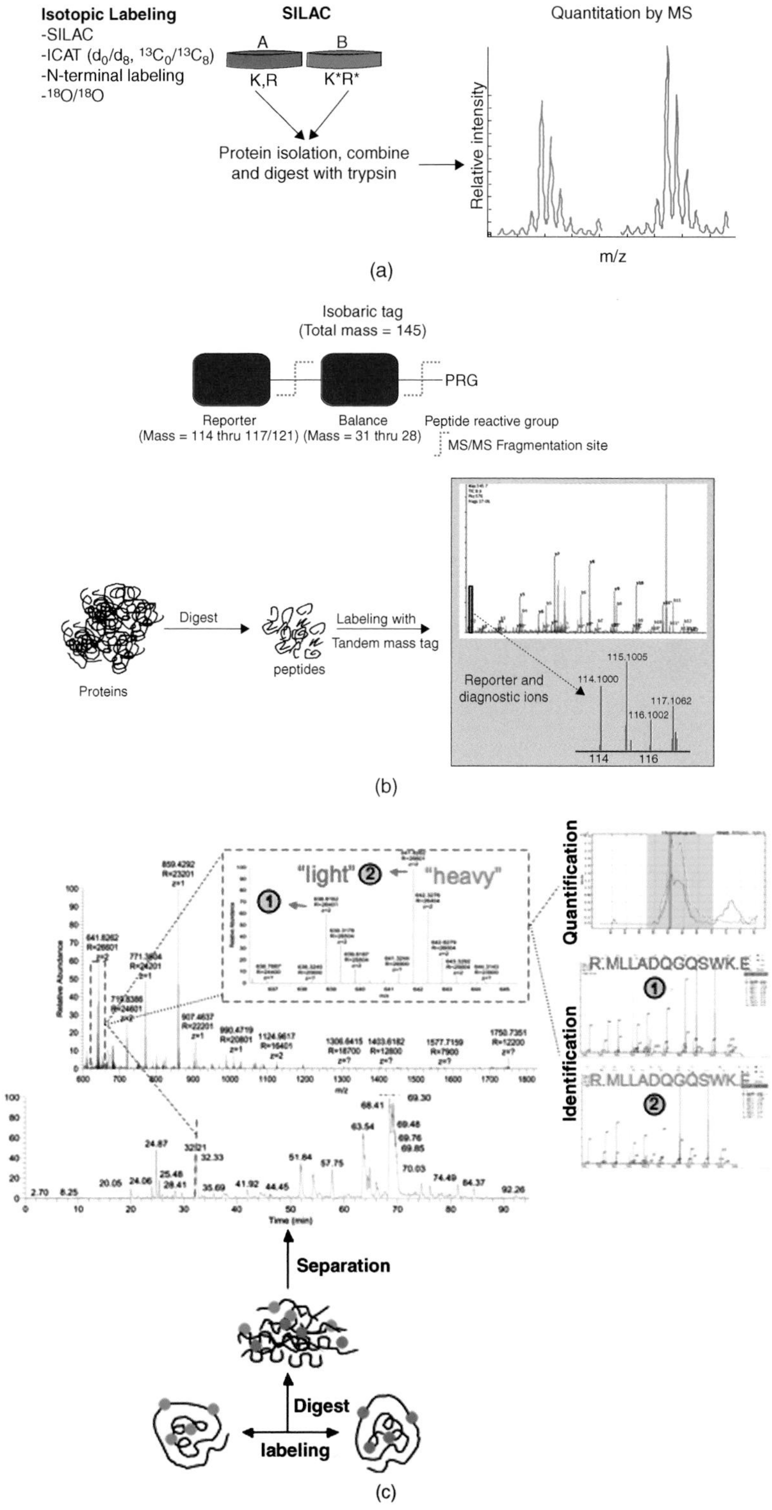

FIGURE 6.2 (a) Quantitation using isotopic labeling. (b) Quantitation using tandem mass tags. (c) Quantitation using chemical labeling strategies. (*See text for full caption.*)

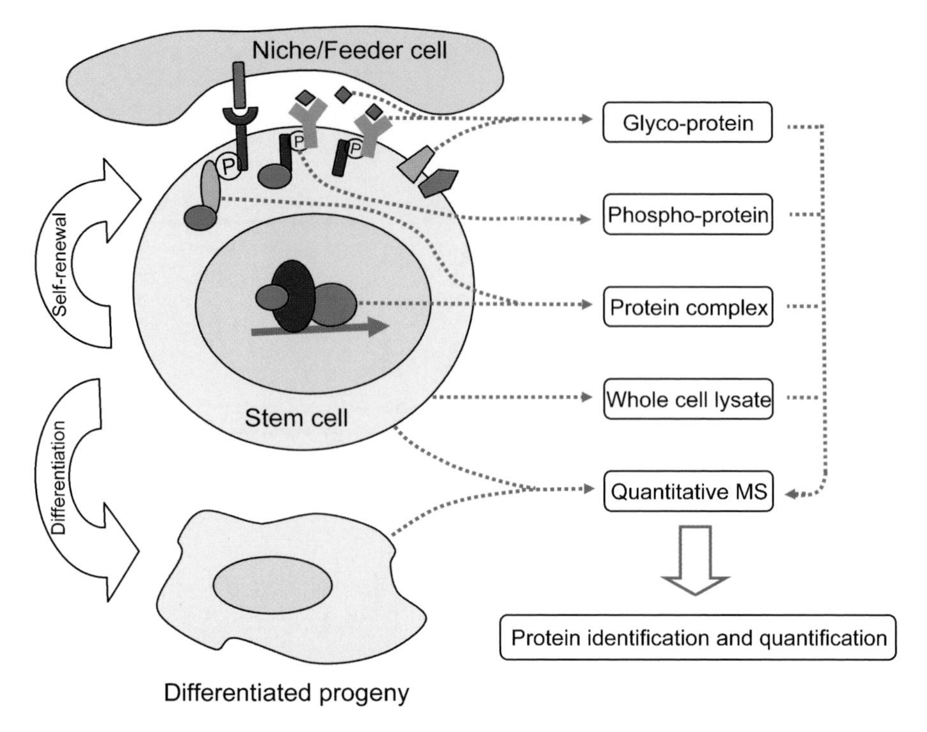

FIGURE 10.1 Schematic representation of proteomic strategies for investigating stem cell biology. (*See text for full caption.*)

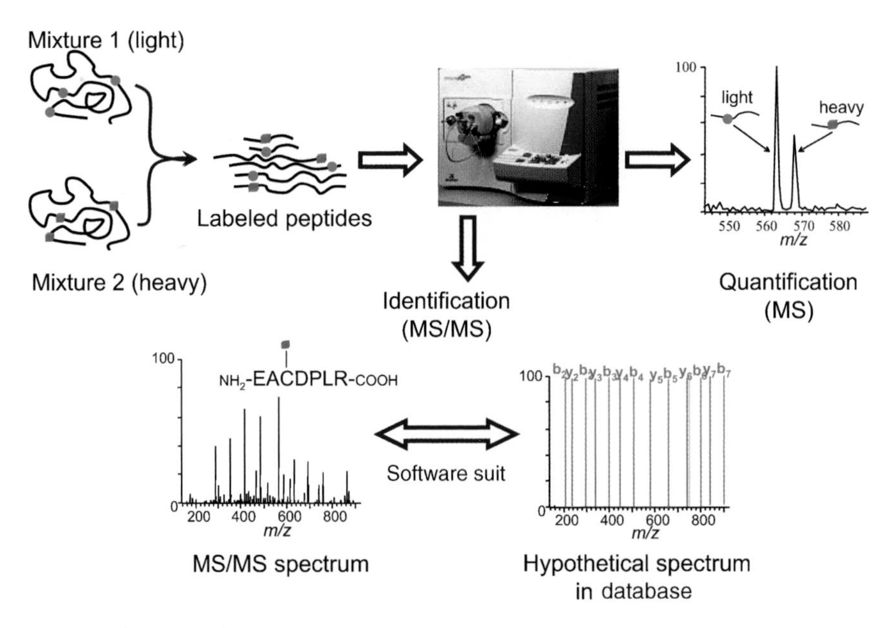

FIGURE 10.2 Quantitative proteomic scheme. (*See text for full caption.*)

The effective use of MS to measure changes in peptide (and protein) levels requires the recognition of several important parameters. The first is that different peptides exhibit widely variable ionization efficiencies. This means that there are no universal standards that can be employed to roughly quantify the various peptides contained within a given sample. Additionally, the absolute signal for a specific amount of a given peptide depends not only on its own ionization efficiency but also on the identities and relative concentrations of other peptides being simultaneously analyzed as a result of ion suppression effects. This requires that measurements of species being compared be made within the same (or an effectively identical) sample. Absolute quantitative measurements of individual peptides can be made by implementing the same stable isolation dilution techniques utilized for numerous decades in pharmacological studies of small molecules [36,37]. Although effective, this technique requires the synthesis of a pure, isotopically labeled version of every species to be quantified. However, given the exploratory nature of many proteomics studies, differential quantitation schemes that instead measure relative changes between specific samples are often employed.

An ideal quantitation scheme would enable the collection of the required data directly during a standard LC/MS/MS analysis without the need for any specific additional preprocessing of the sample. Several such "label-free" quantitation schemes have been described. Spectral counting techniques are based on the positive correlation typically seen in shotgun-type experiments between the number of tandem MS spectra that are assigned as having arisen from a particular protein and the actual abundance of that protein in a given sample [38]. Since its initial formulation, numerous refinements have been suggested in order to better account for differences in factors such as the size and sequence of various proteins [39]. Other techniques attempt to extract spectral or chromatographic peaks (features) from individual analyses, and then compare changes in the signals (extracted-ion chromatograms) of these features across multiple runs [40,41]. Alternatively, the significance of intensity changes in every data point [time and *m/z* (mass/charge) value] can be evaluated rather than first determining what features to compare [42]. Regardless, such feature comparison techniques require the normalization of overall signal intensities between runs as well as highly reproducible sample-handling techniques and chromatographic separations that can readily be aligned between analyses. Additionally, these samples may contain thousands of individual species, many with signals approaching the background signal. Significant improvements in the peak capacity and sensitivity of chromatographic methods combined with the high resolution and mass accuracy of modern mass spectrometers have greatly improved the quality of such analyses. However, despite the significant computational infrastructure required, these "simple" label-free techniques generally produce less accurate quantitative measurements than do other methodologies [43].

In order to obtain more accurate quantitative measurements, the majority of protein-profiling studies employ various stable isotope labeling strategies. Typically, two samples to be compared are individually labeled with different forms of a stable isotopic pair, and the two samples are subsequently combined in equal proportions at some point before the final LC/MS/MS analysis. The result is that each peptide exists as a pair of isotopically labeled species that are identical in every respect except for their masses. Thus, each isotopically labeled peptide effectively serves as its partner's ideal internal standard, and the ratio of the relative heights of the two isotopically labeled species in the MS scan provides quantitative data as to any differential change that occurred in the expression of the protein from which the peptide arose.

One approach for readily incorporating stable isotopes involves growing cells in isotopically enriched media. For example, one group of cells would be cultured in media that contained ^{14}N as the only source of nitrogen atoms, while a second group would be grown in media containing only ^{15}N [44]. Although this methodology efficiently incorporates stable isotopes on a global scale, the determination of which two peptides constitute an isotopically labeled pair is severely complicated by the fact that such pairs possess variable mass differences depending on the number of nitrogen atoms they contain. "Inverse labeling" methodologies have been introduced that address this issue [45], but at the cost of doubling the number of experiments that need to be performed. Additionally, although this methodology has been applied to rodents by restricting diets exclusively to bacteria grown on isotopically enriched media [46], this technique cannot readily be applied to higher organisms or biologically derived samples such as blood or tissue.

Stable isotope labeling with amino acids in cell culture (SILAC) is a more recently introduced variation of the methodology described previously [47]. As shown in Figure 6.2a, this technique involves the growth of cells in different media, but in this case, only certain amino acids exist in isotopically distinct forms. Typically, stable isotope-labeled versions of lysine and arginine are simultaneously employed such that every tryptic peptide (except the one arising from a protein's *C* terminus) bears an isotopic label. However, other amino acids have also been employed for specific applications. For example, ^{13}C-labeled tyrosine was used in an experiment attempting to identify substrates of tyrosine kinases [48]. Importantly, the use of more than two isotopically distinct versions of an amino acid enables multiples samples to be differentially compared. For example, a triple-SILAC strategy employing $^{12}C_6{}^{14}N_4$-Arg, $^{13}C_6{}^{14}N_4$-Arg, and $^{13}C_6{}^{15}N_4$-Arg was used to compare the mechanism of divergent growth factor effects on mesenchymal stem cell differentiation [49]. Although extremely powerful for the study of biological phenomena in cell culture, this technique cannot be readily applied to quantify difference in the tissues or biofluids of animals. Additionally, the metabolic conversion of arginine to proline has been observed in certain cell lines [50], requiring appropriate adjustments or the exclusion of all proline-containing identifications.

Another method for the introduction of stable isotopes involves chemical modification of the samples under study. Typically, these schemes employ chemical reagents that specifically target one of the native reactive moieties of proteins, including thiols [51], carboxylic acids [52], amines [53], the ε-amino group of lysines [54], and the indole group of tryptophan [55]. Alternatively, stable isotopes can also be incorporated into *C*-terminal carboxylic acid functionalities through the action of trypsin in ^{18}O-labeled water [56], or by reaction with "unnatural" functional moieties on proteins that were either artificially introduced [57] or created as a result of a chemical modification, including ketones that result from various oxidation reactions [58] or α,β-unsaturated carbonyls that results from the β-elimination of labile moieties such as phosphoserine groups under basic conditions [59]. Although more expensive, the majority of more recently introduced reagents employ stable isotopes such as ^{13}C, ^{15}N, and/or ^{18}O rather than deuterium, as it has been shown that deuterium-labeled species have slightly different retention times than their hydrogen-containing counterparts under typical reverse-phase chromatography conditions [60]. As exemplified by the prototypical isotope-coded affinity tag (ICAT) [61], many such labeling reagents also incorporate an affinity label such as biotin in order to affect the enrichment of labeled species. However, more recent analogs of such reagents typically also feature a cleavable element that enables more efficient recoveries of the labeled species from the capture agent employed [62]. Although these chemical modification-based approaches are highly versatile in that they can be applied to any type of sample, this versatility must be weighed against possible issues surrounding the efficiency and selectivity of a given reaction.

Figure 6.2c shows the typical experimental workflow for broad quantitative profiling studies that employ stable isotope-labeling techniques. Samples to be compared are individually labeled with different forms of a stable isotopic reagent, and the samples are subsequently combined in equal proportions at some point before the final LC/MS/MS analysis. Other factors being equal, isotope labeling at the protein level is preferred, as this enables the entire analyte to be processed using a single proteolytic digestion. After LC/MS/MS analysis, computer algorithms attempt to match signals in the MS scans that represent the isotopic variants of a single peptide. Although the identification of this species still relies on tandem MS of at least one of the isotopic variants, the differential quantitation of the species is measured by comparing the XICs of each isotopic variant in the MS spectrum. As described previously, such measurements become highly unreliable when a signal of potential interest starts to approach the background signal level.

Improved quantitative measurements can be obtained utilizing the more recently introduced iTRAQ (isobaric tag for relative and absolute quantitation) methodology [63]. This technique employs a series of amine-reactive, isobaric tags that enable quantitative measurements to be taken during tandem mass spectrometry. In comparison to broad-range MS scans, tandem MS scans are performed after the isolation of a narrow *m/z* window. As such, tandem MS spectra exhibit far lower background signals, leading to more accurate quantitative measurements. As shown in Figure 6.2b, the iTRAQ reagents themselves are isobaric, causing the isotopically labeled variants of a given peptide to behave identically during LC/MS. However, on tandem MS, each individual reagent produces a unique reporter ion separated by a single dalton, due to the presence of carbonyl-based mass counterbalance groups. Thus, peptide identification and quantitation are simultaneously measured in the same tandem MS experiment. In addition to improved quantitative measurements, this technique also enables four (or eight in the next-generation implementation soon to be released) different samples to be compared in a single analysis.

As a result of techniques like those described here, MS-based experiments are starting to address one of the most fundamental properties of the proteome—its dynamic nature. For example, the temporal mapping of signaling cascades [64,65] or

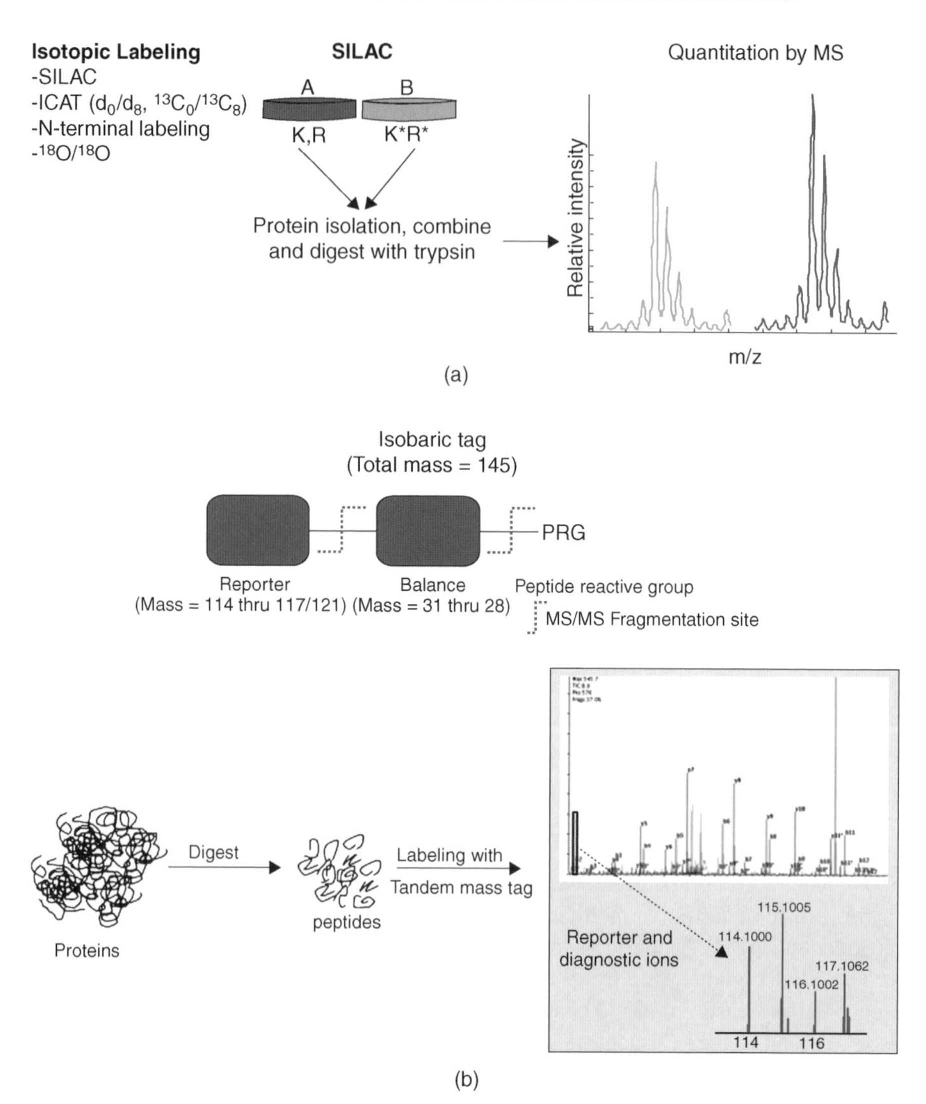

FIGURE 6.2a,b (a) Quantitation using isotopic labeling. In SILAC, two groups of cells are grown in culture media that are identical except that the first medium contains the "light" and the second medium contains the "heavy" form of particular amino acids (K,R). With each cell doubling, the cell population replaces at least half of the original form of the amino acid, eventually incorporating 100% of a given "light" or "heavy" form of the amino acid. The samples are then combined, and identified proteins are quantified from the ratios of the peptide doublets seen in the MS scans. (b) Quantitation using tandem mass tags. An iTRAQ reagent consists of a reporter group (114–117 Da), a balance group (neutral loss of 31–28 Da), and an amine-specific peptide reactive group. The iTRAQ workflow consists of the isolation of proteins from different states to be compared, reduction and alkylation of cysteine residues, and tryptic digestion. The peptide mixtures are then labeled with different iTRAQ reagents. Quantitation is performed in the MS/MS mode by measuring the relative intensity of the reporter groups. (See insert for color representation.)

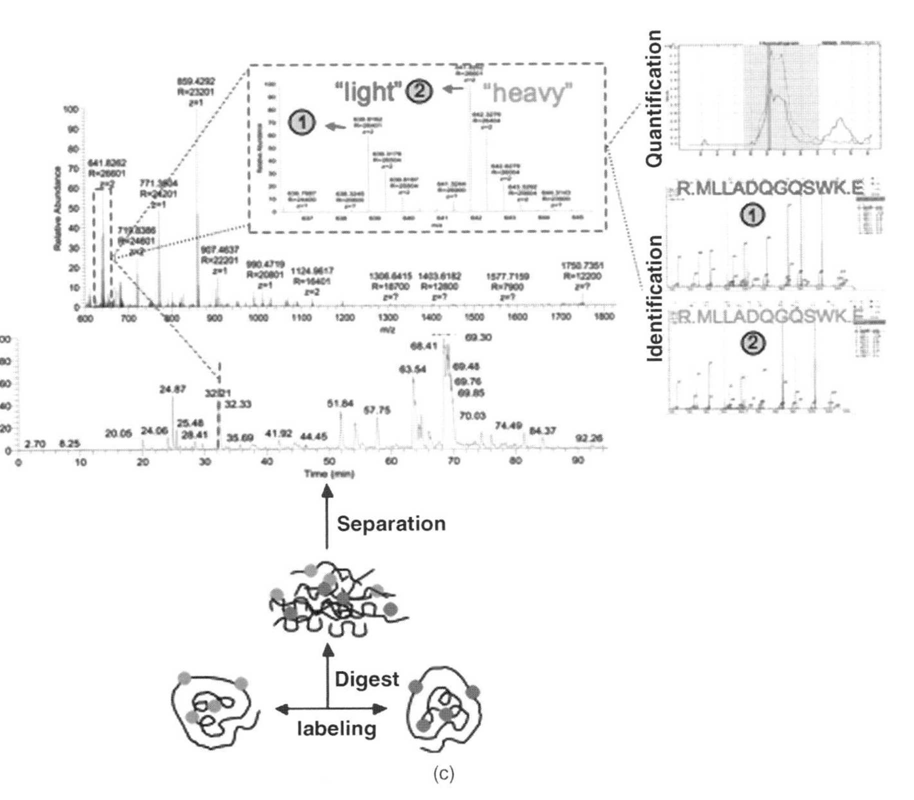

(c)

FIGURE 6.2c (c) Quantitation using chemical labeling strategies. Two protein mixtures representing different cell states are denatured, and reacted with light and heavy reagents (labeling can occur before or after tryptic digestion). Tagged peptides can optionally be enriched, and the peptide mixture is analyzed by LC/MS/MS. The ratio of signal intensities of the tagged peptide pairs is determined by relative abundance. (See insert for color representation.)

changes that occur in the nuclear proteome on treatment with transcription inhibitors have been described [66]. Other studies have measured the dynamics of protein turnover [67] or the identification of newly synthesized proteins [68]. Clearly, such studies will be highly relevant in attempting to understand the underlying mechanisms of stem cell differentiation. It should be noted that isotopic labeling techniques can also be applied to great effect in other applications. For example, isotopic labeling has been used to distinguish true binding partner from nonspecific binders in protein interaction studies [69,70].

6.3 CHARACTERIZING POSTTRANSLATIONAL MODIFICATIONS BY MASS SPECTROMETRY

Although various genomics-based methodologies can provide valuable insights into protein function, only proteomics studies can elucidate the critical role that PTMs play in modulating protein function and activity. To date, well over 200 distinct PTMs have been described (Figure 6.3), and these covalent modifications have been shown to exert profound influences ranging from the modulation of protein three-dimensional structure, subcellular localization, and lifetime to the regulation of processes such as cellular signaling, gene expression, and apoptosis [71]. Combinatorial networks of similar [72] or disparate [73] PTMs provide additional mechanisms for the modulation of biological effects, as exemplified by the regulation of histone function [74]. As such, the ultimate goal of studying PTMs by MS is nothing less than the complete characterization of this highly sophisticated regulatory network. At its basest level, this involves mapping the site of each individual PTM on every gene product. Additionally, both the stoichiometry and the temporal organization of each site of modification need to be determined with respect to other PTMs.

However, the detection and localization of a given PTM represents a far more difficult task than the identification of the presence of a given gene product. For example, tryptic digestion of a protein results in numerous peptides, only a few of which need to undergo successful tandem MS fragmentation and database matching to identify the presence of that protein. By contrast, the existence of a particular PTM can be ascertained only by the successful analysis of the specific peptide that bears the modification. Furthermore, several labile modifications such as phosphorylated serine and threonine residues can produce fragmentation patterns that contain little or no useful sequence information under typical tandem MS conditions. Finally, most regulatory proteins are expressed at only low copy numbers per cell compared to other proteins, and many individual sites of modification are only partially occupied. Therefore, analyses that target a particular PTM often employ selective enrichment techniques as well as highly focused mass spectrometric methodologies. A comprehensive review of these methods for phosphorylation alone would easily fill an entire book, and the different techniques employed are often as unique as the various PTMs they seek to characterize. As such, the following examples were chosen primarily to illustrate the broad scope of this dynamic research area, and by no means should be seen as comprehensive.

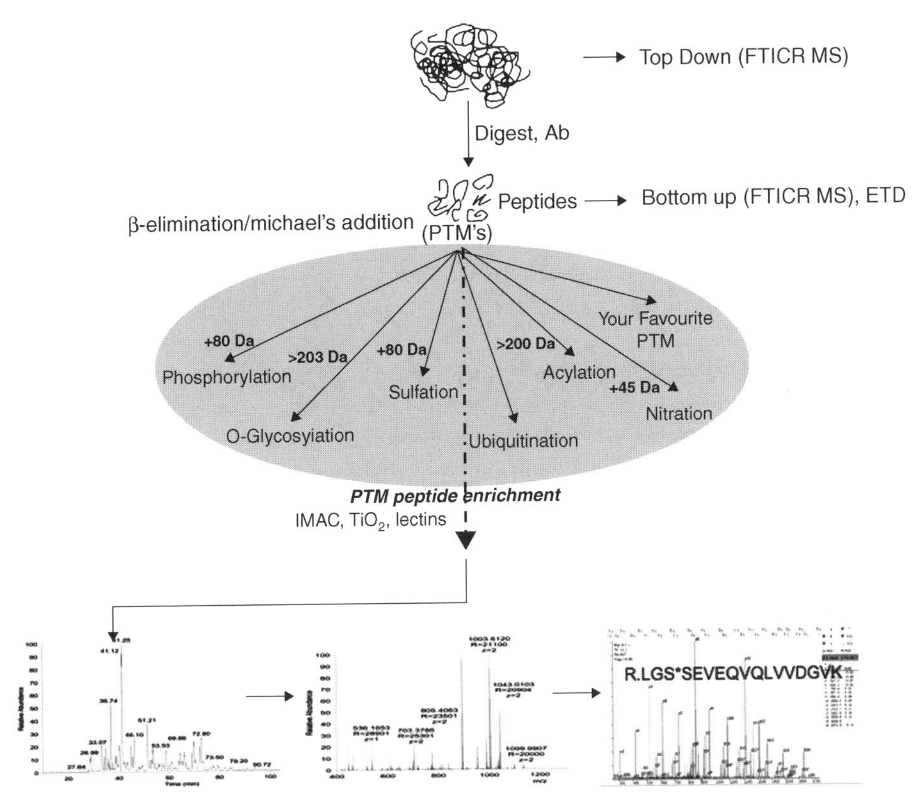

FIGURE 6.3 Posttranslational modifications (PTMs) in proteomics. In a complex protein mixture, large-scale analysis of PTMs is facilitated by PTM-specific protein (e.g., pY antibody enrichment for phosphorylation) and peptide enrichment (e.g., IMAC, TiO_2) prior to MS/MS analysis. Chemistry-based methods for PTM-specific covalent capture of proteins/peptides are also often employed.

Several affinity reagents exist that possess innate selectivities for particular PTMs. For example, both immobilized metal affinity chromatography (IMAC) [75] and titanium dioxide [76] resins have been employed for the enrichment of phosphopeptides. Similarly, various plant proteins called *lectins* have been utilized to enrich glycosylated species with specific glycan structures [77,78]. Although highly effective, these approaches require the discovery/development of an individual reagent for each specific functionality.

As such, a broader approach employs a class of reagents that possess a more general affinity capture moiety. These reagents can affect a specific interaction with a given PTM through a variety of mechanisms. For example, the reagents may selectivity react with a chemical moiety that arises from an intentional modification of the PTM itself. Thus, hydrazine-containing reagents have been employed to selectively react with aldehydes formed by the periodate oxidation of glycosylated peptides [79], while thiol-containing reagents have been utilized in Michael addition reactions with α,β-unsaturated carbonyls that result from the β-elimination of labile moieties such as β-*N*-acetylglucosamine-(*O*-GlcNAc)-modified serines and threonines [80]. Alternatively, the reagent-targeting chemical functionality can enzymatically be added to proteins of interest. For example, azide-containing analogs of sugars [81] and isoprenoids [82] were added to cultured cells, and shown to be incorporated into cellular proteins. The resulting azide-labeled, mucin-like *O*-linked glycosylated and farnesylated proteins (respectively) were isolated using a chemoselective reaction between the incorporated azide groups and a phosphine-functionalized affinity reagent [83]. Although powerful, potential issues of metabolic interconversion as seen in the case of SILAC labeling must be accounted for when employing such a technique. Additionally, a ketone-containing analog of galactose was selectively added to proteins modified with *O*-GlcNac after cell lysis using a modified galactosyltransferase, and the resulting modified species were isolated using a chemoselective reaction between the incorporated ketone moiety and an aminooxy-functionalized affinity reagent [84]. Much like the chemical modification reactions described earlier, this approach is more versatile in that it can be applied to any type of sample.

Typically, the classic biochemical affinity pair biotin–streptavidin has been utilized to effect the isolation of labeled species. However, other affinity enrichment systems have been described. For example, chemoselective ligations such as click chemistry [85] can be used to capture azide-bearing proteins onto an alkyne-functionalized solid support [86]. In another approach, living systems can be transfected so as to directly produce affinity-labeled versions of normal substrates. For example, cell lines were modified to produce only 6His-tagged versions of ubiquitin [87] or SUMO-1 [88]. Subsequently, the proteins modified with these PTMs were enriched from whole-cell lysates using IMAC, and their sites of modification were identified due to the presence of specific mass increases on lysine residues as seen in the tandem MS spectra, such as an increase of 114 Da that represents the "Gly-Gly" tag remaining from ubiquitin after tryptic digestion. Finally, alternative chemical-based affinity systems have also been investigated. For example, species bearing functionalities of interest have been chemically modified with perfluoroalkyl groups, and the unique interplay between the fluorous-derivatized species and the surface of a perfluoroalkyl-silylated stationary phase was subsequently utilized to separate labeled peptides from unlabeled species in a simple solid phase extraction format [89]. Interestingly, the same fluorous derivatization and enrichment methodology can readily be applied for the selective fractionation of both peptide- and small-molecule-based samples.

In addition to a wide range of affinity enrichment techniques, numerous specialized MS scanning and fragmentation methodologies have also been developed to further improve the detection and identification of various PTMs. For example, precursor ion-scanning experiments can be performed using triple quadrupole or more recently introduced hybrid linear ion trap–triple quadrupole (QTrap™) mass spectrometers. In this technique, ions entering the instrument are scanned in Q1 and sequentially fragmented in Q2, while Q3 remains fixed in order to monitor the presence of a particular diagnostic ion. Detection of this characteristic ion indicates that the species that exited Q1 indeed contained the PTM of interest, and thus should be subjected to further study. Typical reporter ions monitored using this technique include −79 *m/z* corresponding to the loss of PO_3^- from phosphopeptides [90], +216 *m/z* corresponding to the loss of the immonium ion of phosphotyrosine [91], and +46 *m/z* corresponding to the loss of the dimethylammonium ion from asymmetrically dimethylated arginine [92].

A conceptually similar approach utilizes the occurrence of a particular dominant neutral loss during tandem MS as an indicator that the originating species contained a specific PTM of interest. For example, phosphoserine and phosphothreonine residues are highly labile under typical tandem MS conditions, and therefore often lead to tandem mass spectra that are dominated by the loss of H_3PO_4 [93]. Neutral losses of HPO_3 are also associated with these species as well as phosphotyrosine residues. The rapid scanning speed of linear ion trap mass spectrometers enables researchers to implement automated MS^3 (MS/MS/MS) strategies in which major ions arising from the loss of H_3PO_4 during tandem MS are automatically subjected to another stage of fragmentation. This third stage of MS often produces useful sequence information, and has resulted in the identification of hundreds of additional phosphopeptides from mammalian cell culture [94]. However, this technique does have several limitations; for example, a neutral loss of H_3PO_4 can manifest itself as different decreases in *m/z* value depending on the charge state of the precursor ion as well as the fact that many sequence-specific phosphorylated residues fail to yield significant neutral loss peaks on tandem MS.

In comparison to large-scale profiling experiments, the phosphorylation profile of an individual pure protein can be mapped in much greater detail using hypothesis-driven multistage MS [95]. Regardless of the instrumentation employed, this technique starts from the premise that each serine, threonine, or tyrosine residue in a protein's sequence may be phosphorylated, and that the resulting phosphopeptide candidates, although numerous, can each be tested experimentally. Accordingly, a computer program calculates several individual tandem MS transitions (i.e., a given *m/z* precursor ion that produces a particular *m/z* fragment ion) that should be unique for each potential site of modification given a particular proteolytic treatment, and then continuously scans these sets of individual transitions. The facile implementation of this technique on a Q-Trap has recently been described and termed multiple-reaction monitoring–initiated detection and sequencing (MIDAS) [96].

Although bottom–up proteomics has proved invaluable for the identification and localization of various PTMs, not surprisingly, its use remains problematic for the study of potential combinatorial interactions and long-range associations between various PTMs. By comparison, top–down experiments seek to obtain high-resolution mass measurements of intact proteins as well as direct fragmentation of those proteins in the gas phase [97]. The complete characterization of an intact protein was first demonstrated using a nonergodic fragmentation technique termed *electron-capture detection* (ECD) [98], and this technique has subsequently been employed to simultaneously map the different PTMs of various histones [99,100]. More recently, an ion–ion analog of ECD, termed *electron transfer disassociation* (ETD), was introduced [101]. Unlike ECD, this technique can readily be implemented on a linear ion trap, and thus is accessible to the great majority of researchers. To date, ETD has shown tremendous potential for the effective characterization of labile PTMs such as phosphorylation [102]. Further, when combined with a second ion–ion reaction termed *proton transfer charge reduction*, ETD has been used to sequence the *N* and *C* termini of proteins (such as the highly modified *N* termini of histones) on a chromatographic timescale [103].

6.4 CONCLUSION

MS has arguably become the single most important technique for the characterization of proteins, while new innovations in methodologies and instrumentation continue to further expand its scope. Although MS-based techniques will undoubtedly continue to play a major role in the exploration of fundamental biological processes, the utility of protein MS is also being actively investigated for more immediate applications such as the discovery and validation of clinically useful biomarkers. Despite the technical progress cited in this review and the great amount of effort currently being expended, the success of MS-based protein biomarker discovery is by no means guaranteed [104]. For example, it has been estimated that serum exhibits a dynamic range of at leaset 10 orders of magnitude in concentration between albumin and the rarest proteins that are currently clinically measured [105]. However, methodology improvements for expanding the capabilities of multiple reaction monitoring assays [106] as well as advances in technologies such as the direct imaging of tissue samples [107] and ion mobility techniques [108] provide hope that even these challenges can one day be addressed.

REFERENCES

[1] Broder S, Venter JC: Sequencing the entire genomes of free-living organisms: The foundation of pharmacology in the new millennium, *Annu Rev Pharmacol Toxicol* **40**:97–132 (2000).

[2] Lockhart DJ, Winzeler EA: Genomics, gene expression and DNA arrays, *Nature* **405**:827–836 (2000).

[3] Rossant J, McKerlie C: Mouse-based phenogenomics for modelling human disease, *Trends Mol Med* **7**:502–507 (2001).

[4] Gygi SP, Rochon Y, Franza BR, Aebersold R: Correlation between protein and mRNA abundance in yeast, *Mol Cell Biol* **19**:1720–1730 (1999).

[5] Pandey A, Mann M: Proteomics to study genes and genomes, *Nature* **405**:837–846 (2000).

[6] Karas M, Hillenkamp F: Laser desorption ionization of proteins with molecular masses exceeding 10,000 daltons, *Anal Chem* **60**:2299–2301 (1988).

[7] Fenn JB, Mann M, Meng CK, Wong SF, Whitehouse CM: Electrospray ionization for mass spectrometry of large biomolecules, *Science* **246**:64–71 (1989).

[8] Henzel WJ, Billeci TM, Stults JT, Wong SC, Grimley C, Watanabe C: Identifying proteins from two-dimensional gels by molecular mass searching of peptide fragments in protein sequence databases, *Proc Natl Acad Sci USA* **90**:5011–5015 (1993).

[9] Shevchenko A, Wilm M, Vorm O, Mann M: Mass spectrometric sequencing of proteins silver-stained polyacrylamide gels, *Anal Chem* **68**:850–858 (1996).

[10] Patterson SD, Aebersold R: Mass spectrometric approaches for the identification of gel-separated proteins, *Electrophoresis* **16**:1791–1814 (1995).

[11] Peng J, Gygi SP: Proteomics: The move to mixtures, *J Mass Spectrom* **36**:1083–1091 (2001).

[12] Hernandez P, Muller M, Appel RD: Automated protein identification by tandem mass spectrometry: Issues and strategies, *Mass Spectrom Rev* **25**:235–254 (2006).

[13] Domon B, Aebersold R: Mass spectrometry and protein analysis, *Science* **312**:212–217 (2006).

[14] Servais AC, Crommen J, Fillet M: Capillary electrophoresis-mass spectrometry, an attractive tool for drug bioanalysis and biomarker discovery, *Electrophoresis* **27**:2616–2629 (2006).

[15] Peters EC, Brock A, Horn DM, Phung QT, Ericson C, Salomon AR, Ficarro SB, Brill LM: An automated LC-MALDI FT-ICR MS platform for high-throughput proteomics, *LC GC Europe* **15**:423–428 (2002).

[16] Chen HS, Rejtar T, Andreev V, Moskovets E, Karger BL: High-speed, high-resolution monolithic capillary LC-MALDI MS using an off-line continuous deposition interface for proteomic analysis, *Anal Chem* **77**:2323–2331 (2005).

[17] Opiteck GJ, Lewis KC, Jorgenson JW, Anderegg RJ: Comprehensive on-line LC/LC/MS of proteins, *Anal Chem* **69**:1518–1524 (1997).

[18] Wall DB, Kachman MT, Gong S, Hinderer R, Parus S, Misek DE, Hanash SM, Lubman DM: Isoelectric focusing nonporous RP HPLC: A two-dimensional liquid-phase separation method for mapping of cellular proteins with identification using MALDI-TOF mass spectrometry, *Anal Chem* **72**:1099–1111 (2000).

[19] Link AJ, Eng J, Schieltz DM, Carmack E, Mize GJ, Morris DR, Garvik BM, Yates JR 3rd: Direct analysis of protein complexes using mass spectrometry, *Nat Biotechnol* **17**:676–682 (1999).

[20] Washburn MP, Wolters D, Yates JR 3rd: Large-scale analysis of the yeast proteome by multidimensional protein identification technology, *Nat Biotechnol* **19**:242–247 (2001).

[21] Blackler AR, Klammer AA, MacCoss MJ, Wu CC: Quantitative comparison of proteomic data quality between a 2D and 3D quadrupole ion trap, *Anal Chem* **78**: 1337–1344 (2006).

[22] Bradshaw RA, Burlingame AL, Carr S, Aebersold R: Reporting protein identification data: The next generation of guidelines, *Mol Cell Proteom* **5**:787–788 (2006).

[23] Syka JE, Marto JA, Bai DL, Horning S, Senko MW, Schwartz JC, Ueberheide B, Garcia B, Busby S, Muratore T, Shabanowitz J, Hunt DF: Novel linear quadrupole ion trap/FT mass spectrometer: performance characterization and use in the comparative analysis of histone H3 post-translational modifications, *J Proteome Res* **3**:621–626 (2004).

[24] Olsen JV, de Godoy LM, Li G, Macek B, Mortensen P, Pesch R, Makarov A, Lange O, Horning S, Mann M: Parts per million mass accuracy on an Orbitrap mass spectrometer via lock mass injection into a C-trap, *Mol Cell Proteom* **4**:2010–2021 (2005).

[25] Jessani N, Cravatt BF: The development and application of methods for activity-based protein profiling, *Curr Opin Chem Biol* **8**:54–59 (2004).

[26] Greenbaum D, Medzihradszky KF, Burlingame A, Bogyo M: Epoxide electrophiles as activity-dependent cysteine protease profiling and discovery tools, *Chem Biol* **7**:569–581 (2000).

[27] Lo LC, Pang TL, Kuo CH, Chiang YL, Wang HY, Lin JJ: Design and synthesis of class-selective activity probes for protein tyrosine phosphatases, *J Proteome Res* **1**:35–40 (2002).

[28] Saghatelian A, Jessani N, Joseph A, Humphrey M, Cravatt BF: Activity-based probes for the proteomic profiling of metalloproteases, *Proc Natl Acad Sci USA* **101**:10000–10005 (2004).

[29] Patricelli MP, Szardenings AK, Liyanage M, Nomanbhoy TK, Wu M, Weissig H, Aban A, Chun D, Tanner S, Kozarich JW: Functional interrogation of the kinome using nucleotide acyl phosphates, *Biochemistry* **46**:350–358 (2007).

[30] Foster LJ, de Hoog CL, Zhang Y, Xie X, Mootha VK, Mann M: A mammalian organelle map by protein correlation profiling, *Cell* **125**:187–199 (2006).

[31] Gavin AC, Bosche M, Krause R, Grandi P, Marzioch M, Bauer A, Schultz J, Rick JM, Michon AM, Cruciat CM, Remor M, Hofert C, Schelder M, Brajenovic M, Ruffner H, Merino A, Klein K, Hudak M, Dickson D, Rudi T, Gnau V, Bauch A, Bastuck S, Huhse B, Leutwein C et al: Functional organization of the yeast proteome by systematic analysis of protein complexes, *Nature* **415**:141–147 (2002).

[32] Willingham AT, Orth AP, Batalov S, Peters EC, Wen BG, Aza-Blanc P, Hogenesch JB, Schultz PG: A strategy for probing the function of noncoding RNAs finds a repressor of NFAT, *Science* **309**:1570–1573 (2005).

[33] Wissing J, Godl K, Brehmer D, Blencke S, Weber M, Habenberger P, Stein-Gerlach M, Missio A, Cotten M, Muller S, Daub H: Chemical proteomic analysis reveals alternative modes of action for pyrido[2,3-d]pyrimidine kinase inhibitors, *Mol Cell Proteom* **3**:1181–1193 (2004).

[34] Haystead TA: The purinome, a complex mix of drug and toxicity targets, *Curr Top Med Chem* **6**:1117–1127 (2006).

[35] Patton WF, Beechem JM: Rainbow's end: The quest for multiplexed fluorescence quantitative analysis in proteomics, *Curr Opin Chem Biol* **6**:63–69 (2002).

[36] Gerber SA, Rush J, Stemman O, Kirschner MW, Gygi SP: Absolute quantification of proteins and phosphoproteins from cell lysates by tandem MS, *Proc Natl Acad Sci USA* **100**:6940–6945 (2003).

[37] Kuhn E, Wu J, Karl J, Liao H, Zolg W, Guild B: Quantification of C-reactive protein in the serum of patients with rheumatoid arthritis using multiple reaction monitoring mass spectrometry and 13C-labeled peptide standards, *Proteomics* **4**:1175–1186 (2004).

[38] Liu H, Sadygov RG, Yates JR 3rd: A model for random sampling and estimation of relative protein abundance in shotgun proteomics, *Anal Chem* **76**:4193–4201 (2004).

[39] Lu P, Vogel C, Wang R, Yao X, Marcotte EM: Absolute protein expression profiling estimates the relative contributions of transcriptional and translational regulation, *Nat Biotechnol* **25**:117–1124 (2007).

[40] Chelius D, Bondarenko PV: Quantitative profiling of proteins in complex mixtures using liquid chromatography and mass spectrometry, *J Proteome Res* **1**:317–323 (2002).

[41] Wang W, Zhou H, Lin H, Roy S, Shaler TA, Hill LR, Norton S, Kumar P, Anderle M, Becker CH: Quantification of proteins and metabolites by mass spectrometry without isotopic labeling or spiked standards, *Anal Chem* **75**:4818–4826 (2003).

[42] Wiener MC, Sachs JR, Deyanova EG, Yates NA: Differential mass spectrometry: A label-free LC-MS method for finding significant differences in complex peptide and protein mixtures, *Anal Chem* **76**:6085–6096 (2004).

[43] Ahn NG, Shabb JB, Old WM, Resing KA: Achieving in-depth proteomics profiling by mass spectrometry, *ACS Chem Biol* **2**:39–52 (2007).

[44] Oda Y, Huang K, Cross FR, Cowburn D, Chait BT: Accurate quantitation of protein expression and site-specific phosphorylation, *Proc Natl Acad Sci USA* **96**:6591–6596 (1999).

[45] Wang YK, Ma Z, Quinn DF, Fu EW: Inverse 15N-metabolic labeling/mass spectrometry for comparative proteomics and rapid identification of protein markers/targets, *Rapid Commun Mass Spectrom* **16**:1389–1397 (2002).

[46] Wu CC, MacCoss MJ, Howell KE, Matthews DE, Yates JR 3rd: Metabolic labeling of mammalian organisms with stable isotopes for quantitative proteomic analysis, *Anal Chem* **76**:4951–4959 (2004).

[47] Mann M: Functional and quantitative proteomics using SILAC, *Nat Rev Mol Cell Biol* **7**:952–958 (2006).

[48] Ibarrola N, Molina H, Iwahori A, Pandey A: A novel proteomic approach for specific identification of tyrosine kinase substrates using [13C]tyrosine, *J Biol Chem* **279**:15805–15813 (2004).

[49] Kratchmarova I, Blagoev B, Haack-Sorensen M, Kassem M, Mann M: Mechanism of divergent growth factor effects in mesenchymal stem cell differentiation, *Science* **308**:1472–1477 (2005).

[50] Ong SE, Kratchmarova I, Mann M: Properties of 13C-substituted arginine in stable isotope labeling by amino acids in cell culture (SILAC), *J Proteome Res* **2**:173–181 (2003).

[51] Ren D, Julka S, Inerowicz HD, Regnier FE: Enrichment of cysteine-containing peptides from tryptic digests using a quaternary amine tag, *Anal Chem* **76**:4522–4530 (2004).

[52] Goodlett DR, Keller A, Watts JD, Newitt R, Yi EC, Purvine S, Eng JK, von Haller P, Aebersold R, Kolker E: Differential stable isotope labeling of peptides for quantitation and de novo sequence derivation, *Rapid Commun Mass Spectrom* **15**:1214–1221 (2001).

[53] Chakraborty A, Regnier FE: Global internal standard technology for comparative proteomics, *J Chromatogr A* **949**:173–184 (2002).

[54] Peters EC, Horn DM, Tully DC, Brock A: A novel multifunctional labeling reagent for enhanced protein characterization with mass spectrometry, *Rapid Commun Mass Spectrom* **15**:2387–2392 (2001).

[55] Matsuo E, Toda C, Watanabe M, Iida T, Masuda T, Minohata T, Ando E, Tsunasawa S, Nishimura O: Improved 2-nitrobenzenesulfenyl method: Optimization of the protocol and improved enrichment for labeled peptides, *Rapid Commun Mass Spectrom* **20**:31–38 (2006).

[56] Brown KJ, Fenselau C: Investigation of doxorubicin resistance in MCF-7 breast cancer cells using shot-gun comparative proteomics with proteolytic 18O labeling, *J Proteome Res* **3**:455–462 (2004).

[57] Xie J, Schultz PG: Adding amino acids to the genetic repertoire, *Curr Opin Chem Biol* **9**:548–554 (2005).

[58] Mirzaei H, Regnier F: Identification and quantification of protein carbonylation using light and heavy isotope labeled Girard's P reagent, *J Chromatogr A* **1134**:122–133 (2006).

[59] Qian WJ, Goshe MB, Camp DG 2nd, Yu LR, Tang K, Smith RD: Phosphoprotein isotope-coded solid-phase tag approach for enrichment and quantitative analysis of phosphopeptides from complex mixtures, *Anal Chem* **75**:5441–5450 (2003).

[60] Zhang R, Sioma CS, Wang S, Regnier FE: Fractionation of isotopically labeled peptides in quantitative proteomics, *Anal Chem* **73**:5142–5149 (2001).

[61] Gygi SP, Rist B, Gerber SA, Turecek F, Gelb MH, Aebersold R: Quantitative analysis of complex protein mixtures using isotope-coded affinity tags, *Nat Biotechnol* **17**:994–999 (1999).

[62] Li J, Steen H, Gygi SP: Protein profiling with cleavable isotope-coded affinity tag (cICAT) reagents: The yeast salinity stress response, *Mol Cell Proteom* **2**:1198–1204 (2003).

[63] Ross PL, Huang YN, Marchese JN, Williamson B, Parker K, Hattan S, Khainovski N, Pillai S, Dey S, Daniels S, Purkayastha S, Juhasz P, Martin S, Bartlet-Jones M, He F, Jacobson A, Pappin DJ: Multiplexed protein quantitation in Saccharomyces cerevisiae using amine-reactive isobaric tagging reagents, *Mol Cell Proteom* **3**:1154–1169 (2004).

[64] Salomon AR, Ficarro SB, Brill LM, Brinker A, Phung QT, Ericson C, Sauer K, Brock A, Horn DM, Schultz PG, Peters EC: Profiling of tyrosine phosphorylation pathways in human cells using mass spectrometry, *Proc Natl Acad Sci USA* **100**:443–448 (2003).

[65] Zhang Y, Wolf-Yadlin A, Ross PL, Pappin DJ, Rush J, Lauffenburger DA, White FM: Time-resolved mass spectrometry of tyrosine phosphorylation sites in the epidermal growth factor receptor signaling network reveals dynamic modules, *Mol Cell Proteom* **4**:1240–1250 (2005).

[66] Andersen JS, Lam YW, Leung AK, Ong SE, Lyon CE, Lamond AI, Mann M: Nucleolar proteome dynamics, *Nature* **433**:77–83 (2005).

[67] Pratt JM, Petty J, Riba-Garcia I, Robertson DH, Gaskell SJ, Oliver SG, Beynon RJ: Dynamics of protein turnover, a missing dimension in proteomics, *Mol Cell Proteom* **1**:579–591 (2002).

[68] Dieterich DC, Link AJ, Graumann J, Tirrell DA, Schuman EM: Selective identification of newly synthesized proteins in mammalian cells using bioorthogonal noncanonical amino acid tagging (BONCAT), *Proc Natl Acad Sci USA* **103**:9482–9487 (2006).

[69] Blagoev B, Kratchmarova I, Ong SE, Nielsen M, Foster LJ, Mann M: A proteomics strategy to elucidate functional protein-protein interactions applied to EGF signaling, *Nat Biotechnol* **21**:315–318 (2003).

[70] Tackett AJ, DeGrasse JA, Sekedat MD, Oeffinger M, Rout MP, Chait BT: I-DIRT, a general method for distinguishing between specific and nonspecific protein interactions, *J Proteome Res* **4**:1752–1756 (2005).

[71] Jensen ON: Interpreting the protein language using proteomics, *Nat Rev Mol Cell Biol* **7**:391–403 (2006).

[72] Holmberg CI, Tran SE, Eriksson JE, Sistonen L: Multisite phosphorylation provides sophisticated regulation of transcription factors, *Trends Biochem Sci* **27**:619–627 (2002).

[73] Khidekel N, Hsieh-Wilson LC: A "molecular switchboard"—covalent modifications to proteins and their impact on transcription, *Org Biomol Chem* **2**:1–7 (2004).

[74] Couture JF, Trievel RC: Histone-modifying enzymes: Encrypting an enigmatic epigenetic code, *Curr Opin Struct Biol* **16**:753–760 (2006).

[75] Ficarro SB, McCleland ML, Stukenberg PT, Burke DJ, Ross MM, Shabanowitz J, Hunt DF, White FM: Phosphoproteome analysis by mass spectrometry and its application to Saccharomyces cerevisiae, *Nat Biotechnol* **20**:301–305 (2002).

[76] Larsen MR, Thingholm TE, Jensen ON, Roepstorff P, Jorgensen TJ: Highly selective enrichment of phosphorylated peptides from peptide mixtures using titanium dioxide microcolumns, *Mol Cell Proteomics* **4**:873–886 (2005).

[77] Yang Z, Hancock WS: Approach to the comprehensive analysis of glycoproteins isolated from human serum using a multi-lectin affinity column, *J Chromatogr A* **1053**:79–88 (2004).

[78] Qiu R, Regnier FE: Comparative glycoproteomics of N-linked complex-type glycoforms containing sialic acid in human serum, *Anal Chem* **77**:7225–7231 (2005).

[79] Zhang H, Li XJ, Martin DB, Aebersold R: Identification and quantification of N-linked glycoproteins using hydrazide chemistry, stable isotope labeling and mass spectrometry, *Nat Biotechnol* **21**:660–666 (2003).

[80] Vosseller K, Hansen KC, Chalkley RJ, Trinidad JC, Wells L, Hart GW, Burlingame AL: Quantitative analysis of both protein expression and serine/threonine post-translational modifications through stable isotope labeling with dithiothreitol, *Proteomics* **5**:388–398 (2005).

[81] Hang HC, Yu C, Kato DL, Bertozzi CR: A metabolic labeling approach toward proteomic analysis of mucin-type O-linked glycosylation, *Proc Natl Acad Sci USA* **100**:14846–14851 (2003).

[82] Kho Y, Kim SC, Jiang C, Barma D, Kwon SW, Cheng J, Jaunbergs J, Weinbaum C, Tamanoi F, Falck J, Zhao Y: A tagging-via-substrate technology for detection and proteomics of farnesylated proteins, *Proc Natl Acad Sci USA* **101**:12479–12484 (2004).

[83] Kohn M, Breinbauer R: The Staudinger ligation—a gift to chemical biology, *Angew Chem Int Ed Engl* **43**:3106–3116 (2004).

[84] Khidekel N, Ficarro SB, Peters EC, Hsieh-Wilson LC: Exploring the O-GlcNAc proteome: Direct identification of O-GlcNAc-modified proteins from the brain, *Proc Natl Acad Sci USA* **101**:13132–13137 (2004).

[85] Kolb HC, Finn MG, Sharpless KB: Click chemistry: Diverse chemical function from a few good reactions, *Angew Chem Int Ed Engl* **40**:2004–2021 (2001).

[86] Duckworth BP, Xu J, Taton TA, Guo A, Distefano MD: Site-specific, covalent attachment of proteins to a solid surface, *Bioconj Chem* **17**:967–974 (2006).

[87] Peng J, Schwartz D, Elias JE, Thoreen CC, Cheng D, Marsischky G, Roelofs J, Finley D, Gygi SP: A proteomics approach to understanding protein ubiquitination, *Nat Biotechnol* **21**:921–926 (2003).

[88] Rosas-Acosta G, Russell WK, Deyrieux A, Russell DH, Wilson VG: A universal strategy for proteomic studies of SUMO and other ubiquitin-like modifiers, *Mol Cell Proteom* **4**:56–72 (2005).

[89] Brittain SM, Ficarro SB, Brock A, Peters EC: Enrichment and analysis of peptide subsets using fluorous affinity tags and mass spectrometry, *Nat Biotechnol* **23**:463–468 (2005).

[90] Carr SA, Huddleston MJ, Annan RS: Selective detection and sequencing of phosphopeptides at the femtomole level by mass spectrometry, *Anal Biochem* **239**:180–192 (1996).

[91] Steen H, Kuster B, Fernandez M, Pandey A, Mann M: Detection of tyrosine phosphorylated peptides by precursor ion scanning quadrupole TOF mass spectrometry in positive ion mode, *Anal Chem* **73**:1440–1448 (2001).

[92] Rappsilber J, Friesen WJ, Paushkin S, Dreyfuss G, Mann M: Detection of arginine dimethylated peptides by parallel precursor ion scanning mass spectrometry in positive ion mode, *Anal Chem* **75**:3107–3114 (2003).

[93] Tholey A, Reed J, Lehmann WD: Electrospray tandem mass spectrometric studies of phosphopeptides and phosphopeptide analogues, *J Mass Spectrom* **34**:117–123 (1999).

[94] Beausoleil SA, Jedrychowski M, Schwartz D, Elias JE, Villen J, Li J, Cohn MA, Cantley LC, Gygi SP: Large-scale characterization of HeLa cell nuclear phosphoproteins, *Proc Natl Acad Sci USA* **101**:12130–12135 (2004).

[95] Chang EJ, Archambault V, McLachlin DT, Krutchinsky AN, Chait BT: Analysis of protein phosphorylation by hypothesis-driven multiple-stage mass spectrometry, *Anal Chem* **76**:4472–4483 (2004).

[96] Unwin RD, Griffiths JR, Leverentz MK, Grallert A, Hagan IM, Whetton AD: Multiple reaction monitoring to identify sites of protein phosphorylation with high sensitivity, *Mol Cell Proteom* **4**:1134–1144 (2005).

[97] Kelleher NL: Top-down proteomics, *Anal Chem* **76**:196A–203A (2004).

[98] Zubarev RA, Kellerher NL, McLafferty FW: Electron capture dissociation of multiply charged protein cations. A nonergodic process, *J Am Chem Soc* **120**:3265–3266 (1998).

[99] Medzihradszky KF, Zhang X, Chalkley RJ, Guan S, McFarland MA, Chalmers MJ, Marshall AG, Diaz RL, Allis CD, Burlingame AL: Characterization of tetrahymena histone H2B variants and posttranslational populations by electron capture dissociation (ECD) Fourier transform ion cyclotron mass spectrometry (FT-ICR MS), *Mol Cell Proteom* **3**:872–886 (2004).

[100] Boyne MT 2nd, Pesavento JJ, Mizzen CA, Kelleher NL: Precise characterization of human histones in the H2A gene family by top down mass spectrometry, *J Proteome Res* **5**:248–253 (2006).

[101] Syka JE, Coon JJ, Schroeder MJ, Shabanowitz J, Hunt DF: Peptide and protein sequence analysis by electron transfer dissociation mass spectrometry, *Proc Natl Acad Sci USA* **101**:9528–9533 (2004).

[102] Chi A, Huttenhower C, Geer LY, Coon JJ, Syka JE, Bai DL, Shabanowitz J, Burke DJ, Troyanskaya OG, Hunt DF: Analysis of phosphorylation sites on proteins from Saccharomyces cerevisiae by electron transfer dissociation (ETD) mass spectrometry, *Proc Natl Acad Sci USA* **104**:2193–2198 (2007).

[103] Taverna SD, Ueberheide BM, Liu Y, Tackett AJ, Diaz RL, Shabanowitz J, Chait BT, Hunt DF, Allis CD: Long-distance combinatorial linkage between methylation and acetylation on histone H3 N termini, *Proc Natl Acad Sci USA* **104**:2086–2091 (2007).

[104] Rifai N, Gillette MA, Carr SA: Protein biomarker discovery and validation: The long and uncertain path to clinical utility, *Nat Biotechnol* **24**:971–983 (2006).

[105] Anderson NL, Anderson NG: The human plasma proteome: History, character, and diagnostic prospects, *Mol Cell Proteom* **1**:845–867 (2002).

[106] Anderson L, Hunter CL: Quantitative mass spectrometric multiple reaction monitoring assays for major plasma proteins, *Mol Cell Proteom* **5**:573–588 (2006).

[107] Chaurand P, Norris JL, Cornett DS, Mobley JA, Caprioli RM: New developments in profiling and imaging of proteins from tissue sections by MALDI mass spectrometry, *J Proteome Res* **5**:2889–2900 (2006).

[108] Valentine SJ, Plasencia MD, Liu X, Krishnan M, Naylor S, Udseth HR, Smith RD, Clemmer DE: Toward plasma proteome profiling with ion mobility-mass spectrometry, *J Proteome Res* **5**:2977–2984 (2006).

7

LARGE-SCALE GENOMIC ANALYSIS OF STEM CELL POPULATIONS

JONATHAN D. CHESNUT AND MAHENDRA S. RAO
Stem Cells and Regenerative Medicine, Invitrogen Corporation, Carlsbad, California

Introduction

Precisely defining the mechanisms of stem cell pluripotency, proliferation, and differentiation will be a challenge for researchers for years to come. Since we have not yet approached an understanding of the differences between stem cell line, nor have we grappled with the effects that various conditions have on stem cells in culture, we must use every relevant tool at our disposal to identify specific characteristics inherent in stem cell populations. This chapter discusses various methods of rapidly obtaining large amounts genomic and epigenetic data from stem cell populations and approaches to data mining for the gold nuggets that will aid in our understanding of the differences between various stem cell lines and help us identify key regulatory genes that are important in their biology and in development of adult tissue.

Cultured embryonic and tissue-specific stem cells hold great promise as research tools to further our understanding of tissue development and eventually as material for drug screening and therapy. This relatively young field is replete with rapid technological development of cell culture and analytical tools. Much has been learned about deriving and culturing embryonic stem cells. However, perhaps with the exception of bone marrow stem cells (and possibly stromal stem cells), very little was known about tissue-specific stem cells in adult tissue. Since the mid-1980s there has been an exponential increase in the number of manuscripts adult stem cell types, including skin, liver, pancreas, brain, lung, intestine, skeletal muscle, cardiac muscle, and cord

Chemical and Functional Genomic Approaches to Stem Cell Biology and Regenerative Medicine
Edited by Sheng Ding

blood [1]. Indeed, their promise has lead to a record number of publications describing the isolation of yet another population of magical cells, neural stem cells, which are present in relatively rare numbers that can be harvested using specialized methods and appear to be able to transform into all major phenotypes in the brain.

Much additional excitement has been generated by the isolation of human embryonic stem cells (hESCs) and by reports of transdifferentiation, cell fusion, somatic nuclear transfer, homologous recombination in somatic stem cells, and the idea that universal stem cells might exist. The rapid development of the field and the multiplicity of reports have lead to a great deal of confusion in the literature. Are there truly such a multiplicity of cells? Or has the same cell be isolated from different regions by slightly different methods?

Stem cells do not generate differentiated progeny directly but rather generate dividing populations of more restricted precursors analogous to the blast cells or restricted progenitors described in the hematopoietic lineages [2–4]. These precursors can divide and self-renew but are located in regions distinct from the stem cell population and can be distinguished from them by the expression of cell surface and cytoplasmic markers [5–7]. Investigators have begun efforts to analyze stem cell populations (Table 7.1) using a variety of techniques with the idea that by understanding these populations and identifying the factors that regulate self-renewal and direct differentiation, one will be able to modulate development or the response of stem cells to environmental signals. Many of these approaches depend on large-scale analytical tools that rely on the comparison of purified populations of cells that differ in either their stage of development or exposure to factors, or carry specific genetic abnormalities.

Part of the problem in classifying stem cell population has been the lack of a clear unambiguous definition of "stemness." This difficulty in arriving at a unified definition has led to a multiplicity of so-called stem cells that may or may not be

TABLE 7.1 Methods Used to Characterize Stem Cell Populations

Cells Characterized	Method Used	References
Neural stem cells and progenitor cells	Microarray (mRNA)	[75]
Neuroepithelial cells	Subtractive suppression hybridization	[76]
Astrocyte-restricted precursors	Immunohistochemistry	[6]
Neural stem cells	Microarray (mRNA)	[77][a]
SSEA1 positive cells	Microarray (mRNA)	[78]
CNS progenitors	Microarray (mRNA)	[79]
Neurosphere-forming cells	MPSS	[80]
hESC	MPSS	[81]
hESC	EST scan	[20]
hESC	SAGE	[82]
hESC	Microarray (mRNA, miRNA)	[14]
hESC, EB	Microarray (mRNA, miRNA)	[66]
MSC, adipocytes, chondrocytes	Microarray (mRNA, miRNA)	[68]

[a]Method due to Clive Svendsen.

distinct. The lack of common standards or definitions, the absence of a common forum for discussion, and the range in the ability to manipulate one's favorite system of stem cells has led a fragmented field bedeviled by controversy. As many scientists enter the stem cell field from different disciplines, they are forced to wrestle with this lack of unity and decide on their own what "stemness" really means.

One thinks of blastocyst-derived embryonic stem cells as the epitome of the stemness phenotype since their pluripotency allows them to give rise to all three embryonal lineages. This is in contrast to adult stem cells, which are termed *multipotent* and are lineage-restricted, essentially giving rise to only a few specific cell types. An example of an opportunity for more clear classification are the adult stem cells that have been isolated from bone marrow, cord blood, and adipose tissue, since, as mentioned above, it is not entirely clear whether these cells represent different cell types or the same cell that has been isolated from different niches. Rigorous classification of this family of cells should go a long way to put these questions to rest.

Another example are the five different stem cell populations found in the nervous system. Neural stem cells (NSCs) reside in the nervous system and give rise to all neurons and supporting cells such as ventricular zone (VZ) and subventricular zone (SVZ) stem cells as well as astrocyte and glial precursor cells [8,9] and appear to be homogenous. Further classification of this family of cells is underway.

For example, we use the term "tissue-specific stem cells" for cells that have a capacity to divide, self-renew, and differentiate into cells characteristic of that particular tissue. "Self-renewal" is taken to mean that when cells divide they give rise to daughter cells whose properties are similar to the cells themselves. Thus, if a parent cell is multipotent and has a particular range of properties, then the daughter cell has the identical range of characteristics (i.e., the cells have self-renewed). In addition, stem cells can differentiate into either one or many kinds of differentiated cells that are characteristic of the tissue from which they have been isolated. In general, prolonged self-renewal (for the lifetime of the organism) or self-renewal sufficient to regenerate the entire tissue or organ has not been required for defining a cell, making this definition far less rigorous.

The promise of stem cell research and understanding what mechanisms control pluripotency, self-renewal, and differentiation lies in gaining a greater knowledge and putting this information to work toward improving the human condition. The challenge will be to assemble a stem cell parts list that can be used to distinguish various classes of cells and to learn how those parts act in concert to regulate stem cell biology. This chapter discusses various methods of genomic and epigenetic analysis that can be used to refine our knowledge of these special cell types on our way to greater applications in basic research in developmental biology as well as their future use as drug discovery tools and therapeutic agents.

7.1 METHODS FOR GENOMIC ANALYSIS

Since 2000 there have been dramatic advances in technology and equally dramatic reduction in costs. Both genomic and proteomic methods are now available at costs that allow an average small laboratory to begin performing such experiments. Modern analytical tools can be used to generate large volumes of information on the status of various stem cell populations (Table 7.1). Using these tools, the amount of material required for such an analysis has also become much less than was previously necessary making it feasible even when the number of stem cells available is limited.

7.1.1 General Considerations

The general workflow for large scale genomic analysis of cell populations is described in Figure 7.1. The overall workflow can easily separated into five stages: sample derivation/isolation, sample verification, determination of analytical method, actual data analysis, and finally independent verification of the results using an orthogonal method.

Here we use terminally differentiated neuronal cells as an example, but cells can be pulled from any point of this or similar pathways for analysis. The key to this step is the isolation of homogeneous populations of cells whenever possible, as the presence of multiple cell types in a sample bound for analysis may confuse the result (and the researcher). Following this, the sample chosen must be validated to ensure that one knows what one has. The first step is simply to validate the RNA sample to be confident that it is pure and of a known quantity. Following this, one must ensure that the cell population from which the RNA was isolated expresses the markers known to identify the cells in question. Here, the stem cell researcher may have only a small set of markers with which to test the cells. These will consist of known transcripts that can be identified by PCR, membrane protein markers confirmed by immunohistochemistry (IHC), or by testing the cells for the proper differentiation potential in culture.

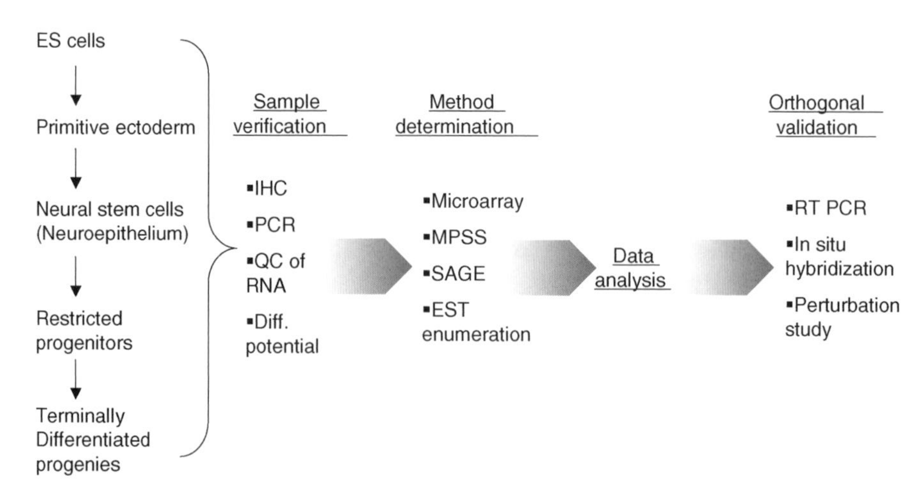

FIGURE 7.1 Flowchart of processes to characterize cell populations using large-scale analysis. Shown is an example of the workflow required for large-scale analysis of terminally differentiated neural cells.

Once the samples are ready for testing, the decision must be made as to which method of analysis is best. The following pages describe in detail various methods and their attributes for particular applications. Finally, the data one gathers from large-scale analysis must be validated using an orthogonal method. This step is often overlooked or minimized in importance but in our opinion is absolutely critical to confirming a biological role of a particular set of genes and furthering our understanding of the pathways in which they participate.

7.1.2 Measures of Identity

One method of classifying cell populations is to analyze subtle differences in their chromosomal DNA. Measures of short tandem repeats (STRs) using as few as nine known loci can be used to identify individual human genomes with the sensitivity of one in hundreds of thousands. This type of analysis allows us to define a genomic fingerprint of each cell line and can be used for early- and late-passage culture as variation in profile of these markers has not been detected with increasing passage number. Similar to this, histocompatibility profiles can be assembled by using a subset of the known HLA markers to type cells by the collection of antigens expressed on their membranes. This type of analysis is exceedingly important in classifying cells for use in the clinic. Mismatching of cells can lead to difficulties in interpretation of data and a failure to perform adequate or appropriate comparative analysis.

7.1.3 Measures of Stability

Having cells in culture and knowing their molecular and proteomic signature is only part of the challenge. Human embryonic stem cells are relatively unstable, and much care is taken to keep them in their pristine, pluripotent state. That said, care must be taken to ensure that genomic and epigenetic changes do not occur within the cells during propagation. A relatively easy and extremely high resolution tool for measuring genomic stability lies in single nucleotide polymorphism (SNP) analysis. Platforms are available that can easily measure nearly 100,000 SNPs in a single cell sample. As SNPs generally occur at approximately every 26 kb (i.e., 26,000 base pairs) in the genome, this analysis enables monitoring chromosomal changes at a relatively fine level. Data from these studies can indicate not only whether changes in chromosomal copy number have occurred but also translocations, deletions, and amplifications at high resolution. Another measure of cellular stability can be gained from SNP analysis of the entire mitochondrial genome. Here we can use established platforms to analyze changes in the nearly 17,000 bp over multiple passages.

7.1.4 Measures of Pluripotency

Various datasets have been compiled using mRNA expression arrays, MPSS, and EST analysis. When comparing global gene expression across hESC lines, correlation coefficients (R^2 values) for genes detected at high confidence generally range within the 0.92–0.94 level. When comparing karyotypically normal hESC lines with abnormal variants or embryonal carcinoma (EC) lines, correlation coefficients can range from 0.92 to as low as 0.88. Further, when comparing hESC embryoid bodies derived from them, the correlation coefficients can decrease to as low as 0.85. These results taken together suggest, not surprisingly, that cells that vary either by acquired abnormality, tissue of derivation, or state of differentiation, carry with them a changing repertoire of highly expressed genes. Of all genes analyzed so far, a set of 105 ES-specific genes have been identified. To this, a set of 194 genes that are upregulated on differentiation to EBs can be compared. As one would like for these profiles to be absolute, some variation is seen across cell lines. Of the 194 differentiation-specific genes identified in EBs, 168 can be detected in low but significant levels in hESCs [10].

All methods rely on having some tests of purity and quality of the sample, require some measure of quality of the methodology use, and require appropriate data-mining tools. One method that researchers in the field have employed to set guidelines is the adaptation of the standards [11], which "minimal information about a microarry experiment" (MIAME) define a set of rules to enable unambiguous interpretation of microarray results and allow design of experiments that can be reproduced faithfully. The checklist that has been developed to help authors and referees to ensure taht the data supporting published microarray results are made available in a format that allows unambiguous interpretation and potential verification. The standards listed cover experiment design, sample origin, extract preparation, and labeling, hybridization procedures, measurement data and specifications, and array design. Besides the MIAME guidelines, several other groups have come forward with frameworks and suggestions for large-scale data mining [12,13].

Use of a Reference Standard To gain the most usefulness from these types of experiments, one should be able to compare like data from multiple experiments. Ideally, one could compare datasets done on multiple days in multiple labs. This poses a potential problem when considering experimental variation between experiments. We have found that the use of an internal reference standard in each experiment allows easy normalization across datasets and enables us to combine data to create large sets, allowing us to fully utilize the power of the techniques described above. As shown in Figure 7.2, one simply has to prepare a large amount of one cell sample and include an aliquot of that same sample in all the experiments one wishes to include in the final dataset. By doing so, this allows us to compare the robustness of each experiment (signal intensity, etc.) and normalize away any other source of variation that may occur. An important point for the stem cell field will be to settle on a single cell type or a subset of cell types that will be an agreed on standard that can be used in various labs. For this, we have proposed the embryonal carcinoma cell line 2102EP as a useful standard [14].

7.2 GENERAL OBSERVATIONS ON THE PROPERTIES OF STEM CELLS

Before we evaluate the differences seen between embryonic stem cells, adult stem cells, and other cells, it is valuable to consider the traits they have in common. A broad

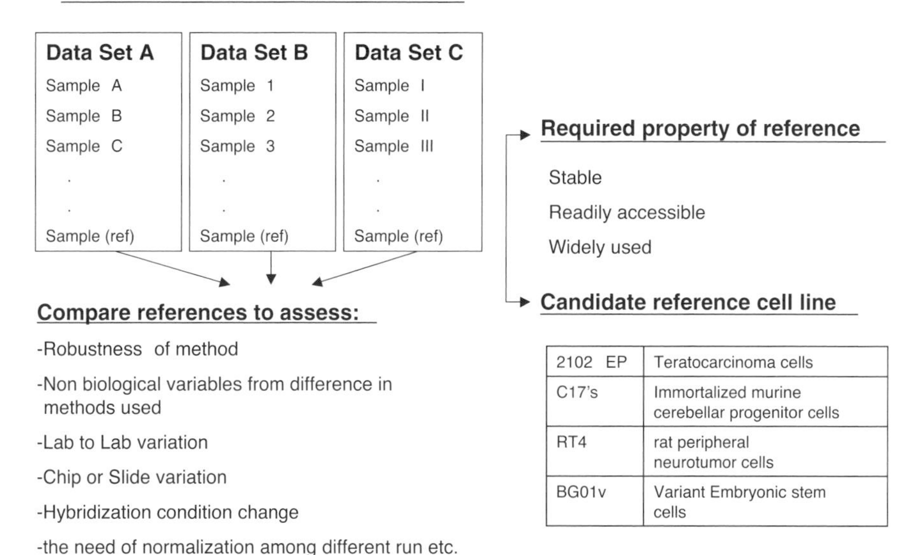

FIGURE 7.2 The use of a reference standard.

analysis of gene expression suggests that stem cells appear similar to other cells in that they synthesize about 10,000–12,000 genes of the approximately 35,000 or 40,000 genes annotated in the RefSeq database. Total RNA levels in stem cells appear similar to those seen in other metabolically active cells (average of 5–10 μg per million cells) but tends to be higher than seen in other somatic cells. While the overall abundance of total RNA is high in stem cells, the distribution of transcripts suggest that most genes are transcribed at relatively low levels of less than 50 transcripts per cell as assessed by MPSS. Transcription factors, growth factors, and other cell-type-specific molecules fall into this category of rare transcripts. These low levels are offset in the total analysis by mitochondrial, ribosomal, and housekeeping genes, which tend to be more abundant. This pattern of gene expression is similar to that seen in most other mitotically active cell types. More specifically, the large majority of expressed genes are shared across cell types with only approximately 20% of genes showing a >10-fold difference between any two cell samples. Mapping of all expressed genes does not show a chromosomal bias, as has been suggested in other stem cell populations [15]. An example from an array experiment performed in our laboratory comparing NSC and APC and OPC is shown in Table 7.2. In this study, we found that the average correlation between NSC cell populations is 0.90 (0.79–0.98) when two different lines grown under different conditions are compared [16]. The correlation is much better when identical samples are compared in two independent sequencing runs (0.99). A major pitfall that can be encountered when comparing datasets is exemplified when identical samples are compared between two methods: SAGE and MPSS or EST and

TABLE 7.2 Correlation (R^2 Score) among NSC, APC, and OPC Samples

Sample	$R^{2\,a}$
NSC/NSC	0.897
NSC/APC	0.649
NSC/OPC	0.750

[a] Average R^2 values of multiple NSC samples compared pairwise with each other or with two APC samples with two OPC samples.

Source: Adapted from Shin et al. [16].

MPSS. In these cases, we found correlation rates were much lower (0.7). A broad analysis of this type of comparison suggests that genes showing expression patterns that are common between two methods are likely to be important. However, the lack of a high correlation when identical samples are analyzed by different methodologies should be taken as a warning that caution must be used when assuming that genes that fall out of consideration, due to lack of agreement between methodologies, are not significant to the analysis.

7.2.1 Species Differences

Not surprisingly, now that multiple gene expression datasets are available for mouse and human ES cells, differences have been identified in their expression profiles. While many key pathways seem to be conserved between mice, humans, and other organisms, many differences have been highlighted as well. For example, Oct3/4 homologues likely do not exist in chicken embryos [17], while LIF signaling, which is critical for ES cell self-renewal in rodents, does not appear to be critical or even required for human ES cells [18]. The low overall correlation between human and rodent ES cells (in one comparison was around 40%) relative to that seen in human-to-human cell comparisons (90% between human ES cell samples) provides additional support for this hypothesis (Wei et al., personal communication).

7.2.2 Lack of a Stemness Phenotype

"Stemness" can be defined as a cell's ability to self-replicate as well as to beget cells representing some or all of the three embryonic lineages. There is considerable debate over the classification cells. Some suggest that the ability to self-renew over the entire lifetime of an organism and their role in substantial contribution to their tissue of origin should define stemness. In this case most, if not all, "stem" cells with the exception of hematopoetic stem cells should be considered progenitor cells. Neural crest stem cells (NCSCs), for example, are a transient population [19], and blastocysts from which embryonic stem (ES) cells are derived do not exhibit prolonged self-renewal throughout the life of the organism [20]. In solid tissues, contribution of stem cells may be regionally restricted such that no single stem cell contributes to a major portion of the organ [21]. We prefer a looser definition of stemness that doesn't restrict cells to lifetime self-renewal and substantial contribution to tissue development. Stem cells classified in this way have been isolated in essentially every tissue of the body. Even with (or especially because of) this more inclusive definition, it is clear that stem cells isolated from various tissues are not similar to each other in their behavior or their gene expression profiles. Further, no common "stemness" gene or pathway has been identified to date (see review by Cai et al. [5]). Comparing datasets with expression in NSC cells does not identify a common subset of genes and suggests that NSC cells are quite different from other somatic stem cells [22–24]. Our results comparing ESC-derived NSC with fetal derived NSC [16] and more limited comparisons [25] have shown clearly that stem cells can be readily distinguished from each other and that stem cells or progenitor cells harvested and different stages of development behave differently.

7.2.3 Allelic Variability

In addition to species and tissue of origin differences, what is clear is that while individual stem cell share properties such as the ability to self-renew in culture and to differentiate into various phenotypes, there will nevertheless be differences between individual cell lines based on the genetic profile of the individual from which they were derived. These allelic differences are not unexpected and merely reflect the diversity of phenotypes seen in the human population. Perhaps what has not been appreciated as much is how much variability this might impose on stem cell behavior, even if care is taken to isolate from the same region at the same developmental time using methods as similar as possible.

When global pairwise comparisons are made between multiple stem cell lines using the same platform, one sees a high reproducibility when the same sample is run repeatedly (correlation coefficients in the range of 0.98–0.99). However, when two samples isolated by the same group at around the same time are run, the differences are much larger (correlation coefficient 0.92–0.94), indicating that a significant number of genes are differentially expressed between two stem cell lines. This sort of difference is consistently seen in all stem cell populations and is generally less than the difference between a cell line and its differentiated progeny and is also less than the difference between a cell line from one species as compared to another [26]. Nevertheless, at the individual gene level such changes can be quite dramatic as is shown in Figure 7.3 [26]. Such allelic variability likely underlies the differences in propagation, self-renewal, and the ability to differentiate into specific phenotypes that have been reported.

7.2.4 Stem Cells Age and Like All Cells, Change When Propagated in Culture

Examination of human cells in culture has shed some light on the cellular basis of aging. When grown in culture, normal human cells will undergo a limited number of divisions before entering a state of replicative senescence in which they remain viable but are unable to divide further [27]. This change has been variously attributed to

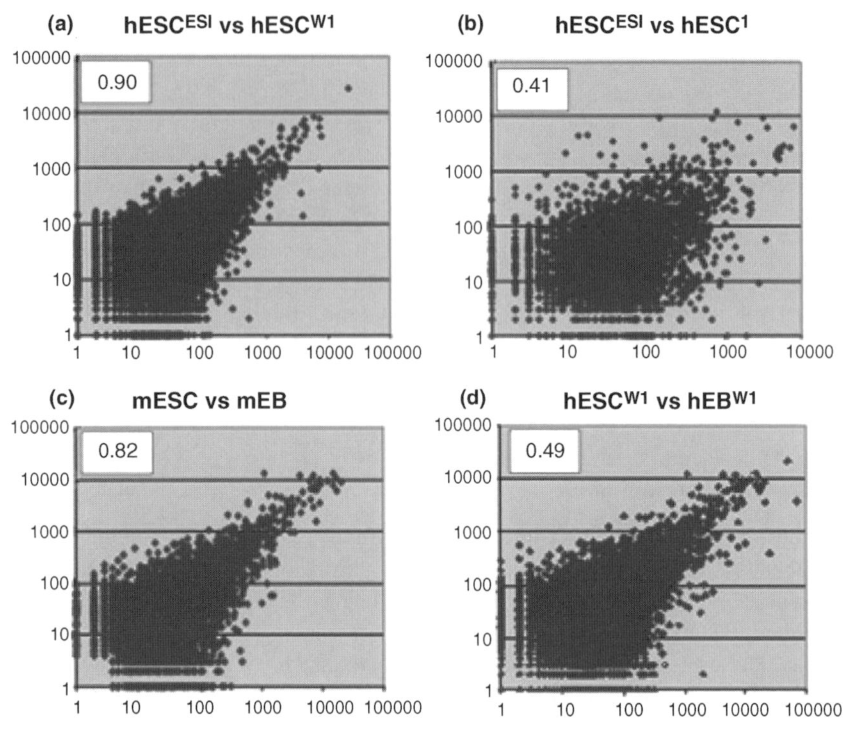

FIGURE 7.3 Gene expression comparisons multiple cell lines. (Adapted from Miura et al. [80].) Scatterplots of gene expression array analysis are shown with the correlation coefficient (R^2) values displayed in the upper left corner of each plot. (a) hESC versus hESC, two different cell lines; (b) human versus mouse ESCs; (c) mouse ESC versus embryoid bodies differentiated therefrom; (d) human ESC versus embryoid bodies differentiated therefrom.

variations in mitochondria, protein changes, loss of genomic integrity, oxidative damage, and progressive loss of DNA repair ability or erosion of telomere ends. Results from our laboratory have shown that ESCs appear to be immortal and can be propagated in continuous culture for a period of at least 2 years. Examination of the expression of immortality-associated genes has suggested that expression of key regulators is important [27]. These include the expression of telomere-associated proteins, expression of some immortality associated genes, high levels of DNA repair enzymes, and the absence of p53 and rb, thus altering cell cycle regulation. Comparing the phenotype of such genes between NSC and ESC shows significant differences. NSC express both p53 and rb and respond to injury by apoptosis, and their patterns of expression of telomere-related genes and immortality-associated genes differ [27]. This appears true for other adult stem cell populations as well [28], suggesting that NSC will not be easily propagated indefinitely in culture. Indeed, more recent results have called this into question. Whittemore and colleagues showed that long-term propagation in culture altered their differentiation ability, while Bailey and colleagues have shown that NSC undergo karyotypic changes as the animal ages providing further confirmation of the hypothesis [29,30]. Our more recent results examining the karyotypic stability of NSC in culture have indicated that NSC are less stable in culture than ESC and can acquire karyotypic changes in as few as 10 passages. Overall large-scale analysis suggests that there are fundamental differences in cell cycle and immortality-associated genes, and these differences suggest that NSC age in culture and will senesce as will other adult stem cell populations.

7.2.5 Cancer Stem Cells Can Be Identified and Appear to Be Related to Tissue-Specific Stem Cell Populations

An exciting finding has been the discovery that many cancers may be propagated by a small number of stem cells present in the tumor mass. This was first described in breast cancers and subsequently in a variety of solid tumors. In the nervous system as well, several reports have suggested that cancer stem cells can be identified and that these cells bear a remarkable similarity to neural stem cells present in early development [31,32]. Likewise, cells resembling glial progenitors have been isolated from some glial tumors [33–35]. We and others have suggested that tumors can perhaps be classified with some thought to the cell of origin on the basis of findings of the kind of lineage relationships that exist between stem and progenitor cells (Figure 7.4) [15]. The advantage of having a detailed database of gene expression from multiple normal cell lines to compare with their transformed counterparts is obvious. One could identify core conserved pathways and key differences and use transforming regimes with normal stem cell populations to develop tools to probe dysregulated pathways. These experiments are not technically difficult and depend on developing an adequate database for mining. Efforts along this path have already been initiated, and several groups have begun to report on the properties of undifferentiated NSC and also to isolate cancer stem cells from appropriate tumors. We anticipate that important insights will be gained by such an analysis.

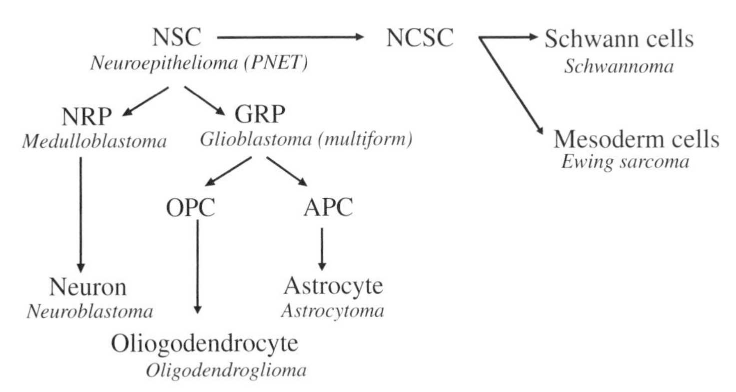

FIGURE 7.4 Stem cell lineage relationships. Stem cell lineages as they relate to known tumor types. (Adapted from Shin et al. [15].)

7.2.6 miRNA Profiling Can Be Used to Type Cells

MicroRNA are short, 20–24-nucleotide noncoding RNAs that have been identified from various sources, including mammalian cells. The sequence of most miRNA are highly conserved across species, with nearly 90% of the currently sequenced human miRNAs identical to mouse and rat and at least 30% homologous to miRNA from Caenorhabditis elegans [36]. miRNAs are thought to negatively regulate gene expression through either direct mRNA cleavage [37–41], mRNA decay by miRNA deadenylation [42,43], or translational repression [44]. To complicate specific mapping of miRNA-binding sites in the transcriptome, it has been determined that in animal cells, at least, translational repression also occurs by annealing of miRNA to mRNA at sites that are imperfectly complementary. Owing to this complexity and lack of a clear understanding of the exact mechanism of miRNA function, identification of target genes regulated by miRNA has been difficult [45]. Nevertheless, the importance of miRNA in several biological processes such as cell growth and apoptosis [46], viral infection [47,48], and human cancer [49–52] is well documented. Several studies have indicated that miRNA might regulate gene expression in more than 30% of the protein-coding genes in humans [53]. The role of miRNA-mediated regulation of stem cell division [54], adipocyte [55], cardiac [56], neural [45,57], and hematopoeitic lineage differentiation [58,59] is well known. More recently, a unique set of miRNA has been shown to be associated with mouse ESC and embryoid body (EB) formation [60–63]. Using Northern blot analysis and cloning, researchers have: identified several miRNAs in human embryonic stem cells, some of which were identical to those previously reported in mouse ES cells [64]. Most of the miRNA identified in this study that were conserved between human and mouse were specific to embryonic stem cells, thereby further suggesting their importance as key regulators.

Analytical tools for gene expression analysis have been available for some time and are widely used in the field. These include computation analysis using sophisticated algorithms to recognize potential microRNA-coding sequences and potential binding sites. Other strategies have including making microRNA chips [65–68]. Lynx Therapeutics has developed sequencing protocols analogous to MPSS to obtain quantitative data on microRNAs made by a particular cell.

Tools for systematic analysis of epigenetic changes in cells have become available, opening the door for broad-scale analysis of another level of transcriptional and transnational regulation. We have used one of these toolsets to analyze microRNA (miRNA) profiles of various hESC cell lines and differentiated cells derived from them [66–67]. We found some variation in miRNA profile between ESC lines but also identified which ones can be correlated with the pluripotent state. Furthermore, as ESCs differentiate to embryoid bodies (EBs), the miRNA profiles change significantly, revealing several differentiation-specific marker candidates.

7.2.7 Epigenetic Profiles May Distinguish Stem Cells from Each Other

The importance of heritable epigenetic remodeling has been highlighted in regulating stem cell proliferation, cell fate determination, and carcinogenesis [69–73]. It has been difficult, however, to study these events on a global scale. The ability to grow a large number of cells and to differentiate them along specific pathways, coupled with the ability to perform such studies in a high-throughput fashion, suggests that this will change in the near future. Global methylation studies can be performed using methylation-specific microarrays [74]. Illumina has described a bead array strategy to study methylation patterns at 1500 loci encompassing regulatory elements in almost 400 genes. These include most genes known to be regulated during early embryonic development and those altered in tumorigenesis. These arrays have been used to examine methylation profiles in cancer stem cells and embryonic stem cells, and experiments assessing NSC are underway.

7.3 COMPILING COMBINED INFORMATION ON CELLS PROVIDES ADDITIONAL INSIGHTS

We are at a point where the gathering of information about stem cells has outpaced our ability to clearly understand the subtleties of how it all fits together. We are generating volumes of gene expression, epigenetic, and proteomic data about cell populations that we are classifying in a sometimes ad hoc manner. This plethora of information presents us with a great opportunity and also a great challenge. We must be able to link the information that results from various levels of regulation in the cell to tell a meaningful story that describes the cell's ability to regulate pluripotency, self-renewal, and differentiation.

First, we can assess the number of genes expressed by their mRNA levels in the cell and correlate the differentially regulated transcripts to specific signaling pathways. Next we can look at how translation of these messages is regulated by small RNAs and potentially other players in the translational machinery. In the end, it will be the final expression levels of active protein and where and with whom they appear in the cell that will tell the tale. Advanced proteomic tools such as SILAC an other mass spectrometry platforms will soon become a critical tool in assaying the cell proteome, giving us yet another level of information for which we must attempt to make sense of the results of the genomic and epigenetic regulation discussed here. These tools are under development and, in many ways, await advances in other ancillary fields before they can be broadly applicable for broad analysis of cellular proteomes.

We can use the data gleaned from the studies described above in several ways. First, without a priori knowledge of the function of molecules that appear to be differentially expressed or chromosomal sites that appear to be differentially modified, we can create short lists of measurable molecules with which to classify stem cells by their particular profiles. This type of approach will be very useful in defining relatedness of specific cell lines to each other and to tissues found in the body. Then, as we get better at linking our datasets in meaningful ways, the molecules that we identify can be linked to specific pathways and confirmed biochemically. In the end, development of advanced bioinformatic and data-mining algorithms will be key to narrowing the field of putative regulatory molecules that are involved in a particular event to allow us to finally perform the biochemical studies necessary for confirmation of the molecular mechanisms involved in various processes.

As we approach these types of studies, the need for novel methods to manipulate stem cells presents itself. Currently there are many challenges in efficiently introducing molecular tools into stem and progenitor cells. For instance, once a particular signaling pathway has been identified for analysis, it is important to do the perturbation experiments to prove the role of the putative regulator in question. Gene transfer to human ESCs, for example, is plagued by poor efficiency and the difficulty in generating clonal populations of stably transfected cells. The use of viral transduction has mitigated this somewhat, allowing efficient and stable delivery of molecular tools such as gene overexpression cassettes, RNAi gene knockdown elements, and lineage reporters. This approach is useful but not ideal as there is little control of the overall copy number of the elements in the engineered cells or where in the genome these elements are inserted. Engineering adult stem and progenitor cells presents another level of difficulty as these populations are limited in their proliferative ability and thus leave a relatively narrow window in which the molecular biologist can work to create engineered cell populations. More recent work in artificial cell immortalization represents a partial solution to this problem, but it is becoming evident that exogenously immortalized adult stem cells acquire variable behaviors in culture that have yet to be defined. Solving these issues will allow controllable and potentially higher-throughput study of multiple putative regulators in various cell types.

Realization of efficient stem cell engineering platforms should allow not only more concise study of regulatory pathways but also, ideally, the ability to use that information to create systems that can be used for drug discovery testing and eventually efficient, directed differentiation of stem cells to primary tissues.

REFERENCES

[1] Lakshmipathy U: Verfaillie C: Stem cell plasticity, *Blood Rev* **19**:29–38 (2005).

[2] Bedi A, Sharkis SJ: Mechanisms of cell commitment in myeloid cell differentiation, *Curr Opin Hematol* **2**:12–21 (1995).

[3] Katsura Y, Kawamoto H: Stepwise lineage restriction of progenitors in lympho-myelopoiesis, *Int Rev Immunol* **20**:1–20 (2001).

[4] Mujtaba T, Piper DR, Kalyani A et al: Lineage-restricted neural precursors can be isolated from both the mouse neural tube and cultured ES cells, *Dev Biol* **214**:113–127 (1999).

[5] Cai J, Weiss ML, Rao MS: In search of "stemness,"*Exp Hematol* **32**:585–598 (2004).

[6] Liu Y, Han SS, Wu Y et al: CD44 expression identifies astrocyte-restricted precursor cells, *Dev Biol* **276**:31–46 (2004).

[7] Kalyani A, Hobson K, Rao MS: Neuroepithelial stem cells from the embryonic spinal cord: Isolation, characterization, and clonal analysis, *Dev Biol* **186**:202–223 (1997).

[8] Rao M: Stem and precursor cells in the nervous system, *J Neurotrauma* **21**:415–427 (2004).

[9] Schubert W, Coskun V, Tahmina M et al: Characterization and distribution of a new cell surface marker of neuronal precursors, *Dev Neurosci* **22**:154–166 (2000).

[10] Liu Y, Shin S, Zeng X et al: Genome wide profiling of hESC, EC and their progenies using bead arrays to develop a base profile of gene expression of pre-August 9th human ES cell lines, *BMC Dev Biology* (2006).

[11] `http://www.mged.org/Workgroups/MIAME/miame.html.`

[12] Verducci JS, Melfi VF, Lin S et al: Microarray analysis of gene expression: considerations in data mining and statistical treatment, *Physiol Genom* **25**:355–363 (2006).

[13] Handl J, Knowles J, Kell DB: Computational cluster validation in post-genomic data analysis, *Bioinformatics* **21**:3201–3212 (2005).

[14] Josephson R, Ording CJ, Liu Y, Shin S, Lakshmipathy U, Toumadje A, Love B, Chesnut JD, Andrews PW, Rao MS, Auerbach JM: Qualification of embryonal carcinoma 2102Ep as a reference for human embryonic stem cell research, *Stem Cells* **25**:437–446 (2007).

[15] Shin S, Chesnut JD, Rao MS: Assessing neural stem cell properties using large scale genomic analysis, *Principles of Developmental Genetics*, Elsevier (2006).

[16] Shin S, Sun Y, Liu Y, Khaner H, Davidson B, Xu X, Reubinoff B, Stice S, Goldman S, Cai J, Zhan M, Rao M, Chesnut J: Whole genome analysis of human neural stem cells derived from embryonic stem cells and isolated from fetal tissues, *Stem Cells* (Feb. 2007).

[17] Soodeen-Karamath S, Gibbins AM: Apparent absence of oct 3/4 from the chicken genome, *Mol Reprod Dev* **58**:137–148 (2001).

[18] Ginis I, Luo Y, Miura T et al: Differences between human and mouse embryonic stem cells, *Dev Biol* **269**:360–380 (2004).

[19] Murphy M, Bartlett PF: Molecular regulation of neural crest development, *Mol Neurobiol* **7**:111–135 (1993).

[20] Brandenberger R, Wei H, Zhang S et al: Transcriptome characterization elucidates signaling networks that control human ES cell growth and differentiation, *Nat Biotechnol* **22**:707–716 (2004).

[21] Song H, Stevens CF, Gage FH: Astroglia induce neurogenesis from adult neural stem cells, *Nature* **417**:39–44 (2002).

[22] Ramalho-Santos M, Yoon S, Matsuzaki Y et al: "Stemness:" Transcriptional profiling of embryonic and adult stem cells, *Science* **298**:597–600 (2002).

[23] Ivanova NB, Dimos JT, Schaniel C et al: A stem cell molecular signature, *Science* **298**:601–604 (2002).

[24] Fortunel NO, Otu HH, Ng HH et al: Comment on "'Stemness:' Transcriptional profiling of embryonic and adult stem cells" and "a stem cell molecular signature," *Science* **302**:393 (2003).

[25] Pevny L, Rao MS: The stem-cell menagerie, *Trends Neurosci* **26**:351–359 (2003).

[26] Wei CL, Miura T, Robson P et al: Transcriptome profiling of human and murine ESCs identifies divergent paths required to maintain the stem cell state, *Stem Cells* **23**:166–185 (2005).

[27] Miura T, Mattson MP, Rao MS: Cellular lifespan and senescence signaling in embryonic stem cells, *Aging Cell* **3**:333–343 (2004).

[28] Brazel CY, Limke TL, Osborne JK et al: Sox2 expression defines a heterogeneous population of neurosphere-forming cells in the adult murine brain, *Aging Cell* **4**:197–207 (2005).

[29] Whittemore SR, Snyder EY: Physiological relevance and functional potential of central nervous system-derived cell lines, *Mol Neurobiol* **12**:13–38 (1996).

[30] Bailey KJ, Maslov AY, Pruitt SC: Accumulation of mutations and somatic selection in aging neural stem/progenitor cells, *Aging Cell* **3**:391–397 (2004).

[31] Singh SK, Clarke ID, Hide T et al: Cancer stem cells in nervous system tumors, *Oncogene* **23**:7267–7273 (2004).

[32] Vescovi AL, Galli R, Reynolds BA: Brain tumour stem cells, *Nat Rev Cancer* **6**:425–436 (2006).

[33] Assanah M, Lochhead R, Ogden A et al: Glial progenitors in adult white matter are driven to form malignant gliomas by platelet-derived growth factor-expressing retroviruses, *J Neurosci* **26**:6781–6790 (2006).

[34] Colin C, Baeza N, Tong S et al: *In vitro* identification and functional characterization of glial precursor cells in human gliomas, *Neuropathol Appl Neurobiol* **32**:189–202 (2006).

[35] Noble M, Gutowski N, Bevan K et al: From rodent glial precursor cell to human glial neoplasia in the oligodendrocyte-type-2 astrocyte lineage, *Glia* **15**:222–230 (1995).

[36] Kim VN, Nam JW: Genomics of microRNA, *Trends Genet* **22**:165–173 (2006).

[37] Bagga S, Bracht J, Hunter S et al: Regulation by let-7 and lin-4 miRNAs results in target mRNA degradation, *Cell* **122**:553–563 (2005).

[38] Yu B, Yang Z, Li J et al: Methylation as a crucial step in plant microRNA biogenesis, *Science* **307**:932–935 (2005).

[39] Yekta S, Shih IH, Bartel DP: MicroRNA-directed cleavage of HOXB8 mRNA, *Science* **304**:594–596 (2004).

[40] Chen X: A microRNA as a translational repressor of APETALA2 in Arabidopsis flower development, *Science* **303**:2022–2025 (2004).

[41] Aukerman MJ, Sakai H: Regulation of flowering time and floral organ identity by a MicroRNA and its APETALA2-like target genes, *Plant Cell* **15**:2730–2741 (2003).

[42] Giraldez AJ, Mishima Y, Rihel J et al: Zebrafish MiR-430 promotes deadenylation and clearance of maternal mRNAs, *Science* **312**:75–79 (2006).

[43] Wu L, Fan J, Belasco JG: MicroRNAs direct rapid deadenylation of mRNA, *Proc Natl Acad Sci USA* **103**:4034–4039 (2006).

[44] Zhang B, Pan X, Anderson TA, MicroRNA: A new player in stem cells, *J Cell Physiol* **209**:266–269 (2006).

[45] Wu L, Belasco JG: Micro-RNA regulation of the mammalian lin-28 gene during neuronal differentiation of embryonal carcinoma cells, *Mol Cell Biol* **25**:9198–9208 (2005).

[46] Cheng AM, Byrom MW, Shelton J et al: Antisense inhibition of human miRNAs and indications for an involvement of miRNA in cell growth and apoptosis, *Nucl Acids Res* **33**:1290–1297 (2005).

[47] Sullivan CS, Ganem D: MicroRNAs and viral infection, *Mol Cell* **20**:3–7 (2005).

[48] Sullivan CS, Grundhoff AT, Tevethia S et al: SV40-encoded microRNAs regulate viral gene expression and reduce susceptibility to cytotoxic T cells, *Nature* **435**:682–686 (2005).

[49] Calin GA, Liu CG, Sevignani C et al: MicroRNA profiling reveals distinct signatures in B cell chronic lymphocytic leukemias, *Proc Natl Acad Sci USA* **101**:11755–11760 (2004).

[50] Calin GA, Sevignani C, Dumitru CD et al: Human microRNA genes are frequently located at fragile sites and genomic regions involved in cancers, *Proc Natl Acad Sci USA* **101**:2999–3004 (2004).

[51] Calin GA, Dumitru CD, Shimizu M et al: Frequent deletions and down-regulation of micro- RNA genes miR15 and miR16 at 13q14 in chronic lymphocytic leukemia, *Proc Natl Acad Sci USA* **99**:15524–15529 (2002).

[52] He L, Thomson JM, Hemann MT et al: A microRNA polycistron as a potential human oncogene, *Nature* **435**:828–833 (2005).

[53] Berezikov E, Guryev V, van de Belt J et al: Phylogenetic shadowing and computational identification of human microRNA genes, *Cell* **120**:21–24 (2005).

[54] Hatfield SD, Shcherbata HR, Fischer KA et al: Stem cell division is regulated by the microRNA pathway, *Nature* **435**:974–978 (2005).

[55] Esau C, Kang X, Peralta E et al: MicroRNA-143 regulates adipocyte differentiation, *J Biol Chem* **279**:52361–52365 (2004).

[56] Zhao Y, Samal E, Srivastava D: Serum response factor regulates a muscle-specific microRNA that targets Hand2 during cardiogenesis, *Nature* **436**:214–220 (2005).

[57] Kuwabara T, Hsieh J, Nakashima K et al: A small modulatory dsRNA specifies the fate of adult neural stem cells, *Cell* **116**:779–793 (2004).

[58] Chen CZ, Li L, Lodish HF et al: MicroRNAs modulate hematopoietic lineage differentiation, *Science* **303**:83–86 (2004).

[59] Chen CZ, Lodish HF: MicroRNAs as regulators of mammalian hematopoiesis, *Semin Immunol* **17**:155–165 (2005).

[60] Yang Z, Wu J: Small RNAs and development, *Med Sci Monit* **12**:RA125–RA129 (2006).

[61] Houbaviy HB, Murray MF, Sharp PA: Embryonic stem cell-specific MicroRNAs, *Dev Cell* **5**:351–358 (2003).

[62] Kanellopoulou C, Muljo SA, Kung AL et al: Dicer-deficient mouse embryonic stem cells are defective in differentiation and centromeric silencing, *Genes Dev* **19**:489–501 (2005).

[63] Tang F, Hajkova P, Barton SC et al: MicroRNA expression profiling of single whole embryonic stem cells, *Nucl Acids Res* **34**:e9(2006).

[64] Suh MR, Lee Y, Kim JY et al: Human embryonic stem cells express a unique set of microRNAs, *Dev Biol* **270**:488–498 (2004).

[65] Krichevsky AM, King KS, Donahue CP et al: A microRNA array reveals extensive regulation of microRNAs during brain development, *Rna* **9**:1274–1281 (2003).

[66] Lakshmipathy U, Love B, Graichen R, Liu Y, Shin S, Maitra A, Rao M, Chesnut J: Micro RNA profiling as a sensitive method to measure differences between cell populations: Comparison of human embryonic stem cells with embryoid bodies, *Methods in Molecular Biology*, Humana Press (2007).

[67] Lakshmipathy U, Love B, Goff LA, Jörnsten R, Graichen R, Hart RP, Chestnut JD: MicroRNA expression pattern of undifferentiated and differentiated human embryonic stem cells, *Stem Cells Dev* **16**:1–14 (2007).

[68] Lakshmipathy U, Love B, Adams C, Rao M, Chesnut J: miRNA profile of human mesenchymal stem cells and their differentiated progeny, in preparation (2007).

[69] Ohgane J, Wakayama T, Senda S et al: The Sall3 locus is an epigenetic hotspot of aberrant DNA methylation associated with placentomegaly of cloned mice, *Genes Cells* **9**:253–260 (2004).

[70] Beaujean N, Taylor J, Gardner J et al: Effect of limited DNA methylation reprogramming in the normal sheep embryo on somatic cell nuclear transfer, *Biol Reprod* **71**:185–193 (2004).

[71] Vignon X, Zhou Q, Renard JP: Chromatin as a regulative architecture of the early developmental functions of mammalian embryos after fertilization or nuclear transfer, *Clon Stem Cells* **4**:363–377 (2002).

[72] Meehan RR: DNA methylation in animal development, *Semin Cell Dev Biol* **14**:53–65 (2003).

[73] Huntriss J, Hinkins M, Oliver B et al: Expression of mRNAs for DNA methyltransferases and methyl-CpG-binding proteins in the human female germ line, preimplantation embryos, and embryonic stem cells, *Mol Reprod Dev* **67**:323–336 (2004).

[74] Maitra A, Arking DE, Shivapurkar N et al: Genomic alterations in cultured human embryonic stem cells, *Nat Genet* **37**:1099–1103 (2005).

[75] Luo Y, Cai J, Liu Y et al: Microarray analysis of selected genes in neural stem and progenitor cells, *J Neurochem* **83**:1481–1497 (2002).

[76] Cai J, Xue H, Zhan M et al: Characterization of progenitor-cell-specific genes identified by subtractive suppression hybridization, *Dev Neurosci* **26**:131–147 (2004).

[77] Wright LS, Li J, Caldwell MA et al: Gene expression in human neural stem cells: Effects of leukemia inhibitory factor, *J Neurochem* **86**:179–195 (2003).

[78] Abramova N, Charniga C, Goderie SK et al: Stage-specific changes in gene expression in acutely isolated mouse CNS progenitor cells, *Dev Biol* **283**:269–281 (2005).

[79] Geschwind DH, Ou J, Easterday MC et al: A genetic analysis of neural progenitor differentiation, *Neuron* **29**:325–339 (2001).

[80] Cai J, Shin S, Wright L et al: Massively parallel signature sequencing profiling of fetal human neural precursor cells, *Stem Cells Dev* **15**:232–244 (2006).

[81] Miura T, Luo Y, Khrebtukova I et al: Monitoring early differentiation events in human embryonic stem cells by massively parallel signature sequencing and expressed sequence tag scan, *Stem Cells Dev* **13**:694–715 (2004).

[82] Richards M, Tan SP, Chan WK et al: Reverse SAGE characterization of orphan SAGE tags from human embryonic stem cells identifies the presence of novel transcripts and antisense transcription of key pluripotency genes, *Stem Cells* **24**(5):1162–1173 (2006).

8

EXPLORING STEM CELL BIOLOGY WITH SMALL MOLECULES AND FUNCTIONAL GENOMICS

JULIE CLARK, YUE XU, SIMON HILCOVE, AND SHENG DING

Department of Chemistry and the Skaggs Institute for Chemical Biology, The Scripps Research Institute, La Jolla, California

Introduction

Stem cells hold great promise for the treatment of many devastating diseases and will also provide new insights into the molecular mechanisms that control developmental and regenerative processes. While stem cells can be expanded and differentiated into different cell types *in vitro*, realization of the therapeutic potential of stem cells will require a better understanding of the signaling pathways that control stem cell fate as well as an improved ability to manipulate stem cell proliferation, differentiation, and reprogramming. In order to gain molecular insights into stem cell biology various chemical and functional genomic approaches have been used to identify and characterize small molecules and genes that can regulate stem cell fate. Such chemical and functional genomic tools will not only provide new insights into stem cell biology but will also facilitate the development of therapeutic agents for regenerative medicine.

Given appropriate conditions, stem cells can self-renew for long periods of time while maintaining the ability to differentiate into various functional cell types in the body [1]. It is these characteristics that not only make stem cells a useful system in

Chemical and Functional Genomic Approaches to Stem Cell Biology and Regenerative Medicine
Edited by Sheng Ding

which to study tissue and organ development but also give them great potential for regenerative medicine. Given the success of well practiced cell-based therapies (e.g., hematopoietic stem cell transplantation for treating hematological diseases and pancreatic islet cell transplantation for type I diabetes), it is conceivable that this approach could be applied to many other serious medical conditions where cells are lost as a result of disease, injury, or aging. Current challenges in cell replacement therapy include the limited source of engraftable stem/progenitor cells and a poor ability to manipulate their functional expansion and differentiation. Alternatively, drug stimulation of the body's own regenerative capabilities (i.e., proliferation, differentiation, migration, or even reprogramming of endogenous cells to replace the damaged cells/structures) may represent a more desirable therapeutic approach to fulfilling the ultimate goal of regenerative medicine. While tremendous efforts have been put into these areas, it is clear that a better understanding of stem cell biology is required before these approaches can be realized.

Chemistry and functional genomics are complementary approaches in stem cell biology; the limitations of one approach are often the strengths of the other. Our understanding of biological processes often develops from discovering or designing ways to perturb a given process and observing the effects of the perturbation. Modulation of gene function can be achieved by overexpression of a cDNA, gene knockdown by RNA interference (RNAi), or chemical regulators. Small molecules offer a number of advantages that include the ability for temporal, tunable, and modular control of specific protein function, while their applications are constrained by their specificity, target characterization (for the ones from phenotypic screens), and inability to saturate biologically active chemical space. In this chapter, more recent developments in the use of small molecules (Figure 8.1) and functional genomic tools will be reviewed in the context of stem cell self-renewal, differentiation, and reprogramming.

8.1 BIOLOGICAL BACKGROUND OF STEM CELLS

Stem cells are classified, on the basis of their origin and differentiation capacity, as either embryonic or adult stem cells [1]. Embryonic stem cells (ESCs) are derived from the inner cell mass of the blastocyst. ESCs can self-renew indefinitely and are pluripotent (possess the ability to differentiate into all cell types in the embryo proper). Adult stem cells, which reside in specific tissues and function as the "reservoir" for cell/tissue renewal during normal homeostasis or tissue regeneration, possess less tumorigenic risks and fewer ethical concerns but have restricted differentiation potential and limited self-renewal capacity.

Stem cell fate is under strict control from both intrinsic and extrinsic factors, and loss of this control has been postulated to be a key step in degenerative and carcinogenic processes [2]. A growing body of evidence suggests that cancer initiation results from an accumulation of oncogenic molecular alterations (intrinsic loss of control) in long-lived stem cells or their progenitors as well as modifications of the surrounding microenvironment (loss of extrinsic control) [3]. Cancer stem cells have been detected

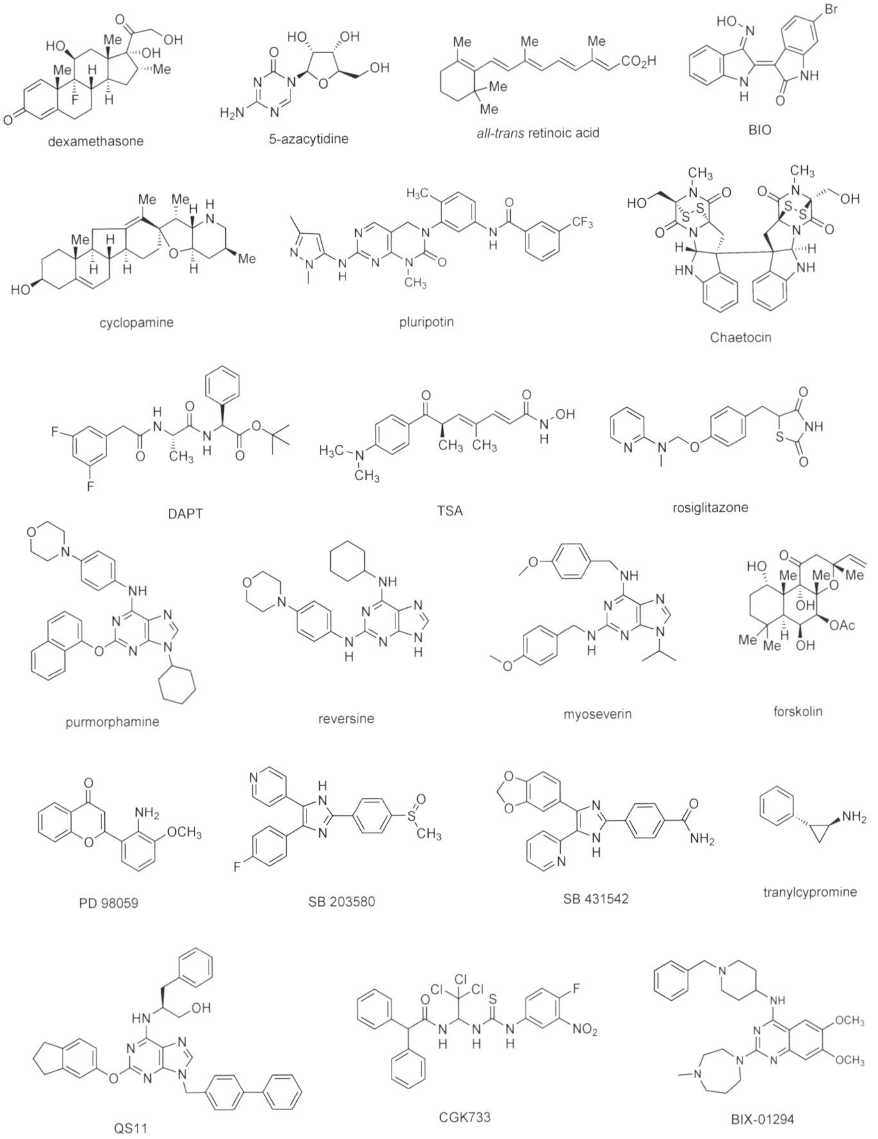

FIGURE 8.1 Chemical compounds regulating stem/progenitor cell fate.

in leukemia, breast, brain, and prostate tumors [4–7]. Therefore, the characterization of a cancer stem cell profile within diverse cancer types, and a better understanding of the biology of its counterpart—the normal stem cell—may open up new avenues for cancer treatment.

Stem cell expansion and differentiation *ex vivo* are commonly controlled by culturing cells in a specific configuration (e.g., an attached monolayer or as suspended aggregates) with cocktails of growth factors and signaling molecules, as well as genetic manipulations. However, most of these conditions either are incompletely defined, or are nonspecific and inefficient at regulating the desired cellular process. Such conditions often result in inconsistency in cell culture and mixed populations of cells that would not be useful in studying specific cellular processes or in cell-based therapies. More efficient and selective conditions for homogeneous stem cell self-renewal and differentiation into specific cell types need to be developed before the various applications of stem cells can be realized. Toward this end, chemical approaches serve as excellent tools to investigate the underlying mechanism and control the specific stem cell fate.

8.2 CHEMICAL APPROACHES TO STEM CELL RESEARCH

Typically, two approaches for using chemical compounds in stem cell biology. A target-based approach involves development of chemical compounds for a specific biological target, followed by the application of these compounds in either cells or organisms for the purpose of linking the produced pharmacological effect to the target's modulation. In a phenotype-based approach, chemical libraries are screened in high-throughput functional assays (in cells or whole organisms) to identify compounds that produce a desired phenotype, followed by elucidation of the molecular targets or pathways that they engage.

The success of both target-based and phenotype-based methods relies heavily on the qualities of the chemical libraries used. Although combinatorial technologies allow the synthesis of a large number of molecules with immense structural diversity, it is impossible to saturate chemical space. Since biological space interacts with only a fraction of chemical space, synthetic attempts to randomly increase the molecular diversity of a chemical library by introducing a high level of structural variability/complexity drastically reduce the library's fitness to a given biological selection/screen, resulting in most molecules being inactive. Furthermore, screening of even larger chemical collections to increase the chance of finding hits would compromise assay functionality, sensitivity, and fidelity for practical reasons. As the diversity and fitness of a library are conflicting goals, the careful design of a chemical library becomes a critical aspect of combinatorial synthesis.

The notion of "privileged structures" describes selected structural motifs that can provide potent and selective ligands for biological targets. Privileged structures typically exhibit a greater tendency of interacting with biological targets and good "drug-like" properties. One of the most straightforward and productive ways to generate "privileged" chemical libraries is to use key biological recognition motifs as the diversity elements for combinatorial synthesis [8]. In this approach, a variety of naturally occurring and synthetic heterocycles that are known to interact with proteins involved in cell signaling (e.g., kinases, cell surface and nuclear receptors, enzymes) are used as the core molecular scaffolds. Then a variety of substitutents can be introduced into each of these scaffolds to create a diverse chemical library. Using this method, a large diverse heterocycle library consisting of over 100,000 discrete small molecules (representing over 30 distinct structural classes) was generated with an average purity of >90%, which has proven to be a rich source of biologically active small molecules targeting various proteins involved in a variety of signaling pathways [8].

The screening method is another important factor for the phenotypic approach. The development of permissive assay conditions with positive and negative controls, if available, should not be taken inconsequentially. While the use of simple reporter systems or enzymatic activity assays would allow higher throughput in cell-based assays, monitoring of more complex phenotypic changes (e.g., cell morphology, multiple biomarker expression and localization, and cell physiology) by high content imaging methods substantially reduces false analysis from primary screens and provides broader assay versatility [64]. As to engineering cell assays (e.g., reporter lines), additional experimental advances with respect to more efficient and precise marking of cells, especially with respect to hES cells, are needed. This would greatly facilitate the use of small molecules as tools to answer fundamental biological questions about development. As opposed to gene perturbation, which is often difficult in ES cells, especially hES cells [65], small molecules offer the additional advantage of being able to elicit a particular phenotype without the need for complicated cell manipulations. Furthermore, informatics on the assays/compounds/genes matrix has greatly facilitated bioactive compound identification and deconvolution of their mechanisms.

8.3 SMALL-MOLECULE REGULATORS OF STEM CELL FATE

8.3.1 Self-Renewal

Self-renewal is the process by which a stem cell divides to generate one (asymmetric division) or two (symmetric division) daughter stem cells with developmental potentials identical to those of the mother cell [9]. The self-renewal ability of stem cells is central to the development and the maintenance of adult tissues.

Self-renewal of stem cells has come to be regarded as a combined phenotypic outcome of cellular proliferation as well as inhibition of differentiation and cell death. Consistent with this notion, while murine embryonic stem (mES) cells are conventionally maintained on inactivated fibroblast feeder cells and with serum, studies have found that the self-renewal of mES cells can be achieved in the absence of feeder cells and serum in a chemically defined media condition by the combined activity of two key signaling molecules: LIF/interleukin-6 (IL6) family members and bone morphogenetic protein (BMP) [10]. LIF activates Stat3 (signal transduction and activation of transcription) signaling through a membrane-bound gp130-LIFR (LIF receptor) complex to promote proliferation and inhibit mesoderm and endoderm differentiations of mES cells via a Myc-dependent mechanism. Activation of gp130 also stimulates extracellular-signal-related kinase (ERK) 1/2 activity. ERK activation is not required for maintenance of pluripotency in mES cells. In fact, ERK activation has been shown to impede self-renewal, and the application of PD098059, the MEK inhibitor [11], promotes self-renewal of mES cells [12,13]. BMP induces expression of Id (inhibitor of differentiation) genes via Smad signaling and also inhibits ERK and p38 MAPK pathways, and consequently inhibit differentiation of mES cells to neuroectoderm and trophectoderm [10,14]. 6-Bromoindirubin-3′-oxime (BIO) [15,16] is another small molecule that promotes self-renewal in both mES and hES cells. BIO, a more recently identified GSK3 (glycogen synthase kinase-3) inhibitor, was used to demonstrate the activation of Wnt/β-catenin pathway also promotes self-renewal [17]. Further studies [18] suggest that BIO may also play a role in stabilizing Myc signaling by suppressing GSK3β-mediated Myc-T58-phosphorylation. It's worth noting that Myc is also regulated at the transcriptional level by Stat3 [18]. Combinations of MEK inhibitor with LIF, GSK3 inhibitor with LIF, or simply three inhibitors for MEK/GSK3/FGFR have been used to sustain self-renewal and to derive various mES cell lines in chemically defined media conditions. On the basis of mechanistic insights to self-renewal, target based kinase inhibitors (ALK-4/5/7 inhibitor, SB431542 [19]), have been applied to study the role of TGF-β/Activin/Nodal signaling in stem cell self-renewal [20,21].

To gain a better understanding of ESC self-renewal, a cell-based phenotypic screen for small molecules that can promote mESC self-renewal was carried out using an established Oct4-GFP reporter mESC line, examining pluripotency marker expression (Oct4) as well as cell morphology (undifferentiated ESCs have compact colony morphology) [22]. A novel synthetic small molecule named *pluripotin* was discovered in this screen that is sufficient to propagate mESCs in the undifferentiated, pluripotent state under chemically defined conditions in the absence of feeder cells, serum, LIF, and BMP [23]. Long-term cultures of mESCs with pluripotin can be differentiated into cells in the three primary germ layers *in vitro*, and can also generate chimeric mice and contribute to the germline *in vivo*. Affinity chromatography using pluripotin-immobilized matrix identified Erk1 and RasGAP as the molecular targets of pluripotin. Additional biochemical and genetic experiments confirm that pluripotin is a dual-function small-molecule inhibitor of both Erk1 and RasGAP, two endogenously expressed proteins with differentiation-inducing activity, and that simultaneous inhibition of both protein activities is necessary and sufficient for pluripotin's effects on mES cells. Because pluripotin's mechanism of action is independent of LIF, BMP, and Wnt signaling, this study suggests that ES cells, and probably other types of stem cells, may possess the intrinsic ability to self-renew; that is, stem cells may autonomously express virtually all gene products that are essential for self-renewal and not require the stimulation of additional gene expression. The endogenous expression of some differentiation-inducing genes at a certain level in the undifferentiated stem cells could cause "spontaneous" differentiation under culture conditions that could not inhibit their negative effects. Therefore, the key to achieve a basic/fundamental self-renewal state of stem cells may be to inhibit the negative effects of endogenously expressed, stemness-inhibitory proteins (e.g., differentiation or cell death–inducing proteins). Not only did the discovery of pluripotin provide a useful chemical tool for ESC expansion and novel insights to the mechanism of ESC self-renewal; it also exemplified the power of small-molecule screens in that it is possible to modulate more than one target by a single small molecule to achieve a desired complex biological phenotype.

Consistent with the self-renewal model, more recent studies on Notch signaling exemplified the contribution of cell survival signaling to stem cell self-renewal. Androutsellis-Theotokis et al. found that activation of Notch promotes fetal neural stem cell (NSC) survival, most likely through activation of AKT, Stat3, and mTor; this survival signal is antagonized by JAK and p38 MAPK, and applying JAK and p38 inhibitors significantly improves survival of NSCs [24]. Interestingly, this mechanism also functions in human ESCs. Moreover, p38 inhibitors have also been shown by another independent study to promote hematopoietic stem cell (HSC) lifespan, further confirming the role of p38 in stem cell self-renewal [25]. While details of the molecular mechanism still need to be elucidated, these inhibitors clearly have had a beneficial effect on stem cell culture.

Most stem cells can divide via either asymmetric or symmetric modes [26]. Since each asymmetric division generates one daughter cell with a stem cell fate (self-renewal) and one daughter cell that differentiates, it was postulated that asymmetric division could be a barrier to stem cell expansion. Therefore, conversion of asymmetric division to symmetric division would promote self-renewal and long-term culture of stem cells. Indeed, this idea was confirmed by utilizing one small molecule, the purine nucleoside xanthosine (Xs) [27]. This small molecule promotes guanine ribonucleotide biosynthesis that reversibly converts cells from asymmetric division kinetics to symmetric division kinetics. It was found that Xs derived stem cell lines exhibit Xs-dependent symmetric kinetics, and this derived stable line shows enhanced self-renewal. This study underscores the importance of balance between the two modes of division, both to stem cell expansion and to the regenerative capacity of adult stem cells.

8.3.2 Lineage-Specific Differentiation

Differentiation is the developmental process by which early, less-specified cells acquire the features of late-stage, more specified cells such as neurons, hepatocytes, or heart muscle cells. Currently, few examples of devised, highly selective and efficient conditions for stem cell differentiation into specific homogeneous cell types have been reported, owing to the lack of understanding of stem cell signaling at the molecular level. Small-molecule phenotypic screens provide another means to generate desired cell types in a controlled manner. A number of small molecules that modulate specific differentiation pathways of embryonic or adult stem cells have been identified by this method.

Neural and Neuronal Differentiation Retinoic acid (RA), forskolin, and HDAC inhibitors have been shown and used to induce neuronal differentiation of hippocampal adult neural progenitor cells [28–30]. However, these factors are either pleiotropic, or of undefined physiological relevance.

Neuropathiazol, a substituted 4-aminothiazole, has been identified from a high-content imaging-based screen of chemical libraries that specifically induces neuronal differentiation of multipotent adult hippocampal neural progenitor cells [31]. More than 90% of the neural progenitor cells treated with neuropathiazol differentiated into neuronal cells as determined by immunostaining with βIII tubulin and the characteristic neuronal morphology. Interestingly, neuropathiazol can also inhibit astroglial differentiation induced by LIF and BMP2, while RA cannot, suggesting that neuropathiazol functions by a different mechanism and has a more specific neurogenic inducing activity. The precise molecular target(s) of neuropathiazol is (are) still unknown, although its (their) identification will surely provide mechanistic insights into neuronal differentiation. Additionally, neuropathiazol can be used as a specific inducer to generate homogeneous neuronal cells, and may also serve as a starting point for development of small molecule drugs to stimulate *in vivo* neurogenesis.

In an elegantly devised approach, functional motor neurons were generated from mESCs by sequential treatments with RA (neuralizing and caudalizing mESCs) and a specific small-molecule agonist (Hh-Ag1.3) of Hedgehog (Hh) signaling (ventralizing the caudalized cells) [32]. This experiment underscores the importance of following developmental progression through sequential induction and combinatorial factors for generating a late-stage cell type from early stem cells.

Differentiation of Mesenchymal Stem/Progenitor Cells Mesenchymal stem cells (MSCs) are capable of differentiating into all of the mesenchymal cell lineages, such as bone, cartilage, adipose tissue, and muscle, and play important roles in tissue repair and regeneration [33]. A number of small molecules have been used to control the differentiation of mesenchymal stem/progenitor cells to a variety of cell types. For example, 5-aza-C (a DNA demethylation agent) can induce C3H10T1/2 cells (a mouse mesenchymal progenitor cell line) to differentiate into myoblasts, osteoblasts, adipocytes, and chondrocytes by enhancing cell differentiation competence via epigenetic modifications [34,35]. Dexamethasone (a glucocorticoid receptor agonist) is a different epigenetic modifier, and its combination with other small molecules, such as ascorbic acid, β-glycerophosphate, isobutylmethylxanthine (IBMX, a nonspecific phosphodiesterase inhibitor), or peroxisome proliferator-activated receptor γ (PPARγ) agonists (such as rosiglitazone) has been widely used to modulate osteogenesis or adipogenesis of MSCs under specific conditions [34,36]. To identify small molecules that selectively induce osteogenesis of MSCs, a high-throughput screen of chemical libraries in C3H10T1/2 cells using an enzymatic assay of alkaline phosphotase (a specific osteoblast marker) led to the identification of a novel synthetic small molecule, purmorphamine [37]. Genomewide expression profiling in conjunction with systematic pathway analysis was used to reveal that the Hh signaling pathway is the primary biological network affected by purmorphamine [38]. Chemical epistasis studies using known Hh pathway antagonists (cyclopamine and forskolin) have confirmed that purmorphamine's mechanism of action on osteogenesis is through specific activation of Hh pathway, and it acts at the level of Smoothened (Smo). More recently, it was shown that purmorphamine directly targets the protein Smoothened (Smo), through competitive displacement assays using fluorescence-tagged cyclopamine [39].

8.3.3 Regeneration

Terminally differentiated, postmitotic mammalian cells are thought to have little or no regenerative capacity, as they are already committed to their final specialized form and function, and have permanently exited the cell cycle. However, it is conceivable that appropriate stimulation of mature cells to reenter the cell cycle and proliferate may provide new therapeutic approaches for treating various degenerative diseases and injuries. The mammalian cardiomyocyte is one such mature cell type that has attracted substantial research efforts toward its regeneration. Using a target-based approach, a p38 MAPK inhibitor, SB203580, and a GSK3 inhibitor, BIO, have been shown independently to promote proliferation in both neonatal and adult cardiomyocytes indicated by BrdU incorporation and histone-3 phosphorylation [40,41]. The proliferation in adult cardiomyocytes was also observed to be associated with transient dedifferentiation of the contractile apparatus. It is interesting to note that activation of canonical Wnt signaling by BIO also promotes proliferation of cardiac progenitor cells. The pancreatic β cell is another highly sought-after cell type for regeneration; the transplantation of which, in conjunction with simultaneous prevention of their immune destruction, may represent a "cure" for type I diabetes. More recent phenotypic screens of large chemical libraries have identified several classes of small molecules that can promote proliferation of human β cells, among which p38 inhibition was identified as the mechanism of action for one class of molecules. A major challenge lying ahead for proliferation of mature cell types is that the process is typically associated with loss of the cell phenotype [e.g., proliferated β cells can undergo epithelial-to-mesenchymal transition (EMT) to acquire fibroblast-like features, and they typically do not redifferentiate back to the mature β cells]. Strategies for inducing functional redifferentiation of the proliferated cells or inhibiting loss of cell identity while proliferating are highly desirable and under intense investigation. It should be noted that a synthetic purine analog, myoseverin, was previously identified from a phenotypic screen, which can induce cleavage of multinucleated myotubes to generate myoblast-like cells, which can proliferate and redifferentiate into myotubes [42,43].

In the mammalian CNS, failure of axonal regeneration is attributed not only to the intrinsic regenerative incompetence of mature neurons but also to the inhibitory actions of CNS myelin and molecules in the glial scar at the lesion sites. In a cell-based screen to identify small molecules that can counteract the inhibitory activity of CNS myelin, several EGFR inhibitors (including PD168393 and AG1478) were found to promote neurite outgrowth of cerebellar granule neurons on an immobilized myelin substrate [44]. More importantly, PD168393 and Erlotinib, an EGFR inhibitor approved by the FDA for the treatment of cancer, were found to be effective *in vivo* in promoting axonal regeneration, providing promise for treatment of brain and spinal cord injury in humans.

8.3.4 Cellular Reprogramming

It was long thought that tissue/organ-specific stem/progenitor cells could give rise only to cells of the same tissue type, but not those of different tissue. In other words, they have irreversibly lost the capacity to generate cell types of other lineages in the body. However, a number of recent reports have demonstrated that tissue-specific stem cells may overcome their intrinsic lineage restriction on exposure to a specific set of *in vitro* culture conditions [45,46]. An extreme example is the reprogramming of a somatic cell to a totipotent state by nuclear transfer cloning, where the nucleus of a somatic cell is transferred into an enucleated oocyte, or the extracts of the oocyte are fused with a somatic cell [47,48]. Although in mammals neither transdifferentiation nor dedifferentiation has yet been identified as a naturally occurring process (except in certain disease states), the discovery of cell plasticity raises the possibility of reprogramming a restricted cell's fate. The ability to dedifferentiate or reprogram lineage-committed cells to multipotent or even pluripotent cells might overcome many of the obstacles associated with using ESCs and adult stem cells in clinical applications (immunocompatibility, cell isolation and expansion, bioethics, etc.).

To identify small molecules that induce reprogramming of lineage-committed myoblasts, a cell-based screen was designed considering the notion that lineage-reversed myoblasts would regain multipotency [49]. Specifically, dedifferentiated myoblasts would acquire the ability to differentiate into (otherwise nonpermitted) mesenchymal cell types under conditions that typically induce differentiation of only multipotent MSCs into adipocytes, osteoblasts, or chondrocytes. A two-stage screening protocol was used in which C2C12 myoblasts were initially treated with compounds to induce dedifferentiation; compounds were then removed and cells were assayed for their ability to undergo osteogenesis on addition of known osteogenic inducing agents [49]. Reversine, a 2,6-disubstituted purine, was found to have the desired dedifferentiation-inducing activity. It inhibits myotube formation, and reversine treated myoblasts can redifferentiate into osteoblasts and adipocytes on exposure to the appropriate differentiation conditions. In addition, the reprogramming effect of reversine on C2C12 cells (as well as some other cell types) can be shown at the clonal level, suggesting that its effect is inductive rather than selective. Furthermore, reversine has also been shown to induce reprogramming of primary mouse and human fibroblasts to a multipotent state, which can be redifferentiated to functional myoblasts and myotubes under myogenic conditions, suggesting that reversine's reprogramming mechanism might be general to different cell types [50]. The cellular targets of reversine were identified through affinity chromatography as mitogen-activated extracellular signal regulated kinase (MEK1) and nonmuscle myosin II heavy chain (NMMII) [37]. Mechanistic studies suggested that inhibition of both target proteins' activities is required, which entails cell cycle G2/M phase staging, cytoskeletal reorganization, and cell signaling modulation to reverse epigenetic modifications. The discovery of reversine also exemplifies that a reprogramming molecule can be identified by a rationally designed phenotypic screen, and that such concepts are readily applicable to other models.

8.3.5 Modulators of Developmental Pathways and Epigenetic Modifiers

Developmental signaling pathways, such as Wnt, Hh, TGF/BMP, and Notch, control embryonic patterning and cell behavior during development and play important roles in regulating tissue homeostasis and regeneration in adulthood. The known chemical modulators of these developmental pathways have been widely used, including GSK3β inhibitors as agonists of canonical Wnt pathway, fumagillin as an antagonist of noncanonical Wnt pathway, purmorphamine as an agonist and cyclopamine as an antagonist of the Hh pathway, DAPT as an antagonist of Notch pathway, and SB431542 as an inhibitor of TGFβ/activin signaling [17,38,51–53].

While activation of Wnt, Hh, and Notch pathways has been involved in various regenerative processes, their aberrant activities are also associated with cancer induction. Strategies to control the activity of these pathways for regeneration while avoiding carcinogenicity would be highly attractive. One possible approach would be to enhance the desired pathway activity by a synergistic agonist, which would be effective only where it is needed (i.e., the pathway activity is inadequate), rather than using a general pathway activator that ectopically activates the signaling. To identify novel compounds and pathways that interact with the canonical Wnt/β-catenin signaling pathway, a reporter-based screen was carried out for molecules that synergistically activate signaling in the presence of Wnt3a. A 2,6,9-trisubstituted purine compound, QS11, was identified to synergize with canonical Wnt ligand both *in vitro* and *in vivo* [54]. Affinity chromatography identified ARF-GAP as a target of QS11. Additional functional studies have confirmed that QS11 inhibits ARF-GAP function, and as a consequence modulates ARF activity and β-catenin localization. Since ARFs play important roles in endocytosis, this study established another link between the endocytosis pathway and Wnt signaling, and provides a useful chemical tool to modulate Wnt pathway and explore novel functions of ARF-GAPs in cell culture and whole organisms.

Epigenetic modifications are central processes in stem cell differentiation and reprogramming that can control specific and heritable gene expression pattern in cells without altering the DNA sequence. Therefore, molecules directly regulating epigenetic machineries and changing the epigenetic status of cells should affect cell fate. Major epigenetic modifications include DNA methylation and histone modifications (acetylation, phosphorylation, methylation, and ubiquitination). 5-Azacytidine and its analogs are widely used DNA demethylation agents, and have been shown to increase cellular plasticity or induce differentiation of certain stem/progenitor cells [55,56]. HDAC inhibitors (e.g., TSA, VPA) have also played essential roles in studying histone acetylation and have been developed for the treatment of cancers. Like all signaling/epigenetic modulators, their effects are context-dependent; they have been shown to enhance self-renewal of HSCs, induce neuronal differentiation of NSCs, or promote myogenesis of muscle cells [57–59]. In addition to these two widely used epig-enetic modifiers, small-molecule inhibitors for protein arginine methyltransferases (PRMTs) [60], histone methyltransferases (HMTs, e.g., Suv39H1 and G9a) [61,62] and histone demethylases (e.g., LSD1) [63], have been identified through various approaches. With ongoing efforts of generating additional, more precise epigenetic chemical probes, these modifiers will no doubt contribute substantially to epigenetic control and stem cell regulation.

8.4 TARGET/PATHWAY IDENTIFICATION

The small-molecule hits identified in phenotypic screens need further characterizations with respect to identifying their specific molecular targets and mechanism of action. A commonly used method to identify chemical compound's molecular targets is affinity chromatography [66]. More recent examples of using this method for target identification of various small molecules that regulate cell fate include myoseverin [43], TWS119 [67], reversine [68], pluripotin [23], and QS11 [54]. Target could also be elucidated with the use of expression libraries and protein microarrays [69]. However, these approaches have not been successfully used for target identification in stem cells.

Gene expression analysis is another technique commonly used for pathway identification. Transcriptome analysis of purmorphamine, a small-molecule inducer of osteogenesis from mesenchymal progenitor cells [37], revealed agonistic activation of the Hedgehog signaling pathway [38]. Development of arrayed cDNA [70] and RNAi libraries [71] on a genome scale may also facilitate target identification from genomic complementation screens. Although the combination of small-molecule and genomic RNAi screens has been successful in the *Drosophila* system [72,73], it remains to be seen whether this can also be applied to target identification in stem cells.

8.5 FUNCTIONAL GENOMICS APPROACHES TO STEM CELL RESEARCH

The two main functional genomics approaches currently applied to stem cell research are gene trap and high-throughput genomic screens (i.e., overexpression of arrayed cDNA libraries or gene knockdown by arrayed RNAi libraries). Gene trap was initially developed for systematic knockout of gene function on a genome scale to identify the roles of mammalian genes during embryonic and postembryonic development, but more recently it has also been modified for functional screens during stem cell development. In contrast to small-molecule or random mutagenesis screens, high-throughput functional genomic screens using arrayed cDNA/RNAi libraries have several advantages: (1) the identities of the hits and primary gene targets are already known; (2) the genome, unlike chemical space, can be saturated; and (3) maintenance/replication of the screen is more convenient and efficient. Given the success of functional screens using small-scale cDNA/RNAi collections [74,75], continuing advancements in high-throughput application to systematically evaluate gene function on a genomewide scale will create an even more powerful tool for the study of stem cell biology.

8.5.1 Gene Trap

Gene trap is a form of random mutagenesis via intragenic insertions designed to perturb gene function [76]. The typical gene trap mechanism relies on a promoterless vector consisting of an upstream 3′ splice site [splice acceptor (SA)], followed by a reporter gene and/or selectable marker and a downstream polyadenylation sequence (polyA). The insertion of the gene trap cassette into an intron of an expressed gene generates a fusion transcript, where the reporter/selectable marker gene is downstream of the endogenous gene's exon(s). Insertion of the polyA induces premature transcription termination, resulting in a nonfunctional version of the endogenous protein fused to the reporter/selectable marker. The gene trap cassette randomly inactivates endogenous genes and provides a DNA tag at the insertion site for easy target identification by rapid amplification of cDNA ends (RACE) [77].

A collection of gene trap mutated mouse embryonic stem cell clones has been generated by To and colleagues [78] and made available to researchers worldwide through the Centre for Modeling Human Disease [79]. The gene trap mutated lines are screened using various *in vitro* differentiation and induction assays. This thorough analysis of gene trap mutations allows researchers to search for insertions in genes expressed under specific *in vitro* conditions or in target cell lineages. Since gene expression can vary under different conditions, Chen's group employed a responder mESC line carrying a dominant selection marker, which could be activated only when an active gene is trapped [80]. This allows for high-throughput screening of trapped lines based on positive marker selection induced by the activation of trapped gene loci on discrete stimuli(s), thereby facilitating functional categorization of trapped genes. Gene trap has been shown to be applicable in both hESCs [81] and adult rodent NSCs [82]. The latter used a β-lactamase reporter enzyme in a retroviral gene trap vector. A nontoxic fluorogenic substrate is cleaved upon activation of the β-lactamase, resulting in an emission wavelength shift. The shift in the fluorescence signal permits the monitoring of gene expression in living NSCs under various treatments [82].

A caveat to this approach lies in the diploid nature of the mouse genome, which requires the generation of homozygous transgenic mice to elicit phenotypic consequences of recessive mutations created by gene trap. Guo and coworkers have developed a sophisticated system to circumvent this problem [83]. Taking advantage of the highly efficient mitotic recombination in *blm*-deficient ESCs, they were able to generate a genomewide library of homozygous mutant cells, allowing for direct phenotypic genetic screens in ESCs by gene trapping. ESC clones carrying traceable insertional mutations can be assayed *in vitro* for gene trap cassette reporter gene activity under discrete cell lineage-specification conditions, allowing for identification of developmentally regulated genes. Compared with traditional targeted gene inactivation based on homologous recombination, gene trap is a simple, unbiased, rapid, and cost-effective approach to studying a broad spectrum of gene inactivation.

8.5.2 High-Throughput Functional Screens

While libraries of pooled cDNA clones have been the primary source of conventional genome-scale screens, the drawback of using such complex libraries is the limitation of using assays with simple readouts (typically a selection method), followed by deconvolution of clone identities. Nanog, a pluripotency-associated master gene, was discovered through this type of method, selecting self-renewing polyclonal ESC population that were "supertransfected" with an episomal expression library and subsequent deconvolution of the clone identity [74,84].

Applying the hypothesis that factors playing important roles in the maintenance of ES cell pluripotency may also have a pivotal function in the induction of pluripotency from differentiated cells, Takanashi and colleagues screened 24 candidate genes with a system where Neo was knocked into the locus of Fbx15 (a pluripotency- specific gene) and induction of pluripotency was detected as resistance to G418 [85]. Initial results showed that transfection with all 24 factors could generate ES-cell-like colonies, and from there they systematically removed factors to reveal that the combination of *Oct3/4, Sox2, c-Myc*, and *Klf4* was sufficient to induce pluripotency in fibroblast cultures, designated induced pluripotent stem (iPS) cells. In addition to exhibiting ES cell morphology and growth properties, iPS cells also expressed ES cell marker genes. Further analysis revealed the ability of iPS cells to form tissues from all three germ layers on subcutaneous transplantation into nude mice as well as the ability to contribute to mouse embryonic development following injection into blastocysts.

Although gain-of-function screens like these are a powerful tool to identify genes that confer a particular cellular phenotype, they may lose important genes during the iterative rounds of selection or miss genes whose overexpression promotes differentiation or cell death. Pritsker's group combined genetic screens with microarray detection methods to allow for high-throughput functional analyses with greater hit sensitivity [70]. In this method, microarray technology is used to monitor the abundance of all gene products from a cDNA library as they function to elicit a given phenotype. They applied this technique to identify genes potentially involved in both differentiation (*JunB* and *c-Fos*) and self-renewal (*Chop-10* and *Akt1*) of mESCs.

8.6 LARGE-SCALE GENETIC APPROACHES

Further advancements in automation and detection technologies have facilitated large-scale high-throughput functional screens with the development of individually arrayed cDNA and RNAi libraries, whose members are spatially separated in different wells of multiwell plates, obviating the need for clone rescue and deconvolution and allowing use of more complex assays for functional screens, such as monitoring dynamic gene regulation or morphological changes.

Functional screens using both cDNA and RNAi libraries have been successfully carried out in mammalian cells [86,87], but the challenge remains for primary cells and stem cells, which are more difficult to transfect and are susceptible to side effects using those methods. Viral delivery methods for functional genomic screens in stem cells have proved to be a credible alternative [88,89]. Michiels and colleagues validated such an approach by identifying BMP4 and BMP7 (two known inducers of osteoblast differentiation) as hits from an osteoblast differentiation screen of human MSCs, transduced with an adenoviral vector-based cDNA library [89]. Zhao's group successfully carried out both osteogenesis and adipogenesis screens in human MSCs (hMSCs) using a 10,000–chemically synthesized siRNA library targeting 5000 human genes, and identified a large number of novel gene hits, which, on silencing, could induce specific osteogenic or adipogenic differentiation [90]. Specifically, the osteogenesis screen yielded 53 candidate suppressors, and 12 of those were further confirmed for their dynamic roles in suppressing osteogenic specification in hMSCs. Furthermore, cAMP was identified to play opposing roles in osteogenesis versus adipogenesis. Further studies on the identified genes will not only help us understand the critical molecular switches between self-renewal and osteogenic/adipogenic differentiations in hMSCs but also facilitate the characterization of the genetic basis of bone diseases as well as the development of new therapies to treat them.

Although high-throughput functional screening is a highly productive and promising technique, single-gene perturbation may not be sufficient to induce a particular biological event. In this case, sensitized screen conditions could be sought out. Furthermore, careful design and validation of constructed libraries and screen assays are essential, especially in the case of RNAi, as off-target effects are currently hard to pursue.

8.7 CONCLUSION

Stem cell research provides tremendous opportunities for understanding human development, regeneration, and diseases, and offers great promise for developing cell-based or drug therapies to treat devastating diseases and injuries. Small molecules have proved useful to probe stem cell biology, to create homogenous self-renewal or differentiation conditions for stem cells that is essential for future cell-based therapy, and to facilitate drug development for controlling endogenous regeneration and treating cancers. However, many challenges remain, including designing better chemical libraries and screening strategies to systematically identify small molecules that regulate the desired cellular process; developing more efficient methods to understand the underlying mechanism; and translating *in vitro* discoveries into approaches for *in vivo* regeneration of desired tissues/organs by small-molecule therapeutics. It is clear that identification of additional small-molecules that control stem cell fate and further technological advancements in functional genomic techniques will contribute to the development of regenerative medicine and stem cell biology.

REFERENCES

[1] Department of Health and Human Services: *Regenerative Medicine 2006*. (cited; available from: `http://stemcells.nih.gov/info/scireport/2006report.htm`).

[2] Pardal R, Clarke MF, Morrison SJ: Applying the principles of stem-cell biology to cancer, *Nat Rev Cancer* **3** (12):895–902 (2003).

[3] Shachaf CM et al: MYC inactivation uncovers pluripotent differentiation and tumour dormancy in hepatocellular cancer, *Nature* **431** (7012):1112–1117 (2004).

[4] Al-Hajj M et al: Prospective identification of tumorigenic breast cancer cells, *Proc Natl Acad Sci USA* **100** (7):3983–3988 (2003).

[5] Bonnet D, Dick JE: Human acute myeloid leukemia is organized as a hierarchy that originates from a primitive hematopoietic cell, *Nat Med* **3** (7):730–737 (1997).

[6] Lapidot T et al: A cell initiating human acute myeloid leukaemia after transplantation into SCID mice, *Nature* **367** (6464):645–648 (1994).

[7] Singh SK et al: Identification of human brain tumour initiating cells, *Nature* **432** (7015):396–401 (2004).

[8] Ding S et al: A combinatorial scaffold approach toward kinase-directed heterocycle libraries, *J Am Chem Soc* **124** (8):1594–1596 (2002).

[9] Molofsky AV, Pardal R, Morrison SJ: Diverse mechanisms regulate stem cell self-renewal, *Curr Opin Cell Biol* **16** (6):700–707 (2004).

[10] Ying QL et al: BMP induction of Id proteins suppresses differentiation and sustains embryonic stem cell self-renewal in col laboration with STAT3, *Cell* **115** (3):281–292 (2003).

[11] Dudley DT et al: A synthetic inhibitor of the mitogen-activated protein kinase cascade, *Proc Natl Acad Sci USA* **92** (17):7686–7689 (1995).

[12] Burdon T et al: Suppression of SHP-2 and ERK signalling promotes self-renewal of mouse embryonic stem cells, *Dev Biol* **210** (1):30–43 (1999).

[13] Yamane T et al: Enforced Bcl-2 expression overrides serum and feeder cell requirements for mouse embryonic stem cell self-renewal, *Proc Natl Acad Sci USA* **102** (9):3312–3317 (2005).

[14] Qi X et al: BMP4 supports self-renewal of embryonic stem cells by inhibiting mitogen-activated protein kinase pathways, *Proc Natl Acad Sci USA* **101** (16):6027–6032 (2004).

[15] Leclerc S et al: Indirubins inhibit glycogen synthase kinase-3 beta and CDK5/p25, two protein kinases involved in abnormal tau phosphorylation in Alzheimer's disease. A property common to most cyclin-dependent kinase inhibitors? *J Biol Chem* **276** (1):251–260 (2001).

[16] Meijer L et al: GSK-3-selective inhibitors derived from Tyrian purple indirubins, *Chem Biol* **10** (12):1255–1266 (2003).

[17] Sato N et al: Maintenance of pluripotency in human and mouse embryonic stem cells through activation of Wnt signaling by a pharmacological GSK-3-specific inhibitor, *Nat Med* **10** (1):55–63 (2004).

[18] Cartwright P et al: LIF/STAT3 controls ES cell self-renewal and pluripotency by a Myc-dependent mechanism, *Development* **132** (5):885–896 (2005).

[19] Laping NJ et al: Inhibition of transforming growth factor (TGF)-beta1-induced extracellular matrix with a novel inhibitor of the TGF-beta type I receptor kinase activity: SB-431542, *Mol Pharmacol* **62** (1):58–64 (2002).

[20] James D et al: TGFbeta/activin/nodal signaling is necessary for the maintenance of pluripotency in human embryonic stem cells, *Development* **132** (6):1273–1282 (2005).

[21] Vallier L, Alexander M, Pedersen RA: Activin/Nodal and FGF pathways cooperate to maintain pluripotency of human embryonic stem cells, *J Cell Sci* ,**118** (Pt 19):4495–4509 (2005).

[22] Ohbo K et al: Identification and characterization of stem cells in prepubertal spermatogenesis in mice small star, filled, *Dev Biol* **258** (1):209–225 (2003).

[23] Chen S et al: Self-renewal of embryonic stem cells by a small molecule.*Proc Natl Acad Sci USA* **103** (46):17266–17271 (2006).

[24] Androutsellis-Theotokis A et al: Notch signalling regulates stem cell numbers in vitro and in vivo, *Nature* **442** (7104):823–826 (2006).

[25] Ito K et al: Reactive oxygen species act through p38 MAPK to limit the lifespan of hematopoietic stem cells, *Nat Med* **12** (4):446–451 (2006).

[26] Morrison SJ, Kimble J: Asymmetric and symmetric stem-cell divisions in development and cancer, *Nature* **441** (7097):1068–1074 (2006).

[27] Lee HS et al: Clonal expansion of adult rat hepatic stem cell lines by suppression of asymmetric cell kinetics (SACK), *Biotechnol Bioeng* **83** (7):760–771 (2003).

[28] Driever PH et al: Valproic acid induces differentiation of a supratentorial primitive neuroectodermal tumor, *Pediatr Hematol Oncol* **21** (8):743–751 (2004).

[29] Palmer TD, Takahashi J, Gage FH: The adult rat hippocampus contains primordial neural stem cells, *Mol Cell Neurosci* **8** (6):389–404 (1997).

[30] Son H et al: Pairing of forskolin and KCl increases differentiation of immortalized hippocampal neurons in a CREB Serine 133 phosphorylation-dependent and extracellular-regulated protein kinase-independent manner, *Neurosci Lett* **308** (1):37–40 (2001).

[31] Warashina M et al: A synthetic small molecule that induces neuronal differentiation of adult hippocampal neural progenitor cells, *Angew Chem Int Ed Engl* **45** (4):591–593 (2006).

[32] Wichterle H et al: Directed differentiation of embryonic stem cells into motor neurons, *Cell* **110** (3):385–397 (2002).

[33] Benayahu D, Akavia UD, Shur I: Differentiation of bone marrow stroma-derived mesenchymal cells, *Curr Med Chem* **14** (2):173–179 (2007).

[34] Grigoriadis AE, Heersche JN, Aubin JE: Differentiation of muscle, fat, cartilage, and bone from progenitor cells present in a bone-derived clonal cell population: Effect of dexamethasone, *J Cell Biol* **106** (6):2139–2151 (1988).

[35] Lassar AB, Paterson BM, Weintraub H: Transfection of a DNA locus that mediates the conversion of 10T1/2 fibroblasts to myoblasts, *Cell* **47** (5):649–656 (1986).

[36] Jaiswal N et al: Osteogenic differentiation of purified, culture-expanded human mesenchymal stem cells in vitro, *J Cell Biochem* **64** (2):295–312 (1997).

[37] Wu X et al: A small molecule with osteogenesis-inducing activity in multipotent mesenchymal progenitor cells, *J Am Chem Soc* **124** (49):14520–14521 (2002).

[38] Wu X et al: Purmorphamine induces osteogenesis by activation of the hedgehog signaling pathway, *Chem Biol* **11** (9):1229–1238 (2004).

[39] Sinha S, Chen JK: Purmorphamine activates the Hedgehog pathway by targeting Smoothened, *Nat Chem Biol* **2** (1):29–30 (2006).

[40] Engel FB et al: p38 MAP kinase inhibition enables proliferation of adult mammalian cardiomyocytes, *Genes Dev* **19** (10):1175–1187 (2005).

[41] Tseng AS, Engel FB, Keating MT: The GSK-3 inhibitor BIO promotes proliferation in mammalian cardiomyocytes, *Chem Biol* **13** (9):957–963 (2006).

[42] Perez OD et al: Inhibition and reversal of myogenic differentiation by purine-based microtubule assembly inhibitors, *Chem Biol* **9** (4):475–483 (2002).

[43] Rosania GR et al: Myoseverin, a microtubule-binding molecule with novel cellular effects, *Nat Biotechnol* **18** (3):304–308 (2000).

[44] Koprivica V et al: EGFR activation mediates inhibition of axon regeneration by myelin and chondroitin sulfate proteoglycans, *Science* **310** (5745):106–110 (2005).

[45] Jang YY et al: Hematopoietic stem cells convert into liver cells within days without fusion, *Nat Cell Biol* **6** (6):532–539 (2004).

[46] Xie H et al: Stepwise reprogramming of B cells into macrophages, *Cell* **117** (5):663–676 (2004).

[47] Hakelien AM et al: Reprogramming fibroblasts to express T-cell functions using cell extracts, *Nat Biotechnol* **20** (5):460–466 (2002).

[48] Lanza RP et al: Generation of histocompatible tissues using nuclear transplantation, *Nat Biotechnol* **20** (7):689–696 (2002).

[49] Chen S et al: Dedifferentiation of lineage-committed cells by a small molecule, *J Am Chem Soc* **126** (2):410–411 (2004).

[50] Anastasia L et al: Reversine-treated fibroblasts acquire myogenic competence in vitro and in regenerating skeletal muscle, *Cell Death Differ* **13** (12):2042–2051 (2006).

[51] Dovey HF et al: Functional gamma-secretase inhibitors reduce beta-amyloid peptide levels in brain, *J Neurochem* **76** (1):173–181 (2001).

[52] Watabe T et al: TGF-beta receptor kinase inhibitor enhances growth and integrity of embryonic stem cell-derived endothelial cells, *J Cell Biol* **163** (6):1303–1311 (2003).

[53] Zhang Y et al: A chemical and genetic approach to the mode of action of fumagillin, *Chem Biol,* **13** (9):1001–1009 (2006).

[54] Qisheng Zhang MBM et al: A small molecule synergist of the wnt/β-catenin signaling pathway, *Proc Natl Acad Sci USA* (in press).

[55] Choi SC et al: 5-azacytidine induces cardiac differentiation of P19 embryonic stem cells, *Exp Mol Med* **36** (6):515–523 (2004).

[56] Tsuji-Takayama K et al: Demethylating agent, 5-azacytidine, reverses differentiation of embryonic stem cells, *Biochem Biophys Res Commun* **323** (1):86–90 (2004).

[57] De Felice L et al: Histone deacetylase inhibitor valproic acid enhances the cytokine-induced expansion of human hematopoietic stem cells, *Cancer Res* **65** (4):1505–1513 (2005).

[58] Hsieh J et al: Histone deacetylase inhibition-mediated neuronal differentiation of multipotent adult neural progenitor cells, *Proc Natl Acad Sci USA* **101** (47):16659–16664 (2004).

[59] Marks PA, Jiang X: Histone deacetylase inhibitors in programmed cell death and cancer therapy, *Cell Cycle* **4** (4):549–551 (2005).

[60] Cheng D et al: Small molecule regulators of protein arginine methyltransferases, *J Biol Chem* **279** (23):23892–23899 (2004).

[61] Greiner D et al: Identification of a specific inhibitor of the histone methyltransferase SU (VAR)3-9, *Nat Chem Biol* **1** (3):143–145 (2005).

[62] Kubicek S et al: Reversal of H3K9me2 by a small-molecule inhibitor for the G9a histone methyltransferase, *Mol Cell* **25** (3):473–481 (2007).

[63] Lee MG et al: Histone H3 lysine 4 demethylation is a target of nonselective antidepressive medications, *Chem Biol* **13** (6):563–567 (2006).

[64] Eggert US, Mitchison TJ: Small molecule screening by imaging, *Curr Opin Chem Biol* **10** (3):232–237 (2006).

[65] Menendez P, Wang L, Bhatia M: Genetic manipulation of human embryonic stem cells: A system to study early human development and potential therapeutic applications, *Curr Gene Ther* **5** (4):375–385 (2005).

[66] Mondal K, Gupta MN: The affinity concept in bioseparation: Evolving paradigms and expanding range of applications, *Biomol Eng* ,**23** (2–3):59–76 (2006).

[67] Ding S et al: Synthetic small molecules that control stem cell fate, *Proc Natl Acad Sci USA* **100** (13):7632–7637 (2003).

[68] Chen S, Takanashi S, Zhang Q, Xiong W, Peters EC, Schultz PG: Reversine induces cellular reprogramming of lineage-committed mammalian cells, *Proc Natl Acad Sci USA* (in press).

[69] Hart CP: Finding the target after screening the phenotype, *Drug Discov Today* **10** (7):513–519 (2005).

[70] Pritsker M et al: Genomewide gain-of-function genetic screen identifies functionally active genes in mouse embryonic stem cells, *Proc Natl Acad Sci USA* **103** (18):6946–6951 (2006).

[71] Moffat J et al: A lentiviral RNAi library for human and mouse genes applied to an arrayed viral high-content screen, *Cell* **124** (6):1283–1298 (2006).

[72] Eggert US et al: Parallel chemical genetic and genome-wide RNAi screens identify cytokinesis inhibitors and targets, *PLoS Biol* **2** (12):e379 (2004).

[73] Perrimon N et al: Drug-target identification in *Drosophila* cells: Combining high-throughout RNAi and small-molecule screens, *Drug Discov Today* ,**12** (1–2):28–33 (2007).

[74] Mitsui K et al: The homeoprotein Nanog is required for maintenance of pluripotency in mouse epiblast and ES cells, *Cell* **113** (5):631–642 (2003).

[75] Takahashi K, Yamanaka S: Induction of pluripotent stem cells from mouse embryonic and adult fibroblast cultures by defined factors, *Cell* **126** (4):663–676 (2006).

[76] Evans MJ, Carlton MB, Russ AP: Gene trapping and functional genomics, *Trends Genet* **13** (9):370–374 (1997).

[77] Niwa H et al: An efficient gene-trap method using poly A trap vectors and characterization of gene-trap events, *J Biochem (Tokyo)* **113** (3):343–349 (1993).

[78] To C et al: The Centre for Modeling Human Disease Gene Trap resource, *Nucl Acids Res* **32** (database issue):D557–D559 (2004).

[79] *Centre for Modeling Human Disease, 2006* (cited; available from `http://www.cmhd.ca/genetrap/index.html`).

[80] Chen YT, Liu P, Bradley A: Inducible gene trapping with drug-selectable markers and Cre/loxP to identify developmentally regulated genes, *Mol Cell Biol* **24** (22):9930–9941 (2004).

[81] Dhara SK, Benvenisty N: Gene trap as a tool for genome annotation and analysis of X chromosome inactivation in human embryonic stem cells, *Nucl Acids Res* **32** (13):3995–4002 (2004).

[82] Scheel JR et al: Quantitative analysis of gene expression in living adult neural stem cells by gene trapping, *Nat Meth* **2** (5):363–370 (2005).

[83] Guo G, Wang W, Bradley A: Mismatch repair genes identified using genetic screens in Blm-deficient embryonic stem cells, *Nature* **429** (6994):891–895 (2004).

[84] Chambers I et al: Functional expression cloning of Nanog, a pluripotency sustaining factor in embryonic stem cells, *Cell* **113** (5):643–655 (2003).

[85] Tokuzawa Y et al: Fbx15 is a novel target of Oct3/4 but is dispensable for embryonic stem cell self-renewal and mouse development, *Mol Cell Biol* **23** (8):2699–2708 (2003).

[86] Huang Q et al: Identification of p53 regulators by genome-wide functional analysis, *Proc Natl Acad Sci USA* **101** (10):3456–3461 (2004).

[87] Paddison PJ et al: A resource for large-scale RNA-interference-based screens in mammals, *Nature* **428** (6981):427–431 (2004).

[88] Berns K et al: A large-scale RNAi screen in human cells identifies new components of the p53 pathway, *Nature* **428** (6981):431–437 (2004).

[89] Michiels F et al: Arrayed adenoviral expression libraries for functional screening, *Nat Biotechnol* **20** (11):1154–1157 (2002).

[90] Yuanxiang Zhao SD: A high throughput siRNA library screen identifies osteogenic suppressors in human mesenchymal stem cells, *Proc Natl Acad Sci USA* (in press).

9

REGENERATION SCREENS IN MODEL ORGANISMS

CHETANA SACHIDANANDAN AND RANDALL T. PETERSON
Cardiovascular Research Center, Massachusetts General Hospital and Harvard Medical School, Boston, Massachusetts

The ability to re-create lost body parts has been a fascination of humans for centuries. The multiheaded monster Hydra in Greek mythology and the *asura*-king Ravana in Hindu mythology attained their invincibility through their ability to regrow a new head every time one was severed. This ability is known as *regeneration*, literally meaning "new birth." In biological terms we define it as the regrowth of a destroyed or amputated organ so as to differentiate it from the "repair" that occurs in cases of minor injury to a small portion of any organ or tissue. The dream of perfect regeneration remains elusive to humans in spite of great advances in medicine and biology in the twentieth century. In fact, we still understand very little about the molecular biology of regeneration compared to early embryonic development. Historically, regeneration studies have focused on organisms that were not genetically tractable, including such examples as hydra, planaria, and amphibians. In the last few decades we have discovered that many of our favorite model organisms are capable of regeneration of various organs and structures. Even human beings are not completely lacking in regenerative capacity. Skeletal muscle, liver, and cornea in humans are capable of regenerating after loss or injury [1]. With the modern tools of genetics, molecular biology, and genomics, we are beginning to make rapid inroads into the mysteries of regeneration.

Chemical and Functional Genomic Approaches to Stem Cell Biology and Regenerative Medicine
Edited by Sheng Ding

Although regeneration appears to recapitulate many events of early embryogenesis, it is unlike development in that regenerating organs or parts of organs have to integrate with existing structures in order to give a functional whole. Most of the reconstruction is carried out by cells that have retained at least some plasticity and pluripotency (stem cells) in a mainly nonplastic postdifferentiation environment. One approach to understanding the problem of regeneration is to study the stem cells that give rise to a number of different cell types in regenerating tissue. These studies have been discussed in detail in other chapters of this book. The goal of such studies is to understand and finally be able to control the different signals that push stem cells down distinct paths of differentiation.

However, regeneration is not a cell autonomous process where the stem cells "singlehandedly" grow and renew the missing organ. It is a complex process involving the crosstalk and interaction between a number of different cell types (Figure 9.1). In the example of a regenerating amphibian limb, the immediate response to amputation is the closure of the wound and the containment of damage [2]. This process involves immediate-early signaling, migration of the epithelial cells, and accumulation of inflammatory and immune response cells to the site of injury. This is followed by the activation of the regenerative cells, that is, dedifferentiation of the growth arrested mesenchymal cells in the proximity of the injury and formation of a blastema [2]. A *blastema* is a collection of undifferentiated or dedifferentiated cells capable of proliferation, differentiation, and reconstruction. The activated cells in a blastema reenter the cell cycle and proliferate to provide cell mass to build the regenerating tissue. Finally, to obtain a functional organ it is essential to have proper differentiation and patterning of newly forming tissue [2]. Therefore, a detailed understanding of stem cell specification, proliferation, and differentiation would tell only part of the story. Understanding the role played by the surrounding cells, including the signaling and regulation from the environmental milieu, is equally important. For this reason, using whole organisms as models for studying regeneration has particular advantages. We can much more easily address effects of cellular interactions, systemic hormones, and stress-induced growth factors on the pluripotent cells in a whole-animal model system. Also, in cases of regeneration of three-dimensional organs (e.g., limb/tail regeneration), model organisms enable assessment of the importance of positional signals on the outcome of the process.

In many cases of regeneration, the regenerative cells are believed to be the product of dedifferentiation of existing differentiated cells of the organ itself, rather than resident or circulating pluripotent stem cells [3,4]. The process of dedifferentiation per se is the subject of intense study because the molecular mechanisms by which these "terminally" differentiated cells downregulate their tissue-specific transcriptional and structural proteins and upregulate cell cycle genes is still not well understood [4,5].

Because regeneration encompasses many disparate processes and because aspects of regeneration differ among various tissues and organisms, no single scientific approach is suitable for all purposes. Choice of model organisms and experimental approach are among the important considerations in selecting a strategy for studying regeneration.

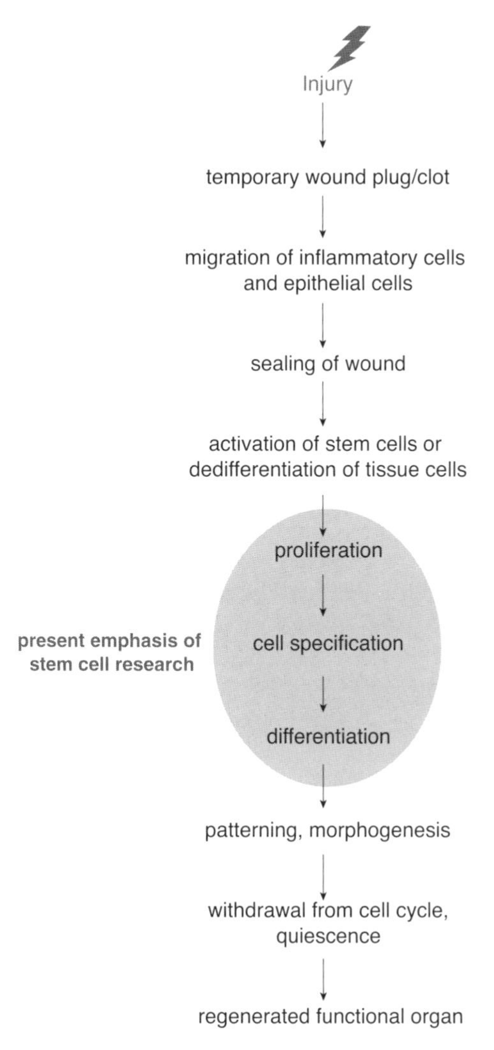

FIGURE 9.1 The present emphases of stem cell research address only a subset of the numerous molecular and cellular events that characterize regeneration.

The selection of strategy to study regeneration presents a number of choices. The first choice is of the model organism, and this must factor in a number of variables, including (1) the ability of the organism to regenerate, (2) the ease of maintenance and manipulation of the organism in the laboratory, (3) the tools available for dissection of the molecular biology of the process in the organism, and (4) the relevance of the organism to the biology of regeneration of other organisms. The second choice is to determine the best approach to study the process. The two basic approaches to a molecular study are (1) the candidate gene approach and (2) large-scale functional genomic screens for new genes. From the existing knowledge of genes and their known roles in development, it is possible to predict the functions of genes in regeneration. This process yields a number of candidates with putative roles in regeneration.

However, the main limitation of this process is that studies are restricted to a known group of genes and their known interactions. In order to expand our knowledge and understanding of the process of regeneration and to approach the problem in an unbiased way, screens are necessary. Screens allow us to identify completely novel genes or novel interactions between known genes in the process of regeneration. Large-scale screens also reveal patterns of gene regulation that would not be apparent in a study focused on a small number of genes.

In this chapter we will attempt to assess the advantages and limitations of selecting different model organisms and experimental approaches in the study of regeneration. We will discuss strategies for exploiting the technologies of molecular biology, genomics, genetics, and chemical biology.

9.1 THE CHOICE OF A MODEL ORGANISM

A number of model organisms with varying abilities to regenerate one or more organs have been used for regeneration studies, but none more so than the *planaria*, a species of free-living flatworms that possess a remarkable ability to regenerate lost body parts; they are able to re-create a whole new organism from a small body part [6]. They are easy to grow and maintain in the laboratory and can rapidly reproduce asexually. They are excellent candidates for regeneration screens, especially with the more recent progress of the planaria genome sequencing and the success of RNA interference (RNAi) demonstrated recently [7]. Reddien and colleagues performed an RNAi screen in planaria and identified 140 genes with roles in regeneration [8]. Although the planaria has been a very useful model system for regeneration screens, it is not yet known how much overlap there is between planarian and vertebrate regeneration. In order to limit the scope of this review we will focus primarily on tools and model systems for vertebrate regeneration.

A widely studied vertebrate model for regeneration is the amphibian. Various amphibians are capable of regenerating some of their organs, including limbs, neurons, spinal cord, eye, and parts of the heart [3]. Urodele amphibians (newts and salamanders) and anuran amphibians (*Xenopus*) have been used for limb and tail regeneration studies for decades. Both are capable of complete limb regeneration in the larval stages, but only urodeles can regenerate as adults [9]. Amphibians have been very useful for studies on regeneration because of the stereotypic progression of events that follows amputation of a limb. The existence of exhaustive studies of the anatomical and morphological changes in a regenerating limb amputee makes the amphibians excellent systems to dissect the significance of spatial and temporal changes occurring in the organ during regeneration. They have been used extensively in studying the molecular pathways involved in these changes, both by candidate gene approaches and by large-scale screens, as will be discussed below.

The ability of teleost fish to regenerate amputated fins was documented many years ago [10]. More recently, there has been growing interest in medaka and zebrafish regeneration, and they have been found to be able to regenerate fins, optic nerve, spinal cord, scales, and heart [11]. Owing to its success as a genetic model for the study of development, the zebrafish has now become a popular model for dissecting pathways of regeneration [11]. In addition to the traditional genetic approaches, innovative small-molecule screens and antisense strategies are also being adopted, making zebrafish a powerful and flexible system when large-scale, phenotype-driven approaches are needed.

Finally, mammalian models for regeneration exist, the most common of which is the mouse. Although mammals are among the least regenerative organisms, mice have the ability to regenerate liver, skin, skeletal muscle, and to a limited extent, some neurons. Although both the quality and quantity of genetic information are very good in the mouse, regeneration screens in adult organisms remain cumbersome. However, there has been significant success in conducting differential gene expression screens in conditions of injury and regeneration. Also, with the ability to generate targeted gene disruptions, candidate gene studies can be very successful [12]. There has also been success in developing and using explant cultures of organ of Corti in ear [13] or brain slices [14] to combine the advantages of culture systems while maintaining tissue environment. Single-myofiber cultures from mouse and rat muscles have also been successfully used to yield models of fiber-associated muscle satellite cells in skeletal muscle regeneration [15].

9.2 EXPERIMENTAL APPROACHES TO ADDRESS THE PROBLEM OF REGENERATION

Since the advent of genetics and molecular biology in the last few decades, a number of molecules have been identified that play important roles in regeneration. The approaches taken to identify and study these genes can be classified into two categories.

9.2.1 Candidate Gene Approaches

Molecular studies of regeneration have relied heavily on candidate gene approaches in the past. Identifying candidate players in the process of regeneration from what is known about development has been a fruitful strategy. Embryonic development and regeneration are very similar processes in that both involve creation of a functional organ through the processes of cell proliferation, specification, differentiation, and patterning. Hence, the signaling cascades and transcriptional regulation pathways involved in embryonic development are good candidates for roles in regeneration. A number of genes with interesting and varied roles in regeneration have been identified by this approach. We shall take the case of the homeobox transcription factors known as *msx* genes in the zebrafish as an example.

Msx genes have been shown to be expressed in the embryonic limb. Various chick and mouse studies have shown that *Msx1* plays a role in the proliferation of undifferentiated cells in the progress zone while *Msx2*, expressed in the regulatory structure known as apical ectodermal ridge (AER), contributes to the function of the AER in developing limb bud [16–18]. Akimenko and colleagues postulated that since *Msx* genes play an important role in the patterning of the embryonic limb bud, they might play a crucial role during the regeneration of fins in zebrafish [19]. They analyzed the expression pattern of the four zebrafish homologs of the mouse *Msx* genes and found that, although undetectable in an adult fin, these genes are upregulated dramatically in the regenerating fin. Moreover, they found that the expression patterns of the four genes are cell-type-specific, with *msxB* and *msxC* expressed in the mesenchyme and *msxA* and *msxD* expressed in the surrounding epidermis in the regenerating fin. The expression of *msxB* was also dependent on the position of amputation of the fin. Although the initial hypothesis for this study was based on the similarities between development and regeneration of the limb/fin, their analysis uncovered clear differences between the expression of *msx* genes in the two processes. Thus, the approach taken by the authors not only reiterated the parallels between development and regeneration but also highlighted crucial differences between them.

Continuing from this study, Nechiporuk and Keating [20] analyzed *msxB* expression further and found that it is expressed in the proliferating cells of the proximal blastema. By the outgrowth phase it becomes partitioned into the nonproliferating distal blastema cells. This study showed that the regeneration blastema is not a homogenous population of cells but rather exhibits proximodistal differences in cell proliferation and patterning.

Following a number of reports that pointed to a role for *msx* genes in the inhibition of differentiation during regeneration [21,22], Raya and colleagues analyzed the expression of *msx* genes in the zebrafish heart [23]. The authors were able to show that *msxB* and *msxC* are upregulated in regenerating zebrafish heart and that the expression of these genes was downstream of *notch1b* and *deltaC*. Thus, by extending the scope of the candidate gene approach, the authors showed that commonalities exist between different regeneration paradigms and also revealed a new key player involved in heart regeneration. The example above illustrates how the selection of candidates for regeneration studies from developmentally regulated genes can lead to increased understanding of the process of regeneration as well as yield new candidates for study.

As is evident from the case described above, taking a candidate gene approach to understanding regeneration can be very rewarding. However, the success of such approaches is limited by the present knowledge of the roles of developmental genes and their interactions. In order to identify hitherto unknown components of regeneration, cleverly designed screens are required. A number of different screens have been performed addressing different aspects of the process of regeneration.

9.2.2 Large-Scale Functional Genomic Screens

Screens for Genes that Are Differentially Expressed during Regeneration As regeneration progresses through its several stages, changes in gene expression occur with temporal and spatial precision. One strategy for discovering mediators of regeneration has been to identify such stage and region-specific genes by comparison of the transcriptomes of two contrasting conditions.

The simplest of differential expression screens are those comparing the gene expression profiles of normal adult tissue and regenerating tissue. This strategy has been adopted in a number of different systems, including a differential display RT-PCR screen in the newt limb [24], a suppression subtraction screen, and a differential display screen in regenerating caudal fin of zebrafish [25], and a suppression-subtraction screen on regenerating rat liver [26]. Such screens have contributed valuable insights into the process of regeneration as a whole. For instance, Lien and colleagues [27] performed a microarray gene expression comparison between normal and regenerating zebrafish heart at various times after injury and found groups of genes that are regulated concurrently. At 3 days postamputation (dpa), they observed that a number of genes involved in wound response and inflammation were upregulated that are probably required for the recruitment of immune response and inflammatory cells to the site of injury. This was followed by the expression of a group of secretory growth factor genes at 3–7 dpa, some of which are involved in the initiation of DNA synthesis and cell proliferation and others in processes such as angiogenesis and vasculogenesis. Finally, at 14 dpa a number of matrix metalloproteinases were upregulated that might be involved in remodeling of the amputated tissue and dissolution of the fibrin clot as the proliferating cardiomyocytes invade the injured area in preparation for regenerative outgrowth. Thus, in addition to generating a number of new candidates for further analysis, this study has revealed patterns of gene expression that suggest a sequence of events important for the proper progression of regeneration. The authors also compared their database of transcripts with differentially expressed transcripts found in a zebrafish fin regeneration screen being conducted in their lab and identified a number of genes in common. These molecules hint at the existence of a universal pattern of gene transcription for regeneration.

While screens such as those described above provide us with insights into the patterns of gene expression following injury and during regeneration, it is important to note that wound response and regeneration are two distinct processes that might not always be linked. For example, most mammalian organs respond to wounding by inflammation, wound closure, and repair, which results in scarring rather than regeneration. Thus, such screens risk identifying genes that participate in the response to wounding without contributing to regeneration. *Xenopus laevis* is a useful model organism to address this issue of specificity. Unlike the urodele amphibians, *Xenopus* lose their capability to efficiently regenerate amputated limbs as they metamorphose. Grow and colleagues conducted a microarray comparison screen on Xenopus limbs that were amputated at either the regeneration-competent stage 53 or the regeneration-incompetent stage 59 [28]. This comparison identified a number of genes that might be responsible for the differences in the regenerative capabilities of the two stages while excluding genes that would be upregulated in response to tissue injury in both cases. The comparison was also extended to two timepoints postamputation: 1 dpa and 5 dpa. On cross-comparison, the authors found that there were far fewer changes between 1 dpa and 5 dpa in stage 59 larvae than in the stage 53 larvae, perhaps accounting for the differences in regenerative capacity at these two stages. They also found a number of immune response genes upregulated at stage 59 that were not present in stage 53. This lends support to the authors' hypothesis that increased immune complexity might contribute to decreased regenerative capacity. Comparisons were also performed between gene expression data from previous screens on organisms as diverse as planaria and mouse, revealing some common signatures of the regenerative process.

A similar broad perspective screen was carried out on *mdx* mice with a mutation in the dystrophin gene. The *mdx* mouse is a model for the debilitating human disease, Duchenne muscular dystrophy (DMD), caused by the loss of dystrophin. While the human patients' muscles suffer repeated cycles of degeneration and inefficient regeneration, finally leading to death of the patient in adolescence, *mdx* mice manage to recover after the initial degenerative episode and grow to adulthood, albeit with increasing disability. Tseng and colleagues [29] carried out a genechip array comparison between adult wild-type mouse muscle and the *mdx* muscle in the recovery phase. They also compared existing gene expression databases of human DMD patient samples with their database. Although these results have to be viewed with caution because of the species, age, load-bearing history, and other differences between the samples, they might be utilized to design further studies to understand why human beings are incapable of recovery from the effects of mutation. Dystrophin is a membrane-associated protein that forms a bridge for communication between actin cytoskeleton, the transmembrane signaling proteins, and the extracellular matrix in the muscle. The authors found that transcripts for some actin-binding proteins (e.g., β-thymosin, calponin, and actin-related protein 2/3) were upregulated in the *mdx* mice, while no changes were observed in the DMD muscles. Further studies will be required to determine whether these differences contribute to or are a consequence of the progressive degeneration of DMD muscles.

Beyond addressing the global differences between two different conditions or two tissues with differing regenerative capacity, focused screens can identify small groups of genes with specific functions. For example, da Silva and colleagues [30] were interested in identifying molecules responsible for imparting proximodistal identities to cells in the newt limb blastema. They took advantage of the ability of retinoic acid (RA) to respecify the limb blastema in a graded and dose-dependent fashion, and performed a cDNA subtractive screen on limb blastema treated with RA or control. The RA-responsive genes were then analyzed for differential expression between proximal and distal blastema. Finally, cell surface proteins were selected because they postulated that the primary proteins determining spatial identity would be involved in cell–cell interactions. Only one gene met these stringent criteria, a GPI-anchored cell surface member of the complement pathway, CD59. They went on to show that inhibiting the activity of CD59 with blocking antibodies causes respecification of the proximodistal identities of the blastema cells in the regenerating newt limb.

The differential gene expression screens described above have been very powerful in identifying genes with crucial roles in regeneration and have provided critical insights into families of genes up- or downregulated at different stages. However, it has to be noted that out of the hundreds of genes identified in these screens, a small percentage are analyzed in any detail. Moreover, only a few of them turn out to be indispensable to the process of regeneration. Functional screens, where genes of interest are selected based on their "essentiality" for different stages of regeneration, have great advantages, as discussed below.

Genetic Screens for Mutations that Prevent Regeneration Screens that assay regeneration of injured or amputated tissue following mutagenesis can specifically address the issue of function, rather than merely expression, in regenerating tissue. Such screens can also identify non-cell-autonomous functions of genes in regeneration. Moreover, obtaining mutant alleles of genes important for regeneration provides us with powerful tools for the molecular dissection of these processes.

In contrast to expression screens, which can be performed on any model organism for which genomic and transcriptome information is available, genetic screens are limited to organisms where mutagenesis and large-scale mutant maintenance in the lab is possible. Among the vertebrate model organisms discussed in this review, the only strong candidates for genetic screens presently are zebrafish and mice. Although theoretically possible, large-scale screens in mice remain unwieldy and expensive, especially for a phenotypelike regeneration assayed in an adult organism. We shall focus our discussion on one of the more recent regeneration screens conducted in zebrafish.

Poss and colleagues [31] performed an ENU (*N*-ethyl-*N*-nitrosourea) mutagenesis screen in zebrafish and parthenogenetically generated diploid fish in order to look for recessive mutant phenotypes. Viable adult mutants were then subjected to caudal fin amputation, and the ones with reduced regeneration efficiency were identified. The screen was designed to identify temperature-sensitive alleles in order to also capture genes with early embryonic phenotypes. Cloning and further studies with these mutants have led to the identification of four genes with mutations that result in reduced or defective regeneration. The genes identified are involved in the proliferative phase of the proximal blastema [22,31], the formation of regeneration epithelium [32], and the survival of blastema cells [33]. The four genes identified by genetic screening are by no means the only genes involved in zebrafish fin regeneration. Carried to saturation, this strategy would undoubtedly identify many more genes essential for regeneration. If the high rate of regeneration genes found in the RNAi screen conducted in planarians holds true for vertebrates (140 of 1065 tested were required for regeneration) [8], a significant proportion of the total transcripts expressed in zebrafish might be required for regeneration. This apparent involvement of a large percentage of the genome in the process of regeneration raises questions about the specificity of the strategy of forward genetic screening for regeneration. Does the set of essential regeneration genes overlap completely with genes required for cell proliferation, survival, patterning, and so on? Or, does a unique population of "instructive" genes exist that play critical roles in sensing a need for and initiating the process of regeneration? (Figure 9.2).

Although it is not clear whether such "regeneration-instructive" genes exist, it is reasonable to hypothesize the existence of a class of genes that mediate the primary response of a regenerative tissue to injury. Such a group of genes or molecular signals would then be responsible for the "launching" of a cascade of events (wound response,

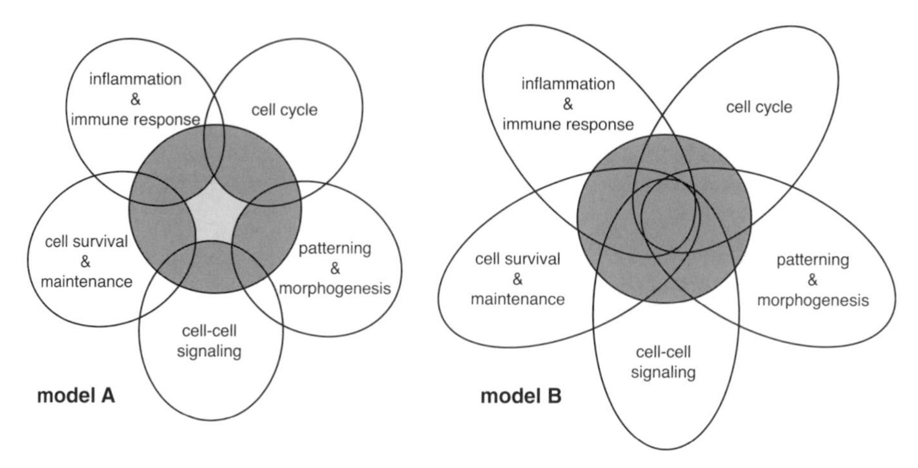

FIGURE 9.2 The total set of genes required for regeneration (gray area) intersects with genes essential for cell proliferation, differentiation, and other cellular processes. Is there a unique set of "regeneration instructive" genes (light gray area) that launch the regenerative process (model A), or do the genes required for routine cellular processes account for all regeneration genes (model B)?

inflammation, proliferation, cell specification and patterning), eventually culminating in regeneration. Because of the large number of essential genes that are required for the progression of regeneration, identification of such "instructive" genes via forward genetic screening with an endpoint of blocked regeneration may be challenging. More sophisticated screens may be required to distinguish between genes essential for the progression of regeneration and those with "instructive" roles in regeneration, if such genes exist.

An additional shortcoming of ENU-induced mutagenesis is that mutants often require a significant investment of time and effort in mapping the mutation to a gene. New tools like retroviral insertion mutagenesis [34] have been employed for studying embryogenesis and could be adapted for regeneration screens. A more recent report on the successful electroporation of antisense morpholinos into regenerating adult zebrafish fin blastemas [35] promises to make reverse genetics approaches like whole-transcriptome morpholino screens feasible for regeneration studies. Successful application of the morpholino antisense technology in other species like Xenopus [36] would make reverse genetic screens possible in these organisms as well.

Screens for Genetic and Molecular Enhancers of Regeneration The efficiency of regeneration is dependent on a number of different factors, including the speed of response to injury, the ability of the cells in the tissue to dedifferentiate, the number of stem cells present, and the presence of a favorable milieu for the function of regulatory/patterning genes. Each of these properties is subject to inhibition and control in a tissue. Genetic screens, where the goal is to obtain mutants with higher efficiency or more rapid regeneration than in wild types, might be able to identify such regulatory mechanisms. The identification of the MRL mouse strain, which possesses increased regenerative capacity, supports the conceptual feasibility of this approach [37,38]. However, attempts at genetic dissection of this phenotype revealed that the trait is quantitatively transmitted and probably conferred by a number of different genetic loci [39,40]. Thus, this example also illustrates why it might be much more difficult to obtain an "enhancing" mutation than deleterious mutations as the number of single genes capable of enhancing regeneration when mutated may be small. The probability of obtaining such mutations may be increased by focusing on individual aspects of regeneration such as activation of stem cells, dedifferentiation of tissue cells, or proliferation of blastemal cells, rather than on the ultimate outcome of regeneration. Although no such studies have been done to date, increasing knowledge of markers of each regeneration stage (e.g., *msx* genes in proliferative blastema) make it feasible to construct reporter genes and reporter lines of fish or frogs for different stages of regeneration. These reporter lines would then facilitate the design of stage-specific screens for regeneration-enhancing mutations.

Compared to obtaining genetic enhancers of regeneration, it is easier to design screens for exogenously applied growth factors or hormones that enhance regeneration. A number of such screens have been successful. For example, in one screen, neurotrophic growth factors and inhibitors of protein kinases (pertussis toxin and GF109203X) were identified as being capable of increasing the number of regenerating fibres in an entorhinalhippocampal slice explant model of nerve cell regeneration [14].

Another strategy for regeneration-enhancing screens that is becoming feasible with the increasing availability of genomic tools is the ectopic gene-expression screen. With the cloning of regeneration-specific and inducible promoters, it should become possible to construct whole-transcriptome expression libraries that would enable screens for genes that enhance the efficiency of regeneration when ectopically expressed in injured tissue.

Chemical Screens for Small Molecules that Affect Regeneration Relative to genetic mutations, small-molecule modulators of protein function hold several advantages for the study of regeneration [41]. Screens for the identification of both chemical suppressors and enhancers of regeneration can be conducted in model organisms. Regenerating organs can be exposed to small molecules at different stages postamputation in order to understand the temporal requirements of the protein activity being modulated. Exogenous application of small molecules also affords precise control over the spatial distribution of the inhibitor/activator on the regenerating tissue being exposed, which can reveal differential requirements of regulatory molecules during regeneration. Different doses of small molecules can also be used for quantitative attenuation or activation of the target molecular pathways, thus enabling modification of the gradients involved in patterning and uncovering differential sensitivity of processes to a regulatory factor. Small-molecule screens therefore have the potential to not only provide candidates for therapeutics but also yield tools for further dissection of the molecular pathways involved in the process of regeneration.

Small molecules have been successfully tested on regeneration, and both positive and negative modulators have been identified. Bayliss and colleagues [42] have demonstrated inhibitory effects of vascular endothelial growth factor (VEGF) antagonists on regenerative angiogenesis in zebrafish fin. On the other hand, an immunosuppressant drug FK506 has been shown to have an independent role in enhancing neural regeneration [43]. Further chemical modification of FK506 by Birge and colleagues [44] led to the identification of a derivative incapable of immunosuppression while retaining the regeneration-enhancing property.

There are independent examples of small-molecule screens conducted on cultured cells. Screening of a purine library for compounds promoting dedifferentiation of cultured muscle cells led to the discovery of myoseverin. Myoseverin is a trisubstituted purine that causes an increase in the efficiency of dedifferentiation of multinucleated myotubes by promoting their reversible fission into viable mononucleatedmyoblasts [45].

Wang and colleagues [46] have demonstrated that regeneration assays need not be limited to visual examination of successful regeneration, which might be limiting or even misleading in some cases. The authors used microarray analysis of RNA from hepatectomized mouse liver as a readout for the efficacy of regeneration following treatments with different doses of a regeneration promoting herbal extract [46]. The profiles of cell cycle genes and immediate early genes from different treatments were then compared and the results confirmed by DNA flow cytometry. From these comparisons they were able to conclude that high doses of the herbal extract upregulated immediate early and cell cycle genes while low doses appeared to have an opposite effect.

As is evident from the discussion above, a number of studies prove the feasibility of obtaining small-molecule modulators of regeneration, and there are separate examples of successful regeneration screens in model organisms. However, it is only recently that a synthesis of the two strategies have led to the design of successful small-molecule screens of regeneration.

High-throughput small-molecule screens for developmental processes have been shown to be feasible in zebrafish [47]. The small size of zebrafish embryos and larvae makes it easy to grow them in small volumes of water in 96-well plates, thus making them amenable to high-throughput screening. They are also very permeable to small molecules, enabling uniform exposure by adding the compounds to the water.

Traditionally, regeneration assays have been done on adult zebrafish, but their larger size makes them less suitable for large-scale compound screens. However, it has been shown that zebrafish larvae undergo fin regeneration through a process very similar to that in adults [48]. This finding was exploited by Tanguay and colleagues [49] to design a larva-stage, large-scale regeneration screen for small molecules that inhibit fin regeneration in zebrafish. From a library of 2000 bioactive molecules, they were able to identify a group of glucocorticoids that attenuated the regenerative ability of zebrafish fin, thus uncovering an important suppressor pathway of regeneration. Such a screen can be easily adapted for identifying regeneration-enhancing compounds as well. These studies have shown the feasibility of hunting for new chemical modulators of regeneration and also identification of potentially new protein targets involved in regeneration.

9.3 CONCLUSIONS

While candidate gene approaches are powerful because they take advantage of existing information about the function of genes in other settings, screens for new genes are essential to extend the study of regeneration into new areas. Differential gene expression screens, genetic screens, and chemical screens provide us with different kinds of information. Differential gene expression screens help us address the question of specificity by identifying genes that are unique to a particular phase of regeneration, while genetic screens address the issue of functionality by identifying only those genes that are required for the progression of regeneration. On the other hand, chemical screens help us discover potential therapeutically relevant molecules and furnish us with tools to modulate gene activities with temporal, spatial, and dose control. Combining the strengths of differential gene expression screens with functional screens should identify novel candidates for better understanding the process of regeneration.

As discussed in the previous sections, regeneration in different animals follows a variety of complex pathways. In order to elucidate these pathways and achieve the goal of improving the efficiency of regeneration, each of these pathways should be approached differently. Studies of chiefly cell-autonomous regeneration paradigms like those followed by blood cells may focus on the biology of the stem cells themselves [e.g., the hematopoietic stem cells (HSCs)]. In cases where the complex interplay of a number of cell types and the effect of environment are crucial (e.g., the regeneration of limb or heart), it may be more productive to adopt whole-organism models. While using cell culture systems as models makes it much easier to perform molecular studies and design large-scale screens, the information gained might not always be relevant to the regenerative process in the animal.

Since 1997 or so, stem cell biologists have been discovering that an intricate combination of differentiation promoting factors and culture conditions are required by the stem cells in culture to steer them down particular differentiation pathways. A complementary approach toward better understanding the non-cell autonomous signals and transcription factors required for the differentiation and patterning of the activated stem cells in the tissue milieu is being taken by regeneration biology. Just as successful regeneration is the result of cooperation and crosstalk between the regenerative stem cells and the injured tissue environment, the future success of regenerative medicine calls for continued cooperation between the fields of stem cell biology and regeneration biology.

REFERENCES

[1] Carlson BM: Some principles of regeneration in mammalian systems, *Anat Rec B New Anat* **287**:4–13(2005).

[2] Han M, Yang X, Taylor G, Burdsal CA, Anderson RA, Muneoka K: Limb regeneration in higher vertebrates: Developing a roadmap, *Anat Rec B New Anat* **287**:14–24 (2005).

[3] Brockes JP, Kumar A: Plasticity and reprogramming of differentiated cells in amphibian regeneration, *Nat Rev Mol Cell Biol* **3**:566–574 (2002).

[4] Odelberg SJ: Cellular plasticity in vertebrate regeneration, *Anat Rec B New Anat* **287**:25–35 (2005).

[5] Tanaka EM: Cell differentiation and cell fate during urodele tail and limb regeneration, *Curr Opin Genet Dev* **13**:497–501 (2003).

[6] Sanchez Alvarado A: Regeneration in the metazoans: Why does it happen? *Bioessays* **22**:578–590 (2000).

[7] Sanchez Alvarado A, Newmark PA: Double-stranded RNA specifically disrupts gene expression during planarian regeneration, *Proc Natl Acad Sci USA* **96**:5049–5054 (1999).

[8] Reddien PW, Bermange AL, Murfitt KJ, Jennings JR, Sanchez Alvarado A: Identification of genes needed for regeneration, stem cell function, and tissue homeostasis by systematic gene perturbation in planaria, *Dev Cell* **8**:635–649 (2005).

[9] Geraudie J, Ferretti P: Gene expression during amphibian limb regeneration, *Int Rev Cytol* **180**:1–50 (1998).

[10] Geraudie J, Singer M: Relation between nerve fiber number and pectoral fin regeneration in the teleost, *J Exp Zool* **199**:1–8 (1977).

[11] Poss KD, Keating MT, Nechiporuk A: Tales of regeneration in zebrafish, *Dev Dyn* **226**:202–210 (2003).

[12] Fernandez MA, Albor C, Ingelmo-Torres M, Nixon SJ, Ferguson C, Kurzchalia T, Tebar F, Enrich C, Parton RG, Pol A: Caveolin-1 is essential for liver regeneration, *Science* **313**:1628–1632 (2006).

[13] Feghali JG, Lefebvre PP, Staecker H, Kopke R, Frenz DA, Malgrange B, Liu W, Moonen G, Ruben RJ, Van de Water TR: Mammalian auditory hair cell regeneration/repair and protection: A review and future directions, *Ear Nose Throat J* **77**:276, 280, 282–275 (1998).

[14] Prang P, Del Turco D, Kapfhammer JP: Regeneration of entorhinal fibers in mouse slice cultures is age dependent and can be stimulated by NT-4, GDNF, and modulators of G-proteins and protein kinase C, *Exp Neurol* **169**:135–147 (2001).

[15] Beauchamp JR, Heslop L, Yu DS, Tajbakhsh S, Kelly RG, Wernig A, Buckingham ME, Partridge TA, Zammit PS: Expression of CD34 and Myf5 defines the majority of quiescent adult skeletal muscle satellite cells, *J Cell Biol* **151**:1221–1234 (2000).

[16] Coelho CN, Krabbenhoft KM, Upholt WB, Fallon JF, Kosher RA: Altered expression of the chicken homeobox-containing genes GHox-7 and GHox-8 in the limb buds of limbless mutant chick embryos, *Development* **113**:1487–1493 (1991).

[17] Reginelli AD, Wang YQ, Sassoon D, Muneoka K: Digit tip regeneration correlates with regions of Msx1 (Hox 7) expression in fetal and newborn mice, *Development* **121**:1065–1076 (1995).

[18] Robert B, Lyons G, Simandl BK, Kuroiwa A, Buckingham M: The apical ectodermal ridge regulates Hox-7 and Hox-8 gene expression in developing chick limb buds, *Genes Dev* **5**:2363–2374 (1991).

[19] Akimenko MA, Johnson SL, Westerfield M, Ekker M: Differential induction of four msx homeobox genes during fin development and regeneration in zebrafish, *Development* **121**:347–357 (1995).

[20] Nechiporuk A, Keating MT: A proliferation gradient between proximal and msxb-expressing distal blastema directs zebrafish fin regeneration, *Development* **129**:2607–2617 (2002).

[21] Koshiba K, Kuroiwa A, Yamamoto H, Tamura K, Ide H: Expression of Msx genes in regenerating and developing limbs of axolotl, *J Exp Zool* **282**:703–714 (1998).

[22] Nechiporuk A, Poss KD, Johnson SL, Keating MT: Positional cloning of a temperature-sensitive mutant emmental reveals a role for sly1 during cell proliferation in zebrafish fin regeneration, *Dev Biol* **258**:291–306 (2003).

[23] Raya A, Koth CM, Buscher D, Kawakami Y, Itoh T, Raya RM, Sternik G, Tsai HJ, Rodriguez-Esteban C, Izpisua-Belmonte JC: Activation of Notch signaling pathway precedes heart regeneration in zebrafish, *Proc Natl Acad Sci USA* **100** (Suppl 1): 11889–11895 (2003).

[24] Simon HG, Oppenheimer S: Advanced mRNA differential display: isolation of a new differentially regulated myosin heavy chain-encoding gene in amphibian limb regeneration, *Gene* **172**:175–181 (1996).

[25] Padhi BK, Joly L, Tellis P, Smith A, Nanjappa P, Chevrette M, Ekker M, Akimenko MA: Screen for genes differentially expressed during regeneration of the zebrafish caudal fin, *Dev Dyn* **231**:527–541 (2004).

[26] Liu ZW, Zhao MJ, Li ZP: Identification of up-regulated genes in rat regenerating liver tissue by suppression subtractive hybridization, *Sheng Wu Hua Xue Yu Sheng Wu Wu Li Xue Bao (Shanghai)* **33**:191–197 (2001).

[27] Lien CL, Schebesta M, Makino S, Weber GJ, Keating MT: Gene expression analysis of zebrafish heart regeneration, *PLoS Biol* **4**:(2006).

[28] Grow M, Neff AW, Mescher AL, King MW: Global analysis of gene expression in Xenopus hindlimbs during stage-dependent complete and incomplete regeneration, *Dev Dyn* **235**:2667–2685 (2006).

[29] Tseng BS, Zhao P, Pattison JS, Gordon SE, Granchelli JA, Madsen RW, Folk LC, Hoffman EP, Booth FW: Regenerated mdx mouse skeletal muscle shows differential mRNA expression, *J Appl Physiol* **93**:537–545 (2002).

[30] da Silva SM, Gates PB, Brockes JP: The newt ortholog of CD59 is implicated in proximodistal identity during amphibian limb regeneration, *Dev Cell* **3**:547–555 (2002).

[31] Poss KD, Nechiporuk A, Hillam AM, Johnson SL, Keating MT: Mps1 defines a proximal blastemal proliferative compartment essential for zebrafish fin regeneration, *Development* **129**:5141–5149 (2002).

[32] Whitehead GG, Makino S, Lien CL, Keating MT: fgf20 is essential for initiating zebrafish fin regeneration, *Science* **310**:1957–1960 (2005).

[33] Makino S, Whitehead GG, Lien CL, Kim S, Jhawar P, Kono A, Kawata Y, Keating MT: Heat-shock protein 60 is required for blastema formation and maintenance during regeneration, *Proc Natl Acad Sci USA* **102**:14599–14604 (2005).

[34] Amsterdam A, Hopkins N: Mutagenesis strategies in zebrafish for identifying genes involved in development and disease, *Trends Genet* **22**:473–478 (2006).

[35] Thummel R, Bai S, Sarras MP Jr, Song P, McDermott J, Brewer J, Perry M, Zhang X, Hyde DR, Godwin AR: Inhibition of zebrafish fin regeneration using*in vivo* electroporation of morpholinos against fgfr1 and msxb, *Dev Dyn* **235**:336–346 (2006).

[36] Heasman J: Morpholino oligos: Making sense of antisense? *Dev Biol* **243**:209–214 (2002).

[37] Clark LD, Clark RK, Heber-Katz E: A new murine model for mammalian wound repair and regeneration, *Clin Immunol Immunopathol* **88**:35–45 (1998).

[38] Heber-Katz E, Leferovich J, Bedelbaeva K, Gourevitch D, Clark L: The scarless heart and the MRL mouse, *Phil Trans Roy Soc Lond B Biol Sci* **359**:785–793 (2004).

[39] Masinde GL, Li X, Gu W, Davidson H, Mohan S, Baylink DJ: Identification of wound healing/regeneration quantitative trait loci (QTL) at multiple time points that explain seventy percent of variance in (MRL/MpJ and SJL/J) mice F2 population, *Genome Res* **11**:2027–2033 (2001).

[40] Yu H, Mohan S, Masinde GL, Baylink DJ: Mapping the dominant wound healing and soft tissue regeneration QTL in MRL x CAST, *Mamm Genome* **16**:918–924 (2005).

[41] Ding S, Schultz PG: Small molecules and future regenerative medicine, *Curr Top Med Chem* **5**:383–395 (2005).

[42] Bayliss PE, Bellavance KL, Whitehead GG, Abrams JM, Aegerter S, Robbins HS, Cowan DB, Keating MT, O'Reilly T, Wood JM et al: Chemical modulation of receptor signaling inhibits regenerative angiogenesis in adult zebrafish, *Nat Chem Biol* **2**:265–273 (2006).

[43] Gold BG: FK506 and the role of immunophilins in nerve regeneration, *Mol Neurobiol* **15**:285–306 (1997).

[44] Birge RB, Wadsworth S, Akakura R, Abeysinghe H, Kanojia R, MacIelag M, Desbarats J, Escalante M, Singh K, Sundarababu S et al: A role for schwann cells in the neuro-regenerative effects of a non-immunosuppressive fk506 derivative, jnj460, *Neuroscience* **124**:351–366 (2004).

[45] Rosania GR, Chang YT, Perez O, Sutherlin D, Dong H, Lockhart DJ, Schultz PG: Myoseverin, a microtubule-binding molecule with novel cellular effects, *Nat Biotechnol* **18**:304–308 (2000).

[46] Wang JY, Chiu JH, Tsai TH, Tsou AP, Hu CP, Chi CW, Yeh SF, Lui WY, Wu CW, Chou CK: Gene expression profiling predicts liver responses to a herbal remedy after partial hepatectomy in mice, *Int J Mol Med* **16**:221–231 (2005).

[47] Peterson RT, Link BA, Dowling JE, Schreiber SL: Small molecule developmental screens reveal the logic and timing of vertebrate development, *Proc Natl Acad Sci USA* **97**:12965–12969 (2000).

[48] Kawakami A, Fukazawa T, Takeda H: Early fin primordia of zebrafish larvae regenerate by a similar growth control mechanism with adult regeneration, *Dev Dyn* **231**:693–699 (2004).

[49] Mathew LK, Sengupta S, Kawakami A, Andreasen EA, Lohr CV, Loynes CA, Renshaw SA, Peterson, Tanguay RL: Unraveling tissue regeneration pathways using chemical genetics, *J Biol Chem.* **282**:35202–35210 (2007).

10

PROTEOMICS IN STEM CELLS

QIANG TIAN

Institute for Systems Biology, Seattle, Washington

W. ANDY TAO

Department of Biochemistry, Purdue University, West Lafayette, Indiana

Stem cells possess the unique properties of self-renewal (to generate at least one daughter cell identical to the mother cell) and differentiation (to give rise to multiple progenies). They hold great promise in regenerative medicine, tissue repair, and tumor biology [1]. A thorough understanding of stem cell biology calls for the constellation of the complete genetic parts list at the levels of RNAs, proteins, and regulatory networks. The complete sequencing of human genome and the genomes of several model organisms coupled with the spawning high-throughput genomic technologies spearheaded by DNA microarrays have led to a dramatically improved molecular description of stem cells and related differentiation programs at the transcriptome level [2–4]. However, until relatively recently, analyses of stem cell markers and signaling molecules during stem cell development at the protein level are conducted mainly at the individual gene scale using conventional protein analysis tools such as Western blotting and flow cytometry. Proteomics is the global study of proteins in a given biological system with the ultimate goal of defining the complete protein complements and its component behaviors. Proteomic research involves delineations of many diverse aspects of protein, including identification, quantification, compartmentalization, interaction with other partners, and posttranslational modifications, and thus is far more diverse and complex than its genomic counterparts.

Chemical and Functional Genomic Approaches to Stem Cell Biology and Regenerative Medicine
Edited by Sheng Ding

Several unique properties of stem cells warrant global proteomic interrogation:

1. The stem cells derived from various tissue origins are often defined by a distinctive panel of cell surface markers, all of which are proteins. Identification of novel protein markers for better definition and further purification of stem cells remain a daunting and utmost important task.
2. The maintenance of self-renewal and multipotency of stem cells depends on a number of protein factors presented either in the sustaining growth media or in the stem cell niche—the cellular microenvironment, which provides support and stimuli for stem cell self-renewal. Most of these factors are secreted protein growth factors or morphogens that can be readily studied using proteomic approaches.
3. The intricate signaling cascade transducing from growth factors to their corresponding stem cell surface receptors, adaptors, and then downstream target proteins involves a series of posttranslational modification events such as protein phosphorylation.
4. The final effectors determining stem cell self-renewal versus differentiation often involve a number of transcription factor and cofactor proteins engaging in dynamic protein–protein interaction complexes in different subcellular compartments.

All of these issues are poised to be addressed by the ever-maturing proteomic technologies (Figure 10.1). Although major challenges exist as to obtaining a sufficient number of stem cells for proteomic analysis, advances in proteomic technologies—especially mass spectrometry (MS)-based approaches—have enabled global capture of protein information during the developmental process of stem cells, contributing to our knowledge of the fundamental mechanisms of stem cell development. In this chapter we will illustrate how novel proteomic technologies can be applied in the identification of protein markers and regulatory proteins leading to a new dimension of information in stem cell fate regulation.

10.1 MASS SPECTROMETRY EMERGES AS THE MAJOR TECHNOLOGY FOR PROTEOMIC RESEARCH

Before the mid-1980s, protein analyses were confined at the single purified protein level using chemical and enzymatic methods. Completion of the Human Genome Project and the successful sequencing of the genomes of several other species have catalyzed the rapid development of powerful protein analysis tools, giving birth to the emerging field of proteomics.

The reason for this is that most genes from individual genomes have been identified, translated into protein sequences and their corresponding proteolytic peptide fragments, creating a database of theoretical masses of all peptide fragments encoded by the genome. Computational programs pioneered by SEQUEST have been developed to permit matching of each experimental peptide MS measurement against those in the theoretical database, thus identifying the gene that encodes the particular peptide [5]. The genes encoding large numbers of peptides from complex protein mixtures can, accordingly, be identified. Mass spectrometry, with its superior sensitivity, accuracy,

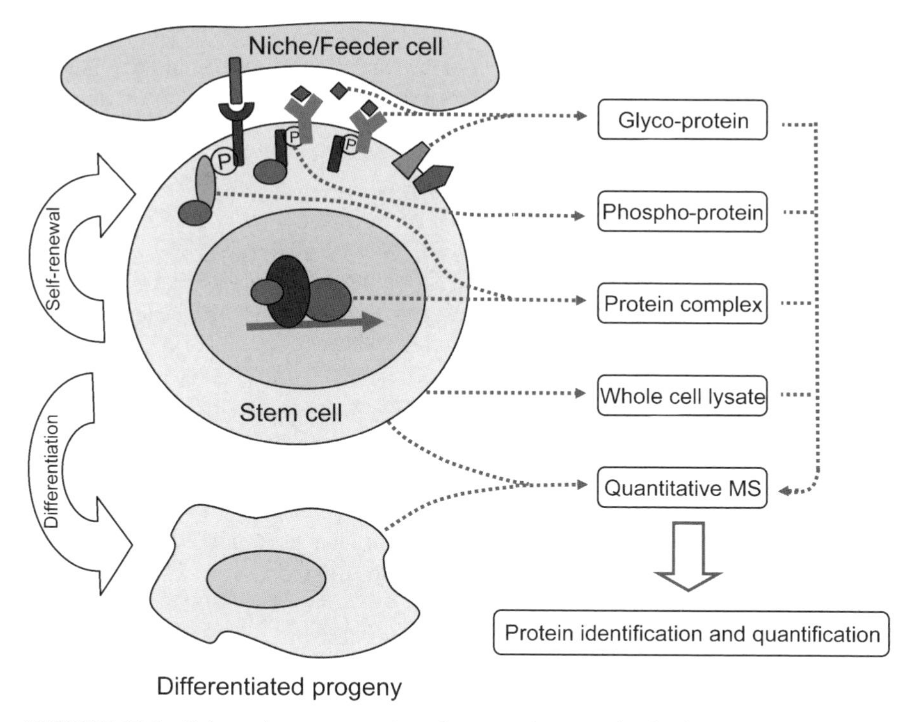

FIGURE 10.1 Schematic representation of proteomic strategies for investigating stem cell biology. Proteins secreted by or expressed on the cytoplasmic membrane of stem cell niche or feeder layer cells, as well as their corresponding receptors on stem cell surface can be isolated using glycoprotein capture reagents. Phosphorylation events after ligand–receptor engagement can be captured using phosphopeptide capture technologies. Protein–protein interaction complex of key signal transduction pathways can be isolated using affinity or immunopurification procedures. In addition, the whole cells can be lyzed or fractionated into subcellular fractions. All of these protein extracts will be subjected to quantitative proteomic analysis using mass spectrometry for comparison with differentiated progenies to identify stem-cell-specific growth factors, surface markers, and regulatory proteins controlling stem cell self-renewal and differentiation. (See insert for color representation.)

and throughput in protein and peptide identification, has emerged to become the most important tool for high-throughput proteomics studies [6]. Tandem mass spectrometry (MS/MS) has the capacity to select precursor peptide ions (first-stage MS) and then fragment them (second-stage MS) and record the resulting spectra through a process called *collision-induced dissociation* (CID). The CID spectra are sequence-specific and, when compared with theoretical spectra generated from sequence databases, can be used for large-scale protein identification.

10.1.1 Current Proteomic Technology Platforms

The analysis of biomolecules by MS has become routine after the introduction of two "soft" ionization techniques in the late 1980s: electrospray ionization (ESI) [7] and matrix-assisted laser desorption/ionization (MALDI) [8]. ESI and MALDI efficiently transfer intact biomolecules into the gas phase in ionized forms. ESI is the major method for proteomics since it can be directly coupled with liquid chromatography (LC) and produces multiply charged ions, which tend to result in more characteristic fragment ions for sequence information in MS/MS. MALDI remains a valuable alternative ionization technique for proteomics and is often used to complement results obtained by ESI. MALDI-MS is very sensitive and more tolerant than ESI to the presence of contaminants.

The MS instruments used for proteomics studies include a number of mass analyzers with different mass accuracy, mass resolution, or detection rate that provide a wide selection window for different analysis purpose. We describe here the respective technical parameters for the most commonly used mass analyzers.

Quadrupole Ion Trap (IT) In IT analyzers, ions are trapped and can therefore be accumulated over time in a physical device. ITMS is well known for its ability to perform multistage mass spectrometry (MS^n) experiments with high sensitivity and fast data acquisition. However, IT analyzers have space-charging effects that limit the ion-trapping capacity, resulting in limited resolution and mass accuracy (typically unit mass accuracy). The introduction of linear ion trap (LIT) analyzers with higher ion-trapping capacities has improved the dynamic range and the overall sensitivity of this technique [9], and LITs have been replacing classical three-dimensional trapping devices. Typically, IT and LIT instruments are operated with data-dependent acquisition to allow high-throughput analyses. The multiple-stage sequential MS/MS capability of IT and LIT, in which fragment ions are iteratively isolated and further fragmented, is very useful for the analysis of posttranslational modifications such as the identification of phosphorylation sites [10].

Fourier Transform Ion Cyclotron Resonance (FTICR) Mass Spectrometry In FTICR, ions are trapped and measured in a cubic magnetic device. Modern FTICR enjoys the highest resolving power of any form of mass analysis to date, and measurements in the low parts per million (ppm) (sub-ppm) range can be achieved. The drawback of FTICR is that the instrument is relatively delicate to maintain and needs frequent calibration. It has now been realized that combining a different mass analyzer to create a hybrid FTICR instrument such as LIT-FTICR has added robustness to this platform and allowed routine generation of low-resolution MS/MS spectra with accurate mass of the precursor ions. FTICR-MS performed on an LIT-ICR hybrid instrument allows true parallel full MS and MS/MS acquisition (not sequential), and it yields high-quality MS data that can be used for quantification.

Time of Flight (TOF) The mass-to-charge ratio (m/z) of an analyte ion in a TOF analyzers is deduced from its flight time through a field-free vacuum tube. Modern TOF analyzers have great resolution and mass accuracy. A resolving power over 10,000 has become routine on many TOF instruments, and with a proper mass calibration protocol, mass accuracies in the low-ppm range are achievable. TOF and MALDI are particularly well matched. The pulsed nature of the ionization event and the fact that desorption from a flat surface removes the "turnaround time" problem are attractive characteristics for mass analysis via TOF [11]. Furthermore, the need for high mass range and the desirability of acquiring the entire mass spectrum from a single ionization event make TOF an obvious choice as a mass analyzer for MALDI. However, today TOF mass analyzers are routely coupled with either ESI or MALDI when the TOF flying tube is geometrically orthogonal to the ion introduction pathway. A single TOF instrument does not have MS/MS capability, and the development of tandem TOF instruments (TOF-TOF) and hybrid TOF instruments such as quadrupole TOF (QTOF) offers such capability, resulting in high resolution and mass accuracy in both MS and MS/MS mode. In the MS mode, one mass analyzer acts as an ion guide to the TOF analyzer where the mass analysis takes place. In the MS/MS mode, the precursor ions are selected in the first mass analyzer and undergo fragmentation through collision-induced dissociation in the collision cell. The product ions are analyzed in the TOF device. High resolution and mass accuracy simplify the identification of peptides via database searches by tightening the search parameters and augmenting the confidence in the results.

Orbitrap Orbitrap was introduced to effect the separation of ions in an oscillating electric field [12,13]. This instrument presents characteristics similar to those of an FT-ICR spectrometer in terms of resolution and mass accuracy but without the burden of an expensive superconducting magnet.

MS-based proteomic strategies can be classified into top–down or bottom–up approaches. In the top–down approach, proteins are directly introduced into mass spectrometer and fragmented for sequence information [14]. Top–down experiments are typically carried out in FTICR instruments, and proteins are fragmented using electron-capture dissociation (ECD) [15]. At the current stage, top–down approach has been used mainly for single-protein analysis and is very useful for identifying posttranslational modifications. The majority of proteomic studies use bottom–up approaches, where protein samples are digested and the resulting peptides are routinely separated on a microscale liquid chromatograph followed by MS analysis. The peptide identity is assigned to the MS/MS spectra through database searching, which is performed according to established guidelines to generate consistent results. A subsequent statistical analysis of the search results is critical to ensure confidence in the identifications [16].

10.1.2 Application of MS/MS for Large-Scale Protein Identification of Stem Cells

One goal of stem cell proteomic research is to define, to the extent limited only by technical sensitivity, all the proteins expressed in certain stem cells to complement the existing transcriptome data and to identify novel protein markers and regulators governing the biological aspects of self-renewal and pluripotency. Given the vast dynamic range of protein expression in a cell—the concentration dynamics extend up to 10^6-fold—and the lack of protein amplification methodologies, any in-depth proteomic analysis of a given cell type using current available proteomic platforms would require a relatively large quantity of starting proteins extracted from a large number of cells, preferably in the magnitude of millions. While this requirement poses some challenge on tissue-specific stem cells that have to be isolated directly from primary tissues, the well-defined *in vitro* culture systems of embryonic stem cells (ESCs) present themselves as better target populations for proteomic interrogation, thus constituting a large portion of the current landscape of stem cell proteomics. Global proteomics has been performed in mouse and human ES cell lines for the purpose of establishing a protein catalog for stem cells and their developmental progenies. A panel of proteins from cell membrane, cytoplasm, and nucleus have been identified that can be used as cell-type benchmarks for population identification and cell differentiation.

Early attempts to define the mouse ESC proteome involve separation of proteins by two-dimensional gel electrophoresis (2D&E), cutting the protein spots from the gel, followed by identification of protein via MALDI MS [17,18]. Since visualization of protein spots by silver staining and manual excision of protein spots are prerequisite by this procedure, the sensitivity and throughput are compromised, resulting in a relatively small number of total proteins of relative higher abundance being identified—24 and 241 proteins, respectively, are identified by these two independent efforts, and none of the known ESC specific proteins are detected in these analyses. More recently, a similar approach was applied on human ESCs. Among the 844 spots analyzed, 685 protein species representing 434 unique proteins were identified [19]. Using an improved 2D LC-MS/MS system, Nagano et al. profiled mouse ESCs by QTOF MS and identified 1790 proteins from tryptic digests of whole-cell lysates [20]. In addition to the identification of hallmark ESC proteins Oct 3/4, Sox2, UTF1, they detected more than 360 potential nuclear and 260 membrane proteins. Of particular interest are the 60 ESC-specific proteins and 41 proteins that have previously been shown to be enriched in stem cells by DNA micrarray analyses [2]. Thirty of them were also identified in the human ESCs by Baharvand and colleagues [19].

Although global proteomic cataloging of undifferentiated stem cell lines lays the groundwork for further exploration of biological function of stem cells, more information can be gained through comparative analysis, in which a differential profile of protein expression between two or more developmental stages is carried out. The presence or absence of specific proteins may represent functional information that impinges on the regulatory signaling pathways. Van Hoof et al. carried out such a research comparing the differential proteome between human and mouse ESCs under both the undifferentiated and differentiated conditions using the more sensitive FTICR mass spectrometer [21]. They identified a large number (ranging from 1532 to 1871) of proteins for each condition, including 191 proteins exclusively expressed in undifferentiated ESCs of both mouse and human but not in the differentiated progenies. These proteins include both ESC benchmarks and uncharacterized proteins. The researches further validated some of the proteins using Western blot, immunofluorescence confocal microscopy, and fluorescence-activated cell sorting, and the results validated the subtractive proteomic approach. However, those proteins that are present both before and after differentiation with quantitative difference in abundance were excluded by this analysis, which can be better addressed by real quantitative proteomic approaches using stable isotope tagging, as will be discussed later.

10.1.3 Identification of Secreted Proteins in Conditioned Media Using MS/MS

Besides the intrinsic protein factors that affect stem cell fate, another important source for stem cell regulators resides in their growing environment, such as the feeder cell layers or the derivative conditioned media that sustain ESC growth. The original feeder layer cells for ESCs are of mouse origin—mouse embryonic fibroblast (MEF). While it is perfectly fitting for growing mouse ESCs, it was later shown that this approach inevitably introduced animal contaminants into human ESC populations, and thus led to searches for new human feeder cell substitutes. Several human cell lines have been established with successful supporting role of hESC growth [22,23]. Some of them have been examined proteomically for secreted proteins to identify potential regulators of hESCs. Lim and Bodnar analyzed MEF conditioned media using a 2D&E MALDI-TOF MS approach and found 136 unique proteins among 828 protein spots [24]. Prowse and colleagues used a combination of 2D LC-MS/MS and 2D&E MALDI-MS/MS to analyze the newly established human feeder cell HNF02-conditioned media, and identified 102 proteins belonging to 15 functional groups, including proteins that are involved in cell adhesion, proliferation, Wnt, and BMP signaling [25]. However, in order to sort out which of these proteins are essential for supporting hESC, one would want to compare the conditioned media of the ESC-supporting cell line with the nonsupporting cell line of the same origin, through quantitative proteomic approaches. Our own laboratory conducted such an analysis using isotope code affinity tags (ICAT; see text below) and two human bone marrow derived feeder cell lines and quantitatively identified more than 60 proteins [23] (Tian et al., unpublished data). A number of proteins identified in the human conditioned media are similar to those identified in the MEFs, including those involved in the Wnt and BMP signaling pathways, which is not surprising because of their role in supporting ESC growth.

10.1.4 MS/MS Coupled with Immunopurification for Identification of Protein Components of Key Signaling Pathways Affecting Stem Cell Fate

In addition to global identification and quantification of protein constituents in stem and progenitor cells, MS-based proteomic technologies have also been applied for the identification of protein factors in specific signaling pathways. The intricate protein interaction network is key to understanding almost any complex biological systems, including stem cells. While the classical yeast two-hybrid screens have long been the method of choice for dissecting protein–protein interaction for more than a decade, the advances in MS and proteomic technologies spawned a new way of evaluating protein interactions by using affinity purification followed by MS analysis. When coupled with classical biochemical purification strategies, the LC-MS/MS-based proteomic technologies have been proved efficient in studying protein complexes. Three steps are involved in such a scheme:

1. The protein of interest is used as bait by tagging to an affinity tag and introduced into appropriate cells.

2. The protein complex is isolated from cell lysates using affinity or immunopurification.
3. The protein components of the complex are analyzed by MS/MS after sodium dodecyl sulfate–polyacrylamide gel electrophoresis (SDS-PAGE), in-gel digestion, or in-solution proteolysis.

Using this approach, Tian and colleagues studied one of the key signal transduction pathways in stem cell development—the Wnt/β-catenin pathway—and identified 12 potential β-catenin interactors in an immune complex containing β-catenin. They confirmed 14-3-3ζ as a novel β-catenin interacting protein. 14-3-3ζ enhances β-catenin-dependent transcription and facilitates its activation by the survival kinase Akt, and stabilizes β-catenin in the cell [26]. More importantly, they demonstrated that 14-3-3ζ, an activated form of Akt, and β-catenin collocalize in intestinal stem cells (ISCs), balancing the antagonizing signaling events between the Wnt and BMP pathways to regulated ISC self-renewal [26–28].

As is the case with most affinity purification–MS/MS approaches, it is always challenging to distinguish between specific and background bindings. Although multiple iterations of MS analysis on rigorously controlled samples can help solve the problem, a better solution is to use quantitative proteomic technique such as the ICAT reagents (see text below), which allows large-scale quantitative comparison of protein mixtures derived from two different states. Ranish and colleagues have demonstrated that it could be used to distinguish between specific and background bindings in the analysis of macromolecule protein complexes [29]. The same strategy can be applied to the analysis of other protein complexes essential for stem cell development, such as the Wnt/β-catenin pathway. However, a major disadvantage is that a fairly large amount of starting materials will be needed given the multiplex of purification procedures.

10.2 QUANTITATIVE PROTEOMIC TECHNOLOGIES AND THEIR APPLICATION IN STEM CELL BIOLOGY

As genomic and proteomic researches enter the era of functional analysis, the quantitative nature of biological processes as exemplified by mRNA expression profiling entail global quantification of proteins. Quantitative proteomics encompass mainly two categories: global function proteomics and targeted function proteomics. Spearheaded by isotope-coded affinity tag (ICAT) technology, which specifically captures cysteine residues [30], there are a number of chemical isotope-tagging strategies for protein quantification, such as *N*-terminal labeling (iTRAQ [31], SPITC [32], etc.) and *C*-terminal labeling (esterification with isotope-labeled methanol [33]). In the meantime, versatile enzymatic isotope labeling techniques have also been developed including *in vivo* labeling through metabolism [34–40], stable isotope labeling of amino acids in culture (SILAC) [41], and enzymatic labeling of ^{16}O–^{18}O at the *C* terminus through the proteases [42–44]. For quantitative analysis of a particular protein or a protein group of specific interest, MS-based methods have also been developed involving the use of visible isotope-coded affinity tag [45], or isotope dilution mass spectrometry [46,47]. In these methods, standard peptides structurally analogous to proteins of interest are isotopically labeled and spiked into sample mixture; the relative ratio between the standard peptides and the targeted peptides in the protein of interest are obtained and the concentration of the standard peptides are used to deduce the absolute quantification of unknown protein [48].

All of these isotope methods are based on the essence that a distinct mass difference is tagged to a pair of (or multiple sets of) peptides by isotopes that have similar chemical properties. The quantification of one peptide relative to the other peptide in the pair is conducted by measuring the difference of signal intensity between the respective peptides with fixed mass difference (Figure 10.2). To circumvent the complexity and cost associated with the design and application of labeling reagents through chemical modifications, a label-free quantitative profiling scheme has also been introduced that makes use of extracted ion chromatograms from MS spectra—with appropriate normalization, the MS signal intensities of analyte ions are linearly correlated with the analyte concentrations, forming the basis of an analytical method for global quantification of differential protein expression in a complex mixture [49–51].

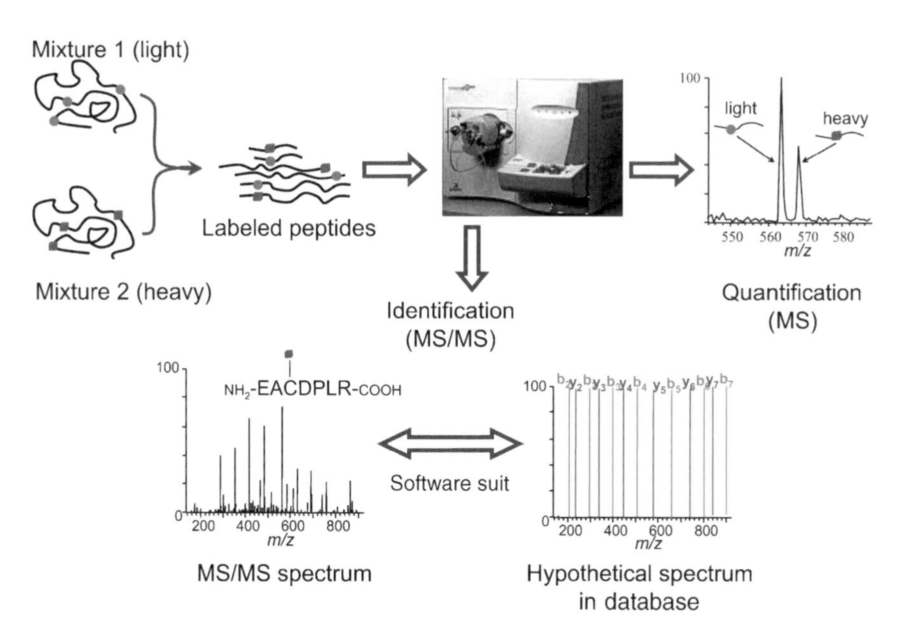

FIGURE 10.2 Quantitative proteomic scheme. The protein or peptide mixtures to be compared are labeled with either light or heavy isotopic reagents, combined and subjected to MS analysis. Protein quantification is achieved through comparison of the intensity of spectra pairs with same chemical property but different mass, whereas identification of protein is accomplished through MS/MS and database search. (See insert for color representation.)

TABLE 10.1 Comparison between Three Widely Used Labeling Methods for Quantitative Proteomics

	ICAT	iTRAQ	SILAC
Labeling method	Chemical derivatization	Chemical derivatization	Metabolic labeling
Labeling amino acid residues	Cysteine	*N*-Terminal and lysine residue	Lysine and arginine
Labeling stage	Labeling on proteins	Labeling on peptides	During cell culture
Quantitation stage	MS	MS/MS	MS
Purification	Yes	No	No
Mass difference of labeled peptides	9 Da	0 Da	6 Da
Multiplex	Two samples simultanously	Up to eight samples simultanously	Typically two samples simultanously
Applications	General	General	Cell culture

[a] Mass difference of the same peptide labeled with light and heavy acid cleavable ICAT.
[b] Mass difference for the same tryptic peptide labeled on Lys and Arg.

We will discuss three major isotopic labeling strategies (ICAT, ITRAQ, and SILAC; see Table 10.1) and their applications in stem and progenitor cell biology in the following paragraphs.

10.2.1 Isotope-Coded Affinity Tag (ICAT) Reagents

The development of a quantitative proteomic technique such as ICAT reagents by Aebersold and colleagues allows large-scale quantitative comparison of protein mixtures derived from two different states [30]. The original ICAT reagent consists of three elements: a thiol-specific reactive group that targets the cysteine residues on a protein or peptide, a linker that incorporates stable isotopes, and an affinity tag (biotin) that is used to isolate ICAT-labeled peptides. The reagent exists in two forms: a heavy form (D8), which contains eight deuteriums in the linker region, and a light form (D0), which contains no deuteriums. To compare the two-protein mixture of interest, each can be labeled with an ICAT reagent, one with the light form, the other with the heavy form. The labeled protein mixtures are then combined and proteolyzed into peptides that are further fractionated by HPLC columns into multiple fractions to reduce the complexity of peptides in each fraction. Cysteine-labeled peptides are purified via affinity avidin columns and subject to microLC-MS/MS analyses. Since the ICAT-labeled peptides are chemically identical with a mass difference of 8, they coelute in the columns as pairs. The relative quantification is determined by analyzing the areas under the curves for the isotopic peptide pairs. More recent improvement of the technology includes using solid phase as support, and the development of cleavable ICAT reagents that remove the ICAT group before MS analysis to significantly increase mass accuracy [52].

The first application of the ICAT-MS/MS approach in stem/progenitor cells was conducted on multipotent hematopoitic progenitor cell lines (EML and MPRO) derived from mouse bone marrow, which represent two distinct hematopoietic stages [53]. These two cell lines provide an excellent model system for studying the molecular control of early myeloid lineage development. The ICAT MS/MS method quantitatively identified 2919 cystein-containing peptides derived from 672 unique proteins, including transcription factors, signal transduction proteins, cell surface proteins, and differentiation markers, as well as various metabolism-related enzymes. Among them are stem cell marker c-Kit and its ligand Kitl, which are highly expressed in the more primitive EML cells. A concurrent DNA microarray analysis on the same pair of cell lines identified more than 500 genes expressed highly in the EML cells and more than 600 genes expressed highly in MPRO cells. The expression ratios of 425 proteins have been mapped to their corresponding mRNA levels. A total of 150 signature genes (those with significant changes at either the protein and/or the mRNA level, $p<0.01$) were identified. Although 113 signature genes (76%) exhibited changes for mRNAs and their cognate proteins in the same direction, only 29 of them changed significantly at both mRNA and protein levels and were thus dubbed correlated genes. In total, 67 genes showed significant change at the mRNA but not the protein level, whereas 52 genes showed significant change at the protein but not the mRNA level. The correlation coefficient between mRNA and protein is 0.64 for the signature genes and 0.59 for all the genes examined. This study suggests that posttranscriptional control is an important regulatory scheme during early hematopoiesis, leading to the hypothesis that a complete view of hematopoiesis requires the integration of multiple high-throughput platforms assessing both transcriptome and proteome. More recently, our own lab applied the ICAT approach to the analysis of human ESC and its derivative liver progenitor cells and quantitatively compared more than 600 proteins in these two populations, including potential novel regulators of hESCs (Tian et al., manuscript submitted).

10.2.2 Isobaric Tagging for Relative and Absolute Quantitation (iTRAQ)

Thiol-specific reagents (e.g., ICAT) have the advantage that they selectively isolate cysteine-containing peptides from a complex sample mixture and therefore reduce the complexity of the sample to be analyzed by MS. The reagents have the disadvantage that proteins and peptides that do not contain any cysteines are transparent to the method and that in general relatively low sequence coverage of the identified proteins is achieved. In an attempt to isotopically label every peptide in a protein digest, *N*-terminal labeling protocols have been described. Among different labeling approaches, iTRAQ is a distinctive approach that quantifies differences in peptide levels at the MS/MS stage [31]. The iTRAQ reagents consist of a reporter group, a balance group, and an amino-reactive group. Peptides of the same sequence labeled with iTRAQ reagents are identical en masse in single MS mode, but produce strong, diagnostic, low-mass MS/MS signature ions (e.g., m/z 114–117), allowing for the quantitation of up to four different samples simultaneously (more recently the reagents were further developed to analyze up to eight samples simultaneously). Any given peptide labeled with different iTRAQ reagents has the same nominal mass. This is an important feature for enhancing the sensitivity of the analysis of low-abundance samples. In addition, quantitative analysis is conducted in the MS/MS experiments with a high signal-to-noise (S/N) ratio, resulting in more accurate measurements that could be readily achieved for peptides of low abundance. Virtually every peptide is quantified using iTRAQ or other amine-specific reagents, and the enormous complexity of the peptide mixtures generated by the digestion of complex protein mixtures makes *N*-terminal peptide tagging challenging for proteomewide analyses. However, the method is in particular appealing for the analysis of relatively simpler complex such as immunopurified protein mixtures.

To characterize the molecular signature of hematopoietic stem cells (HSCs) and their progeny, Unwin is group carried out iTRAQ quantitative proteomics approach, and examined long-term reconstituting hematopoietic stem cells (LSK^+) and non-long-term reconstituting hematopoietic stem cells (LSK^-) [54]. The researchers identified 948 proteins with 145 of them changing in protein expression ratio, 54% could not be matched to transcriptome data on the same cell populations, confirming the importance of posttranscriptional regulation during HSC development. Among the proteins enriched in LSK^+ cells are the proteins involved in glycolytic enzymes, glucose transporter and antioxidant indicating that the LSK^+ cells are adapted for anaerobic environments by a hypoxia-related change in proteins controlling metabolism and oxidative protection, and shift away from anaerobic metabolism as they mature [54]. One advantage of the iTRAQ approach is that no *in vitro* culture is required and thus can be readily applied to freshly isolated stem cells and their progenitor cell populations.

10.2.3 Stable Isotope Labeling with Amino Acids in Cell Culture (SILAC)

Metabolic protein labeling has been used to introduce stable tags into proteins for quantitative proteomics. In SILAC, two groups of cells are grown in culture media identical in all respects except one—the first medium contains the "light" and the second medium contains the "heavy" form of one or two particular amino acids. Through the use of essential amino acids (those not synthesizable by the cell type), the cells are forced to use the labeled or unlabeled form. Thus, with each cell doubling, the cell population replaces at least half of the original form of the amino acid, eventually incorporating 100% of a given light or heavy form of the amino acid. Although different amino acids (e.g., leucine) were used in the past, more recently proteins are typically labeled via the metabolic incorporation of heavy isotope forms of lysine and arginine (Lys-$^{13}C_6$ and Arg-$^{13}C_6$). In most proteomic studies, proteins are digested with trypsin. The resulting tryptic peptides of the identical sequence but labeled with light or heavy Lys and Arg coelute on the chromatography and have the fixed mass difference in MS. The SILAC approach is straightforward and highly efficient—100% of the sample is available for analysis. This allows the experimenter to use any method of protein or even peptide purification (after enzymatic digestion) without introducing error into the final quantitative analysis. The drawback of the approach is that it is applicable only for cell culture experiments and the media with heavy amino acids are relatively expensive.

One successful application of SILAC technology in stem cell research was carried out to distinguish differential usage of signaling proteins in response to distinct growth factors during mesenchymal stem cell (MSC) differentiation [55]. Human MSCs are nonhematopoietic cells that reside within the bone marrow stroma, which can give rise to osteoblasts, adipocytes, and chondrocytes in response to various growth factors. The osteogenic differentiation of MSC relies on epidermal growth factor (EGF), but not platelet-derived growth factor (PDGF). To identify the difference between these two signaling pathways, Kratchmarova and colleagues [55] first grew the hMSCs in media containing normal and two heavier forms of arginine, then treated the cells with EGF and PDGF and immunoprecipitated the cell lysates with anti-phosphortyrosine-specific antibodies to isolate tyrosine phosphorylated protein complexes. The isolated proteins were trypsinized and analyzed using QTOF and LTQ-FT mass spectrometry. The researchers quantitatively identified close to 300 proteins, about 90% of which are shared signaling proteins by both growth factors. Among the proteins that are exclusively activated by PDGF is phosphatidylinositol 3-kinase (PI3K), suggesting a possible role contributing to the unresponsiveness to PDGF. More importantly, when a chemical inhibitor of PI3K was introduced into the MSC culture in the presence of PDGF, a similar level of differentiation was achieved. This research clearly demonstrated the power of quantitative proteomics in delineating molecular basis of stem cell differentiation.

10.2.4 Application of Other Quantitative Proteomics Technologies in Stem and Progenitor Cells

Other than the aforementioned three major quantitative MS approaches, a number of other researchers also employed traditional 2D&E-based approaches to quantitatively measure changes in protein expression in several other stem cell species. For instance, Salasznyk's group: employed quantitative proteomic technique onto mesenchymal stem cells, and studied the differentiation of mesenchymal stem cell after osteogenic induction [56]. Their results brought up an interesting idea: that focusing of gene expression in specific functional clusters as the basis of stem cell differentiation, that is, the specific extracellular matrix–directed osteogenic cell differentiation, was induced by focusing of the expression of genes involved in Ca^{2+}-dependent signaling pathways. Lee and colleagues characterized the adipocytic differentiation from human mesenchymal stem cells using 2D&E MALDI-TOF and identified 32 protein spots with differential expression [57]. Using 2D&E and LC/MS/MS, Wang's group studied the neural differentiation of mouse ESCs and identified 23 proteins with an increase or decrease in expression or phosphorylation after differentiation [58].

10.3 OTHER EMERGING PROTEIN/PEPTIDE-CAPTURING TECHNOLOGIES

As discussed earlier, even with the most powerful MS instruments available, one can identify fewer than 2000 proteins in stem cells, which represents only a small fraction of the whole proteome. This has propelled an extensive search for subproteome-capturing technologies to reduce the proteome complexity which is better suited for subsequent MS analysis. One such technology that is of keen interest to stem cell researchers is the glycoprotein capture method, because many cytoplasmic membrane proteins are glycosylated protein.

10.3.1 *N*-Glycosylated Protein Isolation Technologies

Ruedi Aebersold and colleagues invented a new hydrazine chemistry that reduces the protein complexity by isolation of *N*-linked glycoprotein from a protein mixture [59]. The principle of this method is to oxidize the sugar ring groups of *N*-glycosylated proteins, converting them to aldehyde groups that can be coupled to the amine groups linked to solid-phase beads. All *N*-glycosylated proteins can then be isolated by washing the unattached proteins from the beads. The bead-lined protein samples are digested with trypsin, and the unlinked tryptic peptides washed away—removing perhaps 95% of the protein mass. Then the peptides from the cell membrane lysates or serum are labeled at their amino groups with a light isotope of succinic anhydride (D0, containing no deuteriums), and those from the more differentiated stages are labeled with a heavy isotope (D4, containing four deuteriums) form of succinic anhydride after the epsilon amino groups of the lysine residues have been converted to homoarginine. The peptides are released from the beads by a glycosidase, the peptides derived from two cell populations are mixed together and then fractionated on a reverse-phase column, and the individual fractionated mixtures are deposited in a MALDI microtiter plate and then delivered to the MALDI MS/MS for the comparative quantitative analyses and a determination of the amino acid sequence of the peptide. From these peptide sequences, the encoding genes can be identified.

Another group, led by Toshiaki Isobe, developed a lectin affinity capture technology for the identification of *N*-linked glycoprotein [60]. Glycoproteins were purified on concanavalin A beads, which selectively bind glycoproteins with high-mannose and hybrid-type *N*-glycans. Trypsin digestion generates glycopeptides, which are isolated once again on concanavalin A beads. Eluted glycopeptides are released by PNGase F cleavage in the presence of $H_2{}^{18}O$, resulting in aspartic acid-^{18}O-labeled peptides, followed by two-dimensional liquid chromatography (2DLC)-MS analysis. An advantage of this method is that different lectins can be used to capture proteins with selected glycans; moreover, the noncovalently bound sugar chains are still intact and can be eluted with appropriate monosaccharides for analysis. The approach is not as general as the hydrazine chemistry method. Using a lectin column to select glycopeptides, results in binding of only a subset of *N*-glycans. Because the purification is based on the affinity, peptides derived from very abundant proteins will potentially contaminate the bound glycopeptides. Tagging of during ^{18}O the cleavage is important to avoid false positives due to nonenzymatic deamidation of asparagine to aspartic acid.

The most important reason for applying the glycoprotein capture technology to stem and progenitor cell biology is that many of the membrane proteins are glycoproteins, so one can potentially identify more cell surface markers that otherwise might not be revealed by other peptide capturing technology (e.g., ICAT) alone. This method also achieves the purification of a subset of the complete proteome and accordingly allows in-depth investigation of proteins that would ordinarily be lost because of their inappropriately low abundance. In addition, secreted proteins by stromal cell lines or conditioned media can also be analyzed readily. Both techniques have been applied to the analysis of stem cell membrane proteins. An application of the lectin approach on mouse ESC (D3) identified 324 proteins, 235 of which carried putative signal sequence and transmembrane segments [61]. Among these proteins are cell adhesion molecules; clustering of differentiation-related (CD) molecules; and receptors for LIF, FGF, Wnt, and BMP, some of which have been shown to be critical for ESC self-renewal or differentiation. Our own lab has applied an improved version of hydrazine glycopeptide capture on mouse ESC (E14) and identified more than 200 microsomal proteins, with about 90% of them containing a conserved *N*-glycosylation sites (Tian et al., unpublished data).

10.3.2 Novel Phosphopeptide-Capturing Technology

Protein phosphorylation plays an essential role for the proper transduction of extracellular signals through the membrane receptor/coreceptor complex to the intracellular signaling complexes. It is of enormous interest to capture the phosphorylation events globally during the developmental processes of stem cells. Tao and colleagues developed a novel technology for global analysis of phosphorylated peptides using dendrimer as a soluble polymer support and LC-MS/MS [62]. Three major steps are involved: (1) the peptides are methylated to protect the carboxy group from subsequent reactions; (2) methylated phosphopeptides are subject to a dendrimer conjugation catalyzed by EDC and imidazole, and the subsequent isolation of covalently bound phosphopeptides from nonphosphopeptides by size-selective chromatography; and (3) phosphopeptides are released by acid hydrolysis and analyzed by LC-MS/MS. By employing coupling with an initial immunoprecipitation using antiphophotyrosine antibody and stable isotopic labeling, the researchers identified all known tyrosine phosphorylation sites within the immunoreceptor tyrosine–based motifs (ITAM) of the T-cell receptor CD3 chains on a total of 97 tyrosine phophoproteins. Using the same strategy and purified β-catenin protein, we were able to identify several putative phosphorylation sites on β-catenin by the survival kinase Akt, including serine 552 [26,63]. To confirm the *in vivo* function of this phosphorylation event in the context of intestinal stem cells (ISCs), an antibody that specifically recognizes phosphoserin 552 was generated; immunohistochemistry staining demonstrated that this new form of β-catenin phophorylation was specifically localized in the nucleus of ISCs. Thus, through a combined approach of phosphopeptide capture and mouse genetics, a novel mechanism for controlling self-renewal of ISC was identified [63].

10.4 CONCLUDING REMARKS

The convergence of the burgeoning fields of proteomics and stem cell biology has already generated valuable information regarding stem cell proliferation and differentiation: from large-scale qualitative and quantitative profiling of stem cells and their progenies to the more focused delineation of specific signaling pathways (EGF, PDGF, Wnt/β-catenin, etc.). New regulatory schema governing stem cell development such as posttranscriptional regulation and antioxidant protection have been unveiled; novel components of key stem cell signal transduction complexes have been discovered. The future successful application of proteomic technologies in the stem cell will depend largely on improving the sensitivity of proteomic technologies through improved instrumentation, better subproteome capturing strategies, and the study of interaction between stem cells and the surrounding environmental cues. The combined efforts will result in heightened understanding of stem cell self-renewal and differentiation, leading to potential better therapeutics for regenerative medicine.

REFERENCES

[1] Weissman IL: Translating stem and progenitor cell biology to the clinic: Barriers and opportunities, *Science* **287**:1442–1446 (2000).

[2] Ramalho-Santos M, Yoon S, Matsuzaki Y, Mulligan RC, Melton DA: "Stemness": Transcriptional profiling of embryonic and adult stem cells, *Science* **298**:597–600 (2002).

[3] Park IK, He Y, Lin F, Laerum OD, Tian Q, Bumgarner R, Klug CA, Li K, Kuhr C, Doyle MJ et al: Differential gene expression profiling of adult murine hematopoietic stem cells, *Blood* **99**:488–498 (2002).

[4] Ivanova NB, Dimos JT, Schaniel C, Hackney JA, Moore KA, Lemischka IR: A stem cell molecular signature, *Science* **298**:601–604 (2002).

[5] Yates JR 3rd, Eng JK, McCormack AL: Mining genomes: Correlating tandem mass spectra of modified and unmodified peptides to sequences in nucleotide databases, *Anal Chem* **67**:3202–3210 (1995).

[6] Aebersold R, Goodlett DR: Mass spectrometry in proteomics, *Chem Rev* **101**:269–295 (2001).

[7] Fenn JB, Mann M, Meng CK, Wong SF, Whitehouse CM: Electrospray ionization for mass spectrometry of large biomolecules, *Science* **246**:64–71 (1989).

[8] Karas M, Hillenkamp F: Laser desorption ionization of proteins with molecular masses exceeding 10,000 daltons, *Anal Chem* **60**:2299–2301 (1988).

[9] Schwartz JC, Senko MW, Syka JE: A two-dimensional quadrupole ion trap mass spectrometer, *J Am Soc Mass Spectrom* **13**:659–669 (2002).

[10] Olsen JV, Mann M: Improved peptide identification in proteomics by two consecutive stages of mass spectrometric fragmentation, *Proc Natl Acad Sci USA* **101**:13417–13422 (2004).

[11] McLuckey SA, Wells JM: Mass analysis at the advent of the 21st century, *Chem Rev* **101**:571–606 (2001).

[12] Hardman M, Makarov AA: Interfacing the orbitrap mass analyzer to an electrospray ion source, *Anal Chem* **75**:1699–1705 (2003).

[13] Hu Q, Noll RJ, Li H, Makarov A, Hardman M, Graham Cooks R: The Orbitrap: A new mass spectrometer, *J Mass Spectrom* **40**:430–443 (2005).

[14] Kelleher NL: Top-down proteomics, *Anal Chem* **76**:197A–203A (2004).

[15] Zubarev RA: Electron-capture dissociation tandem mass spectrometry, *Curr Opin Biotechnol* **15**:12–16 (2004).

[16] Keller A, Nesvizhskii AI, Kolker E, Aebersold R: Empirical statistical model to estimate the accuracy of peptide identifications made by MS/MS and database search, *Anal Chem* **74**:5383–5392 (2002).

[17] Guo X, Ying W, Wan J, Hu Z, Qian X, Zhang H, He F: Proteomic characterization of early-stage differentiation of mouse embryonic stem cells into neural cells induced by all-trans retinoic acid *in vitro*, *Electrophoresis* **22**:3067–3075 (2001).

[18] Elliott ST, Crider DG, Garnham CP, Boheler KR, Van Eyk JE: Two-dimensional gel electrophoresis database of murine R1 embryonic stem cells, *Proteomics* **4**:3813–3832 (2004).

[19] Baharvand H, Hajheidari M, Ashtiani SK, Salekdeh GH: Proteomic signature of human embryonic stem cells, *Proteomics* **6**:3544–3549 (2006).

[20] Nagano K, Taoka M, Yamauchi Y, Itagaki C, Shinkawa T, Nunomura K, Okamura N, Takahashi N, Izumi T, Isobe T: Large-scale identification of proteins expressed in mouse embryonic stem cells, *Proteomics* **5**:1346–1361 (2005).

[21] Van Hoof D, Passier R, Ward-Van Oostwaard D, Pinkse MW, Heck AJ, Mummery CL, Krijgsveld J: A quest for human and mouse embryonic stem cell-specific proteins, *Mol Cell Proteomics* **5**:1261–1273 (2006).

[22] Richards M, Fong CY, Chan WK, Wong PC, Bongso A: Human feeders support prolonged undifferentiated growth of human inner cell masses and embryonic stem cells, *Nat Biotechnol* **20**:933–936 (2002).

[23] Cheng L, Hammond H, Ye Z, Zhan X, Dravid G: Human adult marrow cells support prolonged expansion of human embryonic stem cells in culture, *Stem Cells* **21**:131–142 (2003).

[24] Lim JW, Bodnar A: Proteome analysis of conditioned medium from mouse embryonic fibroblast feeder layers which support the growth of human embryonic stem cells, *Proteomics* **2**:1187–1203 (2002).

[25] Prowse AB, McQuade LR, Bryant KJ, Van Dyk DD, Tuch BE, Gray PP: A proteome analysis of conditioned media from human neonatal fibroblasts used in the maintenance of human embryonic stem cells, *Proteomics* **5**:978–989 (2005).

[26] Tian Q, Feetham MC, Tao WA, He XC, Li L, Aebersold R, Hood L: Proteomic analysis identifies that 14-3-3zeta interacts with beta-catenin and facilitates its activation by Akt, *Proc Natl Acad Sci USA* **101**:15370–15375 (2004).

[27] He XC, Zhang J, Tong WG, Tawfik O, Ross J, Scoville DH, Tian Q, Zeng X, He X, Wiedemann LM et al: BMP signaling inhibits intestinal stem cell self-renewal through suppression of Wnt-beta-catenin signaling, *Nat Genet* **36**:1117–1121 (2004).

[28] Tian Q, He XC, Hood L, Li L: Bridging the BMP and Wnt pathways by PI3 kinase/Akt and 14-3-3zeta, *Cell Cycle* **4**:215–216 (2005).

[29] Ranish JA, Yi EC, Leslie DM, Purvine SO, Goodlett DR, Eng J, Aebersold R: The study of macromolecular complexes by quantitative proteomics, *Nat Genet* **33**:349–355 (2003).

[30] Gygi SP, Rist B, Gerber SA, Turecek F, Gelb MH, Aebersold R: Quantitative analysis of complex protein mixtures using isotope-coded affinity tags, *Nat Biotechnol* **17**:994–999 (1999).

[31] Ross PL, Huang YN, Marchese JN, Williamson B, Parker K, Hattan S, Khainovski N, Pillai S, Dey S, Daniels S et al: Multiplexed protein quantitation in Saccharomyces cerevisiae using amine-reactive isobaric tagging reagents, *Mol Cell Proteomics* **3**:1154–1169 (2004).

[32] Lee YH, Han H, Chang SB, Lee SW: Isotope-coded N-terminal sulfonation of peptides allows quantitative proteomic analysis with increased de novo peptide sequencing capability, *Rapid Commun Mass Spectrom* **18**:3019–3027 (2004).

[33] Goodlett DR, Keller A, Watts JD, Newitt R, Yi EC, Purvine S, Eng JK, von Haller P, Aebersold R, Kolker E: Differential stable isotope labeling of peptides for quantitation and de novo sequence derivation, *Rapid Commun Mass Spectrom* **15**:1214–1221 (2001).

[34] Conrads TP, Alving K, Veenstra TD, Belov ME, Anderson GA, Anderson DJ, Lipton MS, Pasa-Tolic L, Udseth HR, Chrisler WB et al: Quantitative analysis of bacterial and mammalian proteomes using a combination of cysteine affinity tags and 15N-metabolic labeling, *Anal Chem* **73**:2132–2139 (2001).

[35] Martinovic S, Veenstra TD, Anderson GA, Pasa-Tolic L, Smith RD: Selective incorporation of isotopically labeled amino acids for identification of intact proteins on a proteome-wide level, *J Mass Spectrom* **37**:99–107 (2002).

[36] Oda Y, Huang K, Cross FR, Cowburn D, Chait BT: Accurate quantitation of protein expression and site-specific phosphorylation, *Proc Natl Acad Sci USA* **96**:6591–6596 (1999).

[37] Shi SD, Hendrickson CL, Marshall AG: Counting individual sulfur atoms in a protein by ultrahigh-resolution Fourier transform ion cyclotron resonance mass spectrometry: Experimental resolution of isotopic fine structure in proteins, *Proc Natl Acad Sci USA* **95**:11532–11537 (1998).

[38] Washburn MP, Ulaszek R, Deciu C, Schieltz DM, Yates JR 3rd: Analysis of quantitative proteomic data generated via multidimensional protein identification technology, *Anal Chem* **74**:1650–1657 (2002).

[39] Zhu H, Hunter TC, Pan S, Yau PM, Bradbury EM, Chen X: Residue-specific mass signatures for the efficient detection of protein modifications by mass spectrometry, *Anal Chem* **74**:1687–1694 (2002).

[40] Zhu H, Pan S, Gu S, Bradbury EM, Chen X: Amino acid residue specific stable isotope labeling for quantitative proteomics, *Rapid Commun Mass Spectrom* **16**:2115–2123 (2002).

[41] Ong SE, Blagoev B, Kratchmarova I, Kristensen DB, Steen H, Pandey A, Mann M: Stable isotope labeling by amino acids in cell culture, SILAC, as a simple and accurate approach to expression proteomics, *Mol Cell Proteomics* **1**:376–386 (2002).

[42] Stewart II, Thomson T, Figeys D: 18O labeling: A tool for proteomics, *Rapid Commun Mass Spectrom* **15**:2456–2465 (2001).

[43] Yao X, Freas A, Ramirez J, Demirev PA, Fenselau C: Proteolytic 18O labeling for comparative proteomics: Model studies with two serotypes of adenovirus, *Anal Chem* **73**:2836–2842 (2001).

[44] Lopez-Ferrer D, Ramos-Fernandez A, Martinez-Bartolome S, Garcia-Ruiz P, Vazquez J: Quantitative proteomics using (16)O/(18)O labeling and linear ion trap mass spectrometry, *Proteomics* **6**:S4–S11 (2006).

[45] Lu Y, Bottari P, Turecek F, Aebersold R, Gelb MH: Absolute quantification of specific proteins in complex mixtures using visible isotope-coded affinity tags, *Anal Chem* **76**:4104–4111 (2004).

[46] Barnidge DR, Goodmanson MK, Klee GG, Muddiman DC: Absolute quantification of the model biomarker prostate-specific antigen in serum by LC-Ms/MS using protein cleavage and isotope dilution mass spectrometry, *J Proteome Res* **3**:644–652 (2004).

[47] Leenheer AP, Thienpont LM: Applications of isotope dilution mass spectrometry in clinical chemistry, pharmacokinetics, and toxicology, *Mass Spectrom Rev* **11**:249–307 (1992).

[48] Buchnall M, Fung KUC, Duncan MW: Practical quantitative biomedical applications of MALDI-TOF mass spectrometry, *J Am Soc Mass Spectrom* **13**:1015–1027 (2002).

[49] Higgs RE, Knierman MD, Gelfanova V, Butler JP, Hale JE: Comprehensive label-free method for the relative quantification of proteins from biological samples, *J Proteome Res* **4**:1442–1450 (2005).

[50] Silva JC, Denny R, Dorschel CA, Gorenstein M, Kass IJ, Li GZ, McKenna T, Nold MJ, Richardson K, Young P et al: Quantitative proteomic analysis by accurate mass retention time pairs, *Anal Chem* **77**:2187–2200 (2005).

[51] Wang W, Zhou H, Lin H, Roy S, Shaler TA, Hill LR, Norton S, Kumar P, Anderle M, Becker CH: Quantification of proteins and metabolites by mass spectrometry without isotopic labeling or spiked standards, *Anal Chem* **75**:4818–4826 (2003).

[52] Zhou H, Ranish JA, Watts JD, Aebersold R: Quantitative proteome analysis by solid-phase isotope tagging and mass spectrometry, *Nat Biotechnol* **20**:512–515 (2002).

[53] Tian Q, Stepaniants SB, Mao M, Weng L, Feetham MC, Doyle MJ, Yi EC, Dai H, Thorsson V, Eng J et al: Integrated genomic and proteomic analyses of gene expression in Mammalian cells, *Mol Cell Proteomics* **3**:960–969 (2004).

[54] Unwin RD, Smith DL, Blinco D, Wilson CL, Miller CJ, Evans CA, Jaworska E, Baldwin SA, Barnes K, Pierce A et al: Quantitative proteomics reveals posttranslational control as a regulatory factor in primary hematopoietic stem cells, *Blood* **107**:4687–4694 (2006).

[55] Kratchmarova I, Blagoev B, Haack-Sorensen M, Kassem M, Mann M: Mechanism of divergent growth factor effects in mesenchymal stem cell differentiation, *Science* **308**:1472–1477 (2005).

[56] Salasznyk RM, Westcott AM, Klees RF, Ward DF, Xiang Z, Vandenberg S, Bennett K, Plopper GE: Comparing the protein expression profiles of human mesenchymal stem cells and human osteoblasts using gene ontologies, *Stem Cells Dev* **14**:354–366 (2005).

[57] Lee HK, Lee BH, Park SA, Kim CW: The proteomic analysis of an adipocyte differentiated from human mesenchymal stem cells using two-dimensional gel electrophoresis, *Proteomics* **6**:1223–1229 (2006).

[58] Wang D, Gao L: Proteomic analysis of neural differentiation of mouse embryonic stem cells, *Proteomics* **5**:4414–4426 (2005).

[59] Zhang H, Li XJ, Martin DB, Aebersold R: Identification and quantification of N-linked glycoproteins using hydrazide chemistry, stable isotope labeling and mass spectrometry, *Nat Biotechnol* **21**:660–666 (2003).

[60] Kaji H, Saito H, Yamauchi Y, Shinkawa T, Taoka M, Hirabayashi J, Kasai K, Takahashi N, Isobe T: Lectin affinity capture, isotope-coded tagging and mass spectrometry to identify N-linked glycoproteins, *Nat Biotechnol* **21**:667–672 (2003).

[61] Nunomura K, Nagano K, Itagaki C, Taoka M, Okamura N, Yamauchi Y, Sugano S, Takahashi N, Izumi T, Isobe T: Cell surface labeling and mass spectrometry reveal diversity of cell surface markers and signaling molecules expressed in undifferentiated mouse embryonic stem cells, *Mol Cell Proteomics* **4**:1968–1976 (2005).

[62] Tao WA, Wollscheid B, O'Brien R, Eng JK, Li XJ, Bodenmiller B, Watts JD, Hood L, Aebersold R: Quantitative phosphoproteome analysis using a dendrimer conjugation chemistry and tandem mass spectrometry, *Nat Methods* **2**:591–598 (2005).

[63] He XC, Yin T, Grindley JC, Tian Q, Sato T, Tao AW, Dirisina R, Porter-Westpfahl KS, Hembree M, Johnson T, Wiedemann LM, Barrett TA, Hood L, Wu H, Li L: PTEN-deficient intestinal stem cells initiate intestinal polyposis, *Nature Genet* (2006).

INDEX

ABI microarrays, 63
A cells (migratory neuroblasts), 39
Actin-binding proteins, 214
Activin A, long-term expression of, 98
Activity-based protein profiling (ABPP), 150
Acute myeloid leukemia (AML), 49–51
Address oligonucleotides, 63
Adenomatous polyposis coli (APC) homologs, 31
Adenomatous polyposis coli tumor suppressor gene, 48
Adenoviral vectors, 96, 98
Adherent growth, 9
Adult neurogenesis, 39
Adult rodent NSCs, gene trap in, 199
Adult stem cell biology, 52–53
Adult stem cell lineages, 33
Adult stem cell niche, 30
  identifying, 32–33
Adult stem cells, 4, 27–58, 171, 188
  fundamental properties of, 28
  genomic mutations of, 47
  properties of, 52
  regulation and disregulation of, 29
Adult stem cell technology, innovation in, 42
Affinity chromatography, 198
Affinity enrichment systems, 158
Affinity labeling, 131–134
Affinity matrix, 133
Affinity purification, 131
Affinity reagents, 157
Affymetrix, 63, 64, 70, 75
Affymetrix dataset, 65
Agilent, 63
Allelic variability, 177
Alopecia, 44–45
α-tubulin hyperacetylation, 122
Alternative mRNA splicing, 75
Alzheimer's disease (AD), 40
Ambion, 74
Amino acids, stable isotope labeling with, 234–235
Amphibian limb, regenerating, 208
Amphibians, in regeneration studies, 210
Amplification technologies, 65
Amplified antisense RNA (cRNA), 60
Analysis of variance (ANOVA), 67
Analytical tools, 171–174
Anticancer compounds, screens for, 73
Anuran amphibians, in regeneration studies, 210
AP1 signaling pathway, functional profiling of, 89
APC/β-catenin destruction complex, 48
Apical ectodermal ridge (AER), 212
Apoptosis, TRAIL-induced, 92–93

Chemical and Functional Genomic Approaches to Stem Cell Biology and Regenerative Medicine
Edited by Sheng Ding

Apratoxin A, 136
ARF-GAP functions, 197
Array-based SNP technologies, 74
Array comparative genomic hybridization (aCGH), 75
Array Express, 71
Array image capture, 67
Arrays, spotted and synthesized, 60
Artificial cell immortalization, 182
Assay conditions, permissive, 191
Assays, long- versus short-term, 95
Astrocyte stem population, 40
Astroglial differentiation, 194
Asymmetric division, 28, 193
ATCC clone library, 85
ATP-binding cassette (ABC) transporter family, 52
Autologous skin transplants, graftable matrices for, 44
Automated microscopy, 90
Automated synthesis, enabling technologies for, 111
Axonal regeneration, 195
5-Aza-C DNA demethylation agent, 194
Azide-containing analogs, 158

Baldness, 44–45
Barluenga, Sofia, vii, 109
Basic helix-loop-helix (bHLH) factors, 7
Basic helix-loop-helix leucine zipper (bHLHLZ) protein family, 125
B cells (SVZ astrocytes), 39
  hematopoietic-lineage, 34
Bcr-Abl-kinase inhibitors, 117–118
Benzodiazepines, 112
Benzopyran library, synthesis of, 119
Benzopyran motif, 118
β-catenin destruction complex, 48
β-catenin interactors, 230
β-catenin phosphorylation, 237
β cells, insulin-producing, 16
β*geo* cassette, 14
bFGF (basic fibroblast growth factor), 8, 9
Bioavailability, predicting, 112
Bioconductor project, 66
Bioinformatic algorithms, development of, 181–182
Biologically active molecules, libraries of, 113
Biological mass spectrometry, protein characterization by, 145–167
Biology, probing with small molecules, 109–144
Biomolecule-compatible ionization techniques, 145
Biotin-streptavidin biochemical affinity pair, 158
Blastema, 208
  cells of, 212
Blastomeres, 1–2
Blood–brain barrier, 41
BMP receptor (BMPr) mutants, 47
Bone-forming osteoblasts, 36
Bone marrow, HSC homeostasis and, 29
Bone marrow transplantation, 27, 28, 34
  workings of, 36–37
Bone morphogenetic protein (BMP), 36, 47, 191, 192. *See also* Serum/BMP
Bone morphogenetic protein-4 (BMP4), 6–7, 8, 200. *See also* BMP receptor (BMPr) mutants
Bone morphogenetic protein-9 (BMP9), 90
Bottom-up approach, 227
Bottom-up proteomics, 147, 159
BrdU DNA base analog, 39
6-Bromoindirubin-3′-oxime (BIO), 192
Bulge activation hypothesis, 43
Bulge cells, 43

C2C12 myoblasts, 196
C57B1/6 embryos, 3–4
Cadherin–catenin interface, 36
cAMP (cyclic AMP), in osteogenesis versus adipogenesis, 201
Cancer. *See also* Tumor entries
  gastrointestinal tract, 48
  initiation of, 188
  therapeutic treatments for, 37
Cancer induction, developmental pathways and, 197
Cancer models, 50
Cancer stem cell model, 51
Cancer stem cell pathways, targeted treatment of, 52
Cancer stem cells, 47, 49–52, 188–189
  relationship to tissue-specific stem cell populations, 179
  therapeutic considerations related to, 51–52
Cancer stem cell theories, 51
Candidate gene approaches, 209, 219
  to regeneration, 211–212
Cap cells, 32
Capillary electrophoresis, 147
Carbohydrate microarrays, 128
Carbohydrate sulfotransferase inhibitors, 117
Cardiomyocytes, 12
  proliferation of, 195
β-Catenin interactors, 230
β-Catenin phosphorylation, 237
C cells (progenitor or transit-amplifying), 39
CD59 pathway, 214
Cdc25A protein phosphatase, 125, 126
Cdc42-dependent functions, 122

CDK1 (cyclin-dependent kinase-1), 93
  inhibition of, 132–133
CDK-related kinases (CRKs), 134
cDNA clone collections/libraries, 60, 199. *See also* cDNA libraries
cDNA clones, 85
cDNA libraries
  biased, 88–89
  development of, 198
cDNAs (complementary DNAs)
  overexpressed, 88–90, 97
  profiling the effects of, 90
  synthesis of, 65
Cell assays, engineering, 191
Cell-autonomous regeneration paradigms, 219
Cell-based therapies, 188
Cell culture, inconsistency in, 190
Cell-cycle-inhibiting proteins, 51
Cell cycle progression, regulating, 93
Cell-cycle-promoting proteins, 51
Cell immortalization, 182
Cell lines, defining relatedness of, 181
Cell plasticity, 196
Cell replacement therapies, 188
  for neurodegenerative diseases, 13–14
Cells
  compiling combined information on, 181–182
  growing in isotopically enriched media, 152
Cell surface marker proteins, 42
Cell surface markers, 37, 39, 49, 74
Cell survival signaling, 193
Cell typing, use of miRNA profiling in, 180
Cellular gene function, exploring via genomic functional profiling, 86–93
Cellular reprogramming, 195–196
CENTA1 protein, 86
Central nervous system (CNS), tissue damage to, 38
Centre for Modeling Human Disease, 199
Centrosomin, 32
Chemical-based affinity systems, 158
Chemical epistasis studies, 194
Chemical isotope-tagging strategies, 230
Chemical labeling, quantitation using, 155
Chemical libraries
  designing better, 201
  privileged, 190
  qualities of, 190
Chemical modulators, of developmental pathways, 197
Chemical screens, 219
  for small molecules that affect regeneration, 217–218
Chemical space, 112, 113
Chemical tags, in encoding methods, 110–112
Chemical technologies, 109–144
Chemoselective ligations, 158
Chemoselective reactions, 127–128
Chemotherapeutic agents, enhancers of, 93
Chesnut, Jonathan D., vii, 169
Chimeric hair follicles, 43
ChIP on chip approaches, 75
Cholesterol metabolism, 118
Chromatin immunoprecipitation, 64
Chromatographic methods, 151
Clark, Julie, vii, 187
Clustering algorithms, 68
CodeLink, 63
Colchicine, 131
Collision-induced dissociation (CID), 225
Colony-forming cells, 49
  hematopoietic, 15
Colony hybridization technique, 60
Columnar cells, 46
CombiMatrix, 64
Combinatorial libraries, 112
  synthesis of, 116
Combinatorial networks, of posttranslational modifications, 156
Combinatorial synthesis, 110–112
Comgenex, 113
Commercial microarrays, 74
  vendors of, 63–64
Comparative analysis, 228
Comparative database-searching techniques, 146
Complementary DNAs. *See* cDNA entries
Computation analysis, 180
Conditioned media, identification of secreted proteins in, 229
Correlation (R^2 score), among NSC, APC, and OPC samples, 176
Cowan, Chad A., vii, 27
CREB signaling, functional profiling of, 89
Cross-species comparisons, 17
Crypt-base columnar cells, 46
C-terminal carboxylic acid functionalities, 152
C-terminal labeling, 230
Cultured cells, small-molecule screens conducted on, 218
CYLD protease, 91
CYP7A1 regulation, 118
Cyst progenitor cells, 31–32
Cytoblot assay, cell-based, 122
Cytokeratin-15 (K15), 44
Cytokines, 7
Cytokine signaling suppressors, 11

Cytotoxic compounds, 135
activity of, 137
novel, 136

Data clustering, 68
Data-dependent analysis schemes, 147, 149
Data-mining algorithms, development of, 181–182
Data repositories, for stem cell microarray experiments, 71
Datasets, comparing, 174, 175–176
"Data warehouses," 70–71
D cells, 40
Dedifferentiation process, 208
Developmental pathways, modulators of, 196–197
Developmental processes, high-throughput small-molecule screens for, 218
Dexamethasone, 194
Differential colony hybridization, 60
Differential gene expression technology, 61–62
Differentially expressed genes, screens for, 212–215, 219
Differentiation
lineage-specific, 193–194
Msx genes in the inhibition of, 212
regulation of, 31–32
Differentiation assays, 99
Differentiation promoting factors, 219
Ding, Sheng, vii, 187
1,3-Dioxane library, 118–121
synthesis and microarraying of, 120
Diseases, stem cells in treating, 187
Diverse molecules, synthesis of libraries consisting of, 113–114
Diversity, measuring and designing, 112–113
Diversity-building reaction, 113–115
Diversity generation, reactions with high potential for, 114
Diversity-oriented synthesis, 115, 122
Diversity space, 112
DNA copy number, tracking changes in, 75
DNA demethylation agents, 194, 197
DNA microarrays, widespread adoption of, 85
DNA sequencing technologies, 60
Dominant neutral loss, during tandem MS, 159
Dopaminergic neurons, 12, 13
Drosophila, niche location and identification in, 32
Drosophila GSC niche, 31–32
Drug approval process, 69
Drug discovery, lead compound optimization in, 73
Duchenne muscular dystrophy (DMD), 214
Dystrophin, 214

Early gene functionalization, in mammalian cells, 84–86
Eberwine amplification scheme, 65
Ectoderm, generation of, 12–14
Ectopic gene-expression screen, 217
Electron-capture detection (ECD), 159
Electron-capture dissociation (ECD), 227
Electron transfer disassociation (ETD), 159–160
Electrospray ionization (ESI), 145, 225–226
elg-19 gene, 136
Elledge–Hannon library, 92
Embryoid bodies (EBs), 9, 12
Embryonic carcinoma (EC) cells, 1, 2, 3
Embryonic germ (EG) cells, 1, 2
Embryonic stem (ES) cells (ESCs), 1–25, 171, 178–179, 188. *See also* ESC entries; ES cell entries; Human embryonic stem cells (hESCs); Murine embryonic stem cells (mESCs)
derivation of, 3–4
key properties of, 4
origin of, 1–3
miRNAs in, 180
as target populations for proteomic interrogation, 228
Embryos, Oct4 deficient, 9
Encoded libraries, 92
Encoding methods, chemical tags in, 110–112
Endoderm derivatives, generation of, 15–16
"Enhancing" mutations, 217
Enteroendocrine cells, 46
Entrez Gene database, 84
Enzymatic isotope labeling techniques, 230
Enzyme profiling, 128
EphB2 cell surface target, 72
Epiblast cells, 2–3
Epicentre, 65
Epidermal growth factor (EGF), 235
Epigenetic changes, tools for systematic analysis of, 180
Epigenetic modifiers, 196–197
Epigenetic profiles, for distinguishing stem cells, 180–181
Epithelial cells, 42
Epithelial stem cell niche, 43–44
Epithelial stem cells, 41–45
as barriers, 48
functions of, 45
identification of, 42–43
skin and intestinal, 41–42
therapeutic considerations related to, 44–45
Epithelial-to-mesenchymal transition (EMT), 195
Epithelium, 41–42

EPO/EPOR binding, 125
Eras-null mouse ES cells, 9
Erg2p target, 136
Erk1 protein, 192
Erlotinib, 195
ESC clones, 199
ES cell-derived hematopoietic progenitors, 15
ES "cell fate" decision, 5
ES cell fate determination
  extrinsic factors in, 8–9
  intrinsic determinants of, 9–11
ES cell identity, establishment of, 11
ES cells, 3. *See also* Embryonic stem (ES) cells (ESCs)
  differentiation of, 12–16
  Nanog-null, 10
  potential applications of, 16–17
  species differences in, 176
ES cell self-renewal, 192
  maintenance of, 5–12
  negative regulators of, 11–12
ESC-specific proteins, 228
ES-like cells, 1
ES-specific genes, 174
N-Ethyl-*N*-nitrosourea (ENU) mutagenesis screen, 215, 216
Ethyl-*N*-nitrosourea (ENU) mutagenesis studies, 71
Exiqon, 74
Exogenously applied growth factor screens, 217
Expressed sequence tags (ESTs), 85
Expression profiling, 86
  using microarrays, 72
External RNA Controls Consortium (ERCC), 69
Extracelluar regulated kinases (ERK), 6
Extracellular-signal-related kinase (ERK), 191–192
Extrusion zone, 45
Ex vivo stem cell culturing, 44

"False discovery rate," 67
False positives/negatives, 94
Familial adenomatous polyposis (FAP), 48
Farnesoid X receptors (FXR), 118
Feature comparison techniques, 151
Feeder cell layers, as a source for stem cell regulators, 229
Feeder cells, 3
Feeder-independent cells, 8
Fenno, Lief, vii, 27
Fetal neural stem cell (NSC) survival, 193
Fexaramine, 118
Fexaramine-induced FXR activation, 118
FGF8 (fibroblast growth factor 8), 13
FGFRs (fibroblast growth factor receptors), 136
53BP1 pathway, 91
Fission process, 46
5-aza-C DNA demethylation agent, 194
FK506 immunosuppressant drug, 133, 218
5-Fluorouracil (5FU), 134, 135
Fluorescent dye, in sample-labeling techniques, 64–65
Fluorescent markers, 98–99
Fluorocarbon tags, 110
Fluorous derivatization and enrichment methodology, 158
Focused libraries, 113
Focused screens, 214
Forkhead transcription factor, 11
Fourier transform ion cyclotron resonance (FTICR) mass spectrometry, 226
FoxD3 transcription factor, in ES cell fate determination, 10–11
FOXO1 translocation, 90
Full-transcriptome exon array, 75
Fumagillin, 131
Functional genomic profiling experiments, weighing factors in, 95–96
Functional genomics approaches, to stem cell research, 198–200
Functional genomic screens, viral delivery methods for, 200
Functional profiling, overexpression-based, 88–90
Functional profiling assays, 88
Functional profiling strategies, considerations when applying, 93–96
Functional screens, 215

G1/S progression, 93
G2/M progression, 93
Gain-of-function screens, 200
Galanthamine, 122
Galanthamine-inspired library, synthesis of, 124
Gastrointestinal cancers, 48–49
Gata4 factor, 11–12
Gata6 factor, 11–12
Gata factors, ES cell self-renewal and, 11–12
Gene expression, miRNA and, 180
Gene expression analysis, 198
  analytical tools for, 180
Gene expression data, steps in analyzing, 66
Gene expression datasets, public, 70–72
Gene expression microarrays, 59
  therapeutic targets and, 72
Gene expression microarray technology, application of, 72

Gene Expression Omnibus, 71
Gene expression pattern, 175
Gene expression profile database, 136
Gene expression profiles, 17, 85–86
  comparing, 213
Gene function, modulation of, 188
Gene Ontology (GO) project, 68
Gene product interaction databases, 68
Gene products, unclassifiable, 85
Genes
  approaches to studying, 145
  Ure2p-dependent, 121
Genetic regeneration enhancers, screens for, 216–217
Genetic screens, for regeneration prevention mutations, 215–216
Gene transfer, 182
Gene trap, 198–199
Gene trap cassette, 199
Genewise queries, 71
Genisphere, 65
Genome-scale analysis, 99
Genome-scale shRNA RNAi libraries, 96
Genome-scale siRNA libraries, 93
Genome sequences, 223
  availability of, 85
Genomewide expression analysis technologies, 59–81
Genomewide expression profiling, 194
Genomic analysis, 17
  methods for, 171–174
Genomic datasets, use in the drug approval process, 69
Genomic functional profiling, 83–107
  advantages of, 86
  application to stem cell biology, 97–99
  exploring cellular gene function via, 86–93
  using RNA interference libraries, 90–93
Genomic functional profiling experiments, methodology of, 87–88
Genomic functional profiling technology, recent developments in, 97–98
Genomic overexpression screen, 136
Genomics revolution, 85
Germline stem cells (GSCs), 29
  in drosophila, 31–32
Glial fibrillary acidic protein (GFAP), 39, 40
Glia precursors, 12
Global gene expression, comparing, 173–174
Global gene expression analysis, 73
Global gene expression profiling, 59
Global methylation studies, 181
Global pairwise comparisons, 177
Global proteomic interrogation, stem cell properties warranting, 224
Global proteomics, 228
Glucagon-like peptide 2 (Glp2), 47
Glycopeptides, 236
Glycoprotein capture technology, 236–237
N-Glycosylated protein isolation technologies, 236–237
"Gly-Gly" tag, 158
Goblet cells, 46
gp130-LIFR (LIF receptor) complex, 191
Graftable matrices, 44
Graft-versus-host disease (GVHD), 37
Granule cells, 40
Granulocyte macrophage colony stimulating factor (GM-CSF), 14
Green fluorescent reporter gene (GFP), 36
GRK4 (G protein-coupled receptor kinase 4), 95
Growth factor screens, 217

H_3PO_4 loss, 159
Hair follicles, 43
Haploinsufficiency, 134
HDAC inhibitors, 197
Heatmaps, 68
Hedgehog signaling pathway, 198
HEK293T cells, 88–90
HeLa cells, 92, 93
Hematopoietic differentiation (mesoderm), 14–15
Hematopoietic lineages, 14
Hematopoietic progenitor cell lines, multipotent, 233
Hematopoietic progenitors, ES cell-derived, 15
Hematopoietic stem cell niche, 29–30, 36–37
Hematopoietic stem cells (HSCs), 27, 34–38. *See also* HSC entries
  benefits of, 37–38
  discovery of, 35
  disregulation of, 29
  human, 36
  identification of, 34–36
  lifespan of, 193
  molecular signature of, 234
  purified, 38
  therapeutic applications of, 37
Hematopoietic system, reconstituting, 37
Heritable epigenetic remodeling, 180–181
Heterocycle library, 191
Hexokinase (HK), 95
Hh pathway antagonists, 194
HIFA (hypoxia inducible factor), 95
High-content phenotypic assays, 137

High-throughput functional genomic screens, 198, 199–200
High-throughput functional screens, large-scale, 200
High-throughput screening (HTS), 94, 112, 137
High-throughput small-molecule screens, 218
High-throughput technologies, 83
Hilcove, Simon, vii, 187
Hippocampal neural progenitor cells, differentiation of, 193–194
His-3 reporter gene, 133, 134
Histocompatibility profiles, 173
Histone deacetylases (HDAC), inhibitors of, 121–122
Histone hyperacetylation, 122
Homozygous mutant cells, genomewide library of, 199
hphtk gene, 14
HSC-activating factors, 36. *See also* Hematopoietic stem cells (HSCs)
HSC transplantation, 37
HSP90 chaperone, 131
Human accelerated region (HAR) RNAs, 91
Human adult neural precursor, 39
Human embryonic stem cells (hESCs), 170, 173, 229
Human ES cell fate determination, extrinsic factors in, 8–9
Human genes, PubMed citations per, 84–85
Human Genome Project, 224
Human MSCs (hMSCs), osteogenesis and adipogenesis screens in, 200–201. *See also* Mesenchymal stem cells (MSCs)
Human umbilical vein endothelial (HUVEC) cells, 90
Hybrid linear ion trap-triple quadrupole (QTrap™) mass spectrometers, 158
Hydrazine-containing reagents, 157
Hydrogen-tritium substitution, 131
Hypothesis-driven multistage MS, 159

ICAT-MS/MS method, 233
Identity, measures of, 173
IL6 family, 6
Illumina, 63, 64, 65, 70, 181
Illumina arrays, 74
Immobilized metal affinity chromatography (IMAC), 157, 158
Immortality-associated genes, 178, 179
Immortal strand theory, 47–48
Immunopurification, MS/MS coupled with, 229–230
Immunoreceptor tyrosine-based motifs (ITAM), 237
Immunosuppressants, 37
Indoloquinolizidines, 126
Information, compiling, 181–182
Information tracking, for microarray experiments, 66–67
Inhibitors, in mouse ES cell fate determination, 7–8
Inner cell mass (ICM), 1, 2
In situ screening, library for, 112
"Instructive" genes, 215–216
Insulin-producing β cells, 16
Insulin-producing pancreatic cells (endoderm), 15–16
Integrative viral vectors, 98
Intestinal crypts, 47–48
Intestinal stem cell niche, 46–47
Intestinal stem cell proliferation, regulation of, 47
Intestinal stem cells, 45–49
  identification of, 46
  molecular markers of, 49
  self-renewal of, 230
  therapeutic considerations related to, 48
Intestinal villi, 45
Intrareplicate expression level, 68
"Inverse labeling" methodologies, 152
Invitrogen, 74
In vitro–generated cells, 12
Ionizing radiation, exposure to, 27
Ischemia, 41
Islet-like structures, derivation of, 15–16
Isobaric tagging for relative and absolute quantitation (iTRAQ), 233–234
Isoindoline library, 125
Isotope-coded affinity tag reagents, 232–233
Isotope coded affinity tags (ICATs), 153, 229
Isotope-coded affinity tag technology, 230
Isotopically enriched media, growing cells in, 152
Isotopic labeling/tagging, 151–152, 156. *See also* Label entries
  quantitation using, 154
  strategies for, 230, 232–235
ITMS (ion trap mass spectrometry), 226
iTRAQ (isobaric tag for relative and absolute quantitation) methodology, 153
iTRAQ reagents, 155

JAK-STAT signal transduction pathway, 31

Keratinocytes, 42
Ketone-containing analog, 158
Key genes, testing the contribution of, 99
Kinase inhibition, 133
Kinase profiling, 128, 130
Kinases, structure-based libraries targeting, 115–118

Knockout embryos, FoxD3, 11
Kyoto Encyclopedia of Genes and Genomes (KEGG), 68

"Label-free" quantitation schemes, 151
Labeling technologies, microarray, 64–65. *See also* Isotopic labeling
Label-retaining cells (LRCs), 48
Label-retaining phagosomes, 46
Label-retaining techniques, 42
Laboratory information management system (LIMS), 67
Large-scale functional genomic screens, 209–210, 212–218
Large-scale genetic approaches, 200–201
Large-scale genomic analysis
  of stem cell populations, 169–186
  workflow stages of, 172
Larva-stage, large-scale regeneration screen, 218
LC-MS analysis, 121. *See also* Liquid chromatography (LC)
LC/MS/MS (liquid chromatography tandem mass spectrometry)
  analysis of peptide mixtures by, 148
  in phosphopeptide-capturing technology, 237
LC-MS/MS-based proteomic technologies, 229–230
LC/MS/MS methodologies, multidimensional, 150
Lead compound optimization, 73
Lectin affinity capture technology, 236
Lectins, 157
Lentiviral shRNA library, 92
Lentiviral vectors, 96, 98
Leukemia inhibitory factor (LIF), in mouse ES cell fate determination, 5–6. *See also* LIF entries
Libraries
  combinatorial, 112
  inhibitor-yielding, 115–127
LIF/interleukin-6 (IL6) family, 191. *See also* Leukemia inhibitory factor (LIF)
LIF signaling, 176
LIF/Stat3 pathway, 8
LIF/Stat3 target gene(s), 6. *See also* Stat3 (signal transducer and activator of transcription 3)
Ligand–protein interaction, 128
Lineage-committed cells, reprogramming, 196
Lineage minus (lin minus), 34
Lineage selection strategy, 13–14
Lineage-specific differentiation, 193–194
Linear ion trap (LIT) analyzers, 226
Linear (two-dimensional) ion traps, 149
Lipid-based transfection, 96
Lipid-mediated transfection, 97
Lipinski's rule-of-five, 112–113
Liquid chromatography (LC), 226. *See also* LC entries; MALDI-LC/MS/MS analysis platforms
LIT-FTICR (linear ion trap–Fourier transform ion cyclotron resonance), 226. *See also* Linear (two-dimensional) ion traps
Liver X receptors (LXR), 118
Long-term assays, 95
Long-term reconstituting hematopoietic stem cells (LSK^+), 234
LTQ FTICR mass spectrometer, 150

Major histocompatability (MHC) immunomatching, 37
MALDI-LC/MS/MS analysis platforms, 147–149. *See also* Liquid chromatography (LC); Mass spectrometry (MS); Matrix-assisted laser desorption/ionization (MALDI)
MALDI-MS, 226
MALDI MS/MS, 236
Mammalian cell lines, use in a resistance screen, 136
Mammalian cells
  early gene functionalization in, 84–86
  regenerative capacity of, 195
Mammalian Genome Consortium, 85
Mammalian models, in regeneration studies, 211
Mammals, immortal strand hypothesis and, 47–48
MAPK pathway, 90. *See also* Mitogen-activated protein kinases (MAPK)
Marelli, Anthony, vii, 83
Marked stem cells, discriminating, 98
MASCOT™, 147
Mass spectrometry (MS), 145–146. *See also* MALDI entries; MS entries; Protein MS
  characterizing posttranslational modifications by, 156–160
  protein identification using, 146–150
  protein quantitation using, 150–156
  in proteomic research, 224–230
  use in measuring changes in peptide levels, 151
Mass-to-charge ratio (*m/z*), 226
Matrix-assisted laser desorption/ionization (MALDI), 145, 225–226. *See also* MALDI entries
mdx mice, in regeneration studies, 214
Medina, Myleen, viii, 83
Mesenchymal cells, BMP-secreting, 47
Mesenchymal stem cells (MSCs). *See also* Human MSCs (hMSCs); Mesenchymal stem/progenitor cells
  differentiation of, 235
  osteogenesis of, 194

Mesenchymal stem/progenitor cells, differentiation of, 194
Mesoderm lineages, 14–15
Messenger RNAs (mRNAs). *See also* mRNA entries
  cDNA synthesis from, 65
  detectability with microarrays, 70
  Gata4, 11
Metabolic interconversion, 158
MetAP2 metalloprotease, 131
Methotrexate (MTX), 133–134
Methylation analysis, 74
Microarray analysis, components of, 69
Microarray analysis methods, maturation of, 69–70
Microarray analysis tools, 65–69
Microarray core facilities, 64
Microarray data, extraction, analysis, and mining of, 60
Microarray designs, improvements in, 70
Microarray detection methods, 200
Microarray experiments
  expense of, 70
  reproducibility of, 69
  software options for, 66
  testing errors in, 67
Microarray gene expression comparison, 213
Microarray Gene Expression Data (MGED) Society, 69
Microarray profiling, 73
Microarray Quality Controls Consortium (MAQC), 69
Microarrays
  applications of, 72–74
  expression profiling using, 72
  global gene expression profiling using, 59
  history of, 59–62
  sample amplification and labeling, 64–65
  stem cell information from, 73–74
Microarray service providers, 64. *See also* Microarray vendors
Microarray technologies, 127–128
  automation of, 70
  evolving, 75
  new and complementary, 74–75
Microarray vendors, 64
  commercial, 63–64
MicroRNA noncoding RNAs, 92
MicroRNAs, 74
Microtubule destabilization, 117
Minimal information about a microarray experiment (MIAME), 174
Minimum expression threshold, 68
miRNA, target genes regulated by, 180
miRNA profiling, use in cell typing, 180
Mitogen-activated extracellular signal regulated kinase (MEK1), 196
Mitogen-activated protein kinases (MAPK), 6
Mitotic spindle, 31
Model organisms
  regeneration screens in, 207–222
  for regeneration studies, 210–211
Modifiers, epigenetic, 196–197
Molecular diversity, 114
Molecular regeneration enhancers, screens for, 216–217
Molecular studies
  approaches to, 209–210
  of regeneration, 211
Molsidomine, 134
Monitoring-initiated detection and sequencing (MIDAS), 159
Monolayer differentiation, 12
Morpholino antisense technology, 216
Motor neurons, generation of, 194
Mouse embryonic fibroblasts (MEFs), 3, 229
  as feeder layers, 8
Mouse embryonic stem cell clones, gene-trap-mutated, 199. *See also* Murine embryonic stem cells (mESCs)
Mouse ES cell fate determination, extrinsic factors in, 5–8
Mouse Gene Prediction Database, The, 71
Mouse genome, diploid nature of, 199. *See also* Null mice; Severe combined immunodeficient (SCID) mice
MptpA protein phosphatase, 125, 126
MptpB phosphatase inhibitor, 127
MptpB protein phosphatase, 125, 126
MRL mouse strain, 216–217
mRNA-level differences, measuring, 67. *See also* Messenger RNAs (mRNAs)
mRNA levels, 233
  measurement of, 74
  number of genes expressed by, 181
mRNA sequences, unhybridized, 60
mRNA splicing, 75
MS^3 (MS/MS/MS) strategies, 159. *See also* Mass spectrometry (MS); Tandem mass spectrometry (MS/MS)
MS analyzers, 147
MS-based protein biomarker discovery, 160
MS-based proteomic strategies, 227
MS-based studies, evaluating, 149–150
MS scanning and fragmentation methodologies, 158
msxB expression, 212
Msx genes, 211–212

Multidimensional protein identification technology (MuDPIT), 149
Multiple sclerosis (MS), genes expressed in, 72
Multipotent adult stem cells, 28
Multistage mass spectrometry, 226
Murine embryonic stem cells (mESCs), 191, 194. *See also* Mouse embryonic entries
  long-term cultures of, 192
Mushashi-1, 47
Mutant alleles, 215
Myc-Max dimerization, 125
Myc signaling, 192
Myelin, inhibitory activity of, 195
Myoblasts, reprogramming, 196
Myoseverin, 117, 195, 198, 218
Myotube fission, 117

Nanog, 83, 199
  in ES cell fate determination, 10
  long-term expression of, 98
Natural-product-based libraries, 113
Natural products, total synthesis of, 113
N-CoR nuclear corepressor, phosphorylation status of, 99
ncRNAs, 91–92
Neomycin B library, 115
Neural cells, embryonic stem cell-derived, 13–14
Neural crest stem cells (NCSCs), 176
Neural differentiation (ectoderm), 12–14, 193–194
Neural precursors, 14, 39
Neural stem cells (NSCs), 38–41, 170, 171, 177, 178–179. *See also* NSC entries
  biology of, 41
  identification of, 38–39
  potency of, 39
  therapeutic considerations related to, 40–41
Neurodegenerative diseases, cell replacement therapies for, 13–14
Neurogenesis, adult, 39
Neuronal differentiation, 193–194
Neurons
  generation of, 13
  regenerative incompetence of, 195
Neuropathiazol, 193–194
NF-κB activation, 91
NIC-60 cancer cell line, 136
Niche, 29–34
  adult stem cell, 32–33
  drosophila GSC, 31–32
  epithelial stem cell, 43–44
  hematopoietic stem cell, 36–37
  intestinal stem cell, 46–47
  molecular mechanisms of, 33–34
  regulation of proliferation in, 30
  regulation of stem cell fate in, 31–32
Nimblegen, 63, 64
Noggin, 40, 47
Noncoding RNAs, 91–92
Nonmuscle myosin II heavy chain (NMMII), 196
Notch1 pathway, 36
Notch signaling, 193
Novel phosphopeptide-capturing technology, 237–238
NRON ncRNA, 91–92
NSC cell populations, 175. *See also* Neural stem cells (NSCs)
NSC-mediated repair, exogenous, 41
NSC transplantation, 41
N-terminal labeling, 230
Nuclear factor of activated T cells (NFAT), 91–92
Nuclear transfer cloning, 196
Nucleic acid libraries, 87–88
Nucleic acids, amplified, 65
NuGen proprietary amplification scheme, 65
Null mice, Gata4 and Gata6, 11
Nurr1 transcription factor, 13
Nurse cells, 30–31

Oct3/4 homologues, 176
Oct4-GFP reporter, 192
Oct4 transcription factor, 83
  in ES cell fate determination, 9–10
  long-term expression of, 98
Off-target activity, 94–95
Oligodendrocytes, derivation of, 13
Oligonucleotide sequences, 134
Oligonucleotide spotting/synthesizing methods, 60
Olomoucine, 116
Oncogenesis, cellular changes related to, 49
Oncogenic kinases, 131
Oncogenic pathways, 48
  nurse cells and, 31
Oncogenic targets, 51
129 embryos, 3–4
Optical encoding system, 117
Orbitrap, 227
Organ rejection, 37
Organs, regeneration of, 207
Orth, Anthony P., viii, 83
Osteogenesis, modulation of, 194
Osteosarcoma cells, 136
Overexpressed cDNAs, profiling the function of, 88–90
Overexpression
  of Nanog, 10
  of Socs3, 11

Overexpression approach, versus RNAi-mediated approach, 95–96
Overexpression experiments, false positives in, 94

p38 inhibition, 195
p38 inhibitors, 193
p42/p44 MAP kinase inhibition, 133
p53 function, genes critical for, 91
p53 signaling, functional profiling of, 89
Pancreatic β cells, proliferation of, 195
Pancreatic cells, generation of, 15–16
Pancreatic duodenal homeobox 1 (Pdx1), 15–16
Paneth cells, 45–46
Parabiosis experiments, 36
Parallel synthesis methods, 110
Parathyroid hormone (PTH), 36
Parkinson's disease (PD), 40
Pathologies, protein–protein interactions and, 125
Pathway-building software packages, 68
Pax4 transcription factor, 15
PD098059 MEK inhibitor, 192
PD168393 EDFR inhibitor, 195
Peptide mass mapping, 146
Peptide mixtures, analysis by reverse-phase LC/MS/MS, 148
Peptides
  genes encoding, 224
  ICAT-labeled, 232
  important parameters of, 151
  solid-phase synthesis of, 110
  separation of, 147
Peroxisome proliferator-activated receptor γ (PPARγ) agonists, 194
Perturbation experiments, 182
Petasis reaction, 115
Peters, Eric C., viii, 145
Peterson, Randall T., viii, 207
Pharmacological approach, 109
Pharmacotherapy
  predicting patient response to, 73
  realistic targets for, 71
Phenotype-based approach, 190
  screening method for, 191
Phenylalanine hydroxylase (PAH) activity, 94
3-Phosphoinositide-dependent protein kinase (PDK1), 128
Phosphatase inhibitors, 125–127
Phosphatases, 93
Phosphatidylinositol 3-kinase (PI3K), 235
Phosphoinositide 3-kinase (PI3K), 15
Phosphoinositol-3 kinase (PI3K), 6
Phosphopeptide-capturing technology, novel, 237–238
Phosphopeptides, enrichment of, 157
Phosphorylation profile, 159
Photocrosslinking, 128, 131
Pianowski, Zbigniew, viii, 109
Pkc1 identification, 135
Planaria, in regeneration studies, 210
Plaque screening, 60
Plasmodium falciparum, purine activity against, 117
Platelet-derived growth factor (PDGF), 235
PLK1 (polokinase-1), 93
Pluripotency
  induction of, 200
  measures of, 173–174
  Oct4 and, 9–10
Pluripotent ES cells, 4
Pluripotent stem (iPS) cells, 2, 200
Pluripotin, 8, 192, 198
Polysialic acid form of neural cell adhesion molecule (PSA-NCAM), 39–40
Posttranslational modifications (PTMs)
  characterizing by mass spectrometry, 156–160
  detection and localization of, 156
  in proteomics, 157
Precursor ion-scanning experiments, 158–159
Precursors, 170
Preimplantation blastocyst stage, 1
Privileged scaffold, libraries based on, 118
"Privileged structures," 190–191
Proapoptotic factors, 89–90
Profiling assays, complex, 96
Progenitor cancer pathway, 51
Progenitor cells, 28
Proliferation markers, 38–39
Proof-of-concept study, 90, 92
Protease profiling, 128, 130
Protein(s)
  enforced overexpression of, 90
  mass spectrometric characterization of, 137–138, 145–167
  tryptic digestion of, 156
Protein expression levels, assessing changes in, 150
Protein function, small-molecule modulators of, 217
Protein identification, defined, 149
Protein kinase A (PKA), 128
Protein kinase inhibitors, 217
Protein MS, 150. *See also* Mass spectrometry (MS); Protein quantitation
Protein/peptide-capturing technologies, emerging, 236–238
Protein phosphorylation, 237

Protein–protein interactions
small-molecule antagonists of, 125
synthesis of libraries targeting, 126
Protein quantitation, using mass spectrometry, 150–156
Protein trafficking, 122
Proteolytic activity, measuring differences in, 128
Proteome, dynamic nature of, 153–156
Proteomic research, mass spectrometry in, 224–230
Proteomics
bottom-up, 159
posttranslational modifications in, 157
in stem cells, 223–242
Proteomic strategies, 225
MS-based, 227, 229
Proteomic technologies
advances in, 224
successful application of, 238
Proteomic technology platforms, 225
Proton transfer charge reduction, 160
PTH/PTHrP receptor (PPR), 36
PTH-related protein (PTHrP), 36
PTP1B protein phosphatase, 125, 126
Public gene expression datasets, 70–72
PubMed citations, 84–85
Purine-dependent enzymes, structure-based libraries targeting, 115–118
Purines, combinatorial libraries based on, 115–117
Purmorphamine, transcriptome analysis of, 198
Puromycin, 99
Purvalanol B, 117
Purvalanols, 132–133
targets of, 134

QPCR (quantitative polymerase chain reaction), 70
QS11 purine compound, 197, 198
QTrap™ (hybrid linear ion trap-triple quadrupole) mass spectrometers, 158
Quadrupole ion trap (IT) analyzers, 226
Quadrupole TOF (QTOF), 227
Quantitative profiling studies, experimental workflow for, 153, 154
Quantitative proteomics, labeling methods for, 232
Quantitative proteomic scheme, 231
Quantitative proteomics technologies, 230–232
alternative, 235
Quantitative trait locus (QTL), 71

Radicicol, 131
Rao, Mahendra S., viii, 169
RasGAP protein, 192
Rate-limiting components, detecting, 94
Reagent-targeting chemical functionality, 158
Reddy, Venkateshwar A., viii, 145
Reference standard, use of, 174, 175
Regeneration, 195
approaches to, 208–210
candidate gene approaches to, 211–212
chemical screens for small molecules that affect, 217–218
choice of model organism for, 209, 210–211
complexity of, 208
complex pathways in, 219
efficiency of, 216
experimental approaches to, 211–218
molecular and cellular events characterizing, 209
screens for differentially expressed genes during, 212–215
screens for genetic and molecular enhancers of, 216–217
total set of genes required for, 216
versus wound response, 213–214
Regeneration assays, 218
"Regeneration-instructive" genes, 215–216
Regeneration prevention mutations, genetic screens for, 215–216
Regeneration screens, 207–222
Regeneration stages, markers of, 217
Regeneration studies, 207
Regenerative capabilities, drug stimulation of, 188
Regenerative cells, 208
Regenerative medicine, 169
Reporter genes, 217
Research, proteomic, 224–230. *See also* Stem cell research
Resident intestinal stem cells (ISCs), 45
Responder mESC line, 199
Retinoic acid (RA) responsive genes, 214
Retroviral insertion mutagenesis, 216
Retroviral vectors, 96, 98
Reverse genetics approaches, 216
Reverse-phase HPLC (high performance liquid chromatography), 147
Reverse-phase LC/MS/MS, 148
Reverse transcription-polymerase chain reaction, 16
Reverse-transfection approach, 88
Reversine, 196, 198
Rf tags, 110–112
Rho GTPase, secramine and, 122
Ribosomal assembly inhibition, 134, 135
RNA, microarray analysis of, 218
RNA expression, DNA microarrays for measuring, 85

RNAi-based genomic functional profiling, 90–91
RNAi-based profiling experiments, false positives in, 94–95
RNAi Consortium, The (TRC), 92
RNAi libraries, development of, 198
RNAi-mediated approach, versus overexpression approach, 95–96
RNA interference (RNAi) libraries, genomic functional profiling using, 90–93
RNA interference mechanisms, 87
RNAi screens
  large-scale, 91
  in planaria, 210
RNA samples, partially degraded, 65

Sachidanandan, Chetana, viii, 207
Sample information software packages, 65–66
Samples
  chemical modification of, 152
  correlation (R^2 score) among, 176
Schofield, Robert, 29–30
Schulze, Eric N., 1
Screening strategies, designing better, 201
Screens
  for anticancer compounds, 73
  chemical, 217–218, 219
  for differentially expressed genes, 212–215, 219
  focused, 214
  functional genomic, 198, 199–200, 209–210, 212–218
  for genes differentially expressed during regeneration, 212–215
  genomic overexpression, 136
  high-throughput, 94, 112, 137, 198, 199–200, 218
  in situ, 112
  osteogenesis and adipogenesis, 200–201
  phenotype, 191, 193
  plaque, 60
  regeneration, 207–222
  RNAi, 91, 210
  small-molecule, 193
  synthetic lethality, 134
Sebaceous gland, 42
Secramine, 122
  discovery of, 124
Secreted proteins, identifying in conditioned media, 229
Self-renewal, 171, 176, 187, 191–193, 223
  regulation of, 31–32
Sengstaken, Crystal L., viii, 1
Separation strategies, multidimensional, 149
Sequencing protocols, 180
Sequest™, 147, 224
Serial transplantation, 28
Serum/BMP, in mouse ES cell fate determination, 6–7
Severe combined immunodeficient (SCID) mice, 36, 49
SHDC activator, 88
SHH (sonic hedgehog) transcription factor, 13
Short-hairpin RNAs (shRNAs), 87. *See also* shRNA libraries
  virally encoded, 99
Short tandem repeats (STRs), measures of, 173
Short-term assays, 95
Shotgun-based protein identification studies, 150
"Shotgun" proteomics studies, 147
Shp2 protein phosphatase, 125, 126
shRNA libraries, 87, 91–92, 96, 99. *See also* Short-hairpin RNAs (shRNAs)
Shumate, Deanna, viii, 83
Signaling pathway protein components, identification of, 229–230
Signaling pathways, 49
  developmental, 196–197
Signal transducer and activator of transcription (STAT), 6
"Silenced" region of chromosomes, detection of, 17
Single-gene perturbation, 201
Single nucleotide polymorphism analysis, 173
Single-nucleotide polymorphisms (SNPs), identification of, 74
siRNA libraries, 92–93. *See also* Small interfering RNAs (siRNAs)
  genome-scale, 93
siRNA transfection protocols, 97, 98
Skeletal diversification reactions, 115
Skin cancers, 48
Skin grafting, 44
Skin stem cells, 42, 43, 45
  molecular markers of, 44
Small interfering RNAs (siRNAs), 87. *See also* siRNA entries
  synthetic, 96
Small-molecule antagonists, of protein–protein interactions, 125
Small-molecule FXR agonists, 118
Small-molecule inhibitors, 137, 197
Small molecule libraries, 109, 137
Small-molecule microarrays, 110, 128
Small-molecule phenotypic screens, 193
Small-molecule regulators, of stem cell fate, 191–197

Small molecules, 188, 201
benefits of, 191
bioavailability of, 112
immobilizing, 128, 129
probing biology with, 109–144
target identification of, 198
Smoothened (Smo), 194
Socs3 cytokine signaling suppressor, ES cell self-renewal and, 11
Software, for microarray experiments, 66
Solid-phase synthesis, of peptides, 110
Solution-phase libraries, 125
Somatic cell reprogramming, 196
Somatic stem cells (SSCs), 29
SOURCE database, 71
Southern blots, 59
Sox1 transcription factor, 13–14
Sox2 transcription factor, 13–14
in ES cell fate determination, 10–11
Species differences, in ES cells, 176
Spectral counting techniques, 151
Sperm progenitor cell, 31
Spinal cord injury (SCI), 40
Spinal insult, 40–41
Spleen colony-forming units (CFU-S), 29
"Split and mix" synthesis, 110
"Split-pool" synthesis, 110, 117
"Spontaneous" differentiation, 192
Squamous layer, 42
Stability, measures of, 173
Stable isotope-labeling strategies/techniques, 151–152, 153
Stable isotope labeling with amino acids in cell culture (SILAC), 152, 155, 181, 230, 234–235
Stat3 (signal transducer and activator of transcription 3), 6, 136, 191
Stat3/mouse ES cell lines, 8
StemBase, 71
Stem cell antigen1 (Sca1)-positive cells, 34–36
Stem cell biology
application of genomic functional profiling to, 97–99
convergence with proteomics, 238
genome-scale analysis of, 99
proteomic strategies for investigating, 225
quantitative proteomic technologies in, 230–232
Stem Cell Database, 71
Stem cell differentiation, 171
Stem cell division, miRNA-mediated regulation of, 180
Stem cell engineering platforms, 182
Stem cell fate, 188
small-molecule regulators of, 191–197
Stem cell lineage relationships, 179
Stem cell lines, global pairwise comparisons of, 177, 178
Stem cell microarray experiments, data repositories for, 71
Stem cell mobilization, for medical applications, 33–34
Stem cell populations
analysis of, 170
large-scale genomic analysis of, 169–186
methods used to characterize, 170
obtaining genomic and epigenetic data from, 169
Stem cell proliferation
disregulation of, 32
injury as a regulator of, 43
self-renewal aspect of, 4
Stem cell proteomic research, goal of, 227
Stem cell regulation, 52–53
injury in, 43
Stem cell research, 83, 201
chemical approaches to, 190–191
functional genomics approaches to, 198–200
SILAC technology in, 235
Stem cells. *See also* Adult stem cell entries; Cancer stem cell entries; Embryonic stem (ES) cells (ESCs); Epithelial stem cell entries; Hematopoietic stem cell entries; Human embryonic stem cells (hESCs); Human MSCs; Intestinal stem cell entries; Mesenchymal stem cells (MSCs); Mouse embryonic stem cell clones; Murine embryonic stem cells (mESCs); Neural crest stem cells (NCSCs); Neural stem cells (NSCs); Pluripotent stem (iPS) cells; Resident intestinal stem cells (ISCs); Subventricular zone stem cells; Tissue-specific stem cell entries; Ventricular zone stem cells
aging of, 177–179
allelic variability in, 177
biological background of, 188–190
cancer and, 49–51
categories of, 97
common traits in, 174–175
engineering, 182
expansion and differentiation of, 189–190
gathering information about, 181–182
improved molecular description of, 223
MS/MS for large-scale protein identification of, 227–228
mutations in, 52
novel methods to manipulate, 182
promise associated with, 169–170

properties of, 174–181
proteomics in, 223–242
RNA levels in, 175
self-renewal of, 191–193
therapeutic potential of, 187
tissue-specific, 196
transducing, 98–99
unique properties of, 224
"Stemness," defined 170–171, 176–177
Stemness phenotype, lack of, 176–177
Stem/progenitor cell fate, chemical compounds regulating, 189
Stereochemical diversity, 115
Streptozotocin (STZ)-induced diabetic mice, 15
Stroke, ischemia from, 41
Structure-based libraries, 115–118
Study data, use of, 181–182
Sub-genome-scale overexpression cDNA libraries, 96
Subgenomic synthetic siRNA libraries, 93
Subgranular zone (SGZ), 39
Subgranular zone A (SGZ-A), 40
Subgranular zone niche, 40
Subproteome-capturing technologies, 236
Substrates, affinity-labeled versions of, 158
Subventricular zone (SVZ), 39
Subventricular zone niche, 39–40
Subventricular zone stem cells, 171
"Supertransfection," 200
SVZ germinal area, 39
Symatlas, 71
Symmetric division, 28, 193
Synthesis, combinatorial, 110–112
Synthetic lethality screens, 134
Synthetic siRNAs, advantages of, 96

T7-RNA polymerase, 65
Tamoxifen analog library, 115
Tandem mass spectrometry (MS/MS), 147, 225. *See also* MS^3 (MS/MS/MS) strategies
coupled with immunopurification, 229–230
dominant neutral loss during, 159
for large-scale protein stem cell identification, 227–228
identification of secreted proteins using, 229
Tandem mass tags, quantitation using, 155
Tandem MS scans, 153
Tandem TOF instruments (TOF-TOF), 227
Tan group study, 69
Tao, W. Andy, viii, 223
Target-based approach, 190
Target based kinase inhibitors, 192
Targeted analyses, decision-driven, 147
Target identification, 128–136
genomic approaches to, 135
selected molecules used for, 132
Target/pathway identification, 198
Teleost fish, in regeneration studies, 210–211
Teratocarcinomas, 3
Terminally differentiated cells, 51
The Mouse Gene Prediction Database, 71
Therapeutic applications, of hematopoietic stem cells, 37
Therapeutic considerations
for cancer stem cells, 51–52
for epithelial stem cells, 44–45
for intestinal stem cells, 48
for neural stem cells, 40–41
Therapeutics, identifying mechanism of action of, 73
The RNAi Consortium (TRC), 92
Thiol-containing reagents, 157
Thiol-specific reagents, 233
Three-dimensional organs, regeneration of, 208
Thymadine marker, 38
Tian, Qiang, viii, 223
Time of flight (TOF) analyzers, 226–227
"Tissue gene expression" databases, 70, 71
Tissue maintenance, adult stem cells and, 27–28
Tissue/organ-specific stem/progenitor cells, 195–196
Tissue-specific stem cell populations, relationship to cancer stem cells, 179
Tissue-specific stem cells, 171, 196
Top-down approach, 227
Top-down experiments, 159
Totipotent blastomeres, 1–2
TRAIL-induced apoptosis, 92–93
Transcription factors, 175
Transduction, tracking, 98
Transfected cells, assays in, 96
Transfection efficiencies, 96, 97
Transforming growth factor β (TGFβ) superfamily, 7
Transit-amplifying (TA) cell, 44
Transplantable growth matrices, 44
TRC lentiviral shRNA library, 99
Trichostatin A, 122
Trichostatin A-inspired library, synthesis of, 123
Triple-SILAC strategy, 152
Tritiated thymadine, 38, 39
Trophectoderm differentiation, 9
Tubacin, 122
Tubulin, 131
α-Tubulin hyperacetylation, 122

Tumor-initiating cell, 49
Tumor suppressors, 91
Tunicamycin, 134
2DGE MALDI-TOF, 235
Two-dimensional gel electrophoresis (2DGE), 146, 150, 228
Two-dimensional liquid chromatography (2DLC)-MS analysis, 236
2D LC-MS/MS experiment, 149
Two-hybrid method (Y2H), 133

Ubiquitin, 158
Ugi reaction, 114–115
Unbiased genomic approaches, with unlabeled molecules, 134–136
Unbiased libraries, 118–121
Unique-clone array experiments, 94
Universal stem cells, 170
Unlabeled molecules, unbiased genomic approaches with, 134–136
Ure2p-controlled genes, 121
Uretupamine A, 121
Uretupamine B, 121
Uretupamines, discovery of, 118–121
Urodele amphibians, in regeneration studies, 210

Vascular endothelial growth factor (VEGF), 40, 218
Ventricular zone (VZ), 39
Ventricular zone stem cells, 171
VE-PTP protein phosphatase, 125, 126
VHR protein phosphatase, 125, 126
Viral libraries, 99
Virally encoded RNAi libraries, 92
Viral transduction, 182
Viral tropisir, 96
Viral vectors, 97
Viral vector systems, 97–98
Virtual dynamic library, 112

Walker, John R., viii, 59
Welch, Genevieve, viii, 83
Whole organisms, for studying regeneration, 208
"Whole-transcriptome" arrays, 63
Williams three-component reaction, 115
Winssinger, Nicolas, viii, 109
Wnt/β-catenin pathway, 192, 230
Wnt pathway, 47, 197
Wnt signaling, functional profiling of, 89
World Drug Index, 112
Wound response, versus regeneration, 213–214

Xanthosine (Xs), 193
Xenopus laevis, in regeneration studies, 213
Xu, Yue, ix, 187

Yeast three-hybrid (Y3H), 133–134
Ying, Qi-Long, ix, 1

Zebrafish, in regeneration studies, 210–211, 212, 213, 215, 218
Zhang, Jia, ix, 83